Mathematical and Statistical
Developments of Evolutionary Theory

NATO ASI Series

Advanced Science Institutes Series

A Series presenting the results of activities sponsored by the NATO Science Committee, which aims at the dissemination of advanced scientific and technological knowledge, with a view to strengthening links between scientific communities.

The Series is published by an international board of publishers in conjunction with the NATO Scientific Affairs Division

A Life Sciences	Plenum Publishing Corporation
B Physics	London and New York
C Mathematical and Physical Sciences	Kluwer Academic Publishers
D Behavioural and Social Sciences	Dordrecht, Boston and London
E Applied Sciences	
F Computer and Systems Sciences	Springer-Verlag
G Ecological Sciences	Berlin, Heidelberg, New York, London,
H Cell Biology	Paris and Tokyo

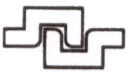

Mathematical and Statistical Developments of Evolutionary Theory

edited by

Sabin Lessard

Département de mathématiques et de statistique,
Université de Montréal,
Montréal, Canada

Proceedings of the NATO Advanced Study Institute and
Séminaire de Mathématiques Supérieures on
Mathematical and Statistical Developments of Evolutionary Theory
Montréal, Canada
August 3–21, 1987

Kluwer Academic Publishers

Dordrecht / Boston / London

Published in cooperation with NATO Scientific Affairs Division

Proceedings of the NATO Advanced Study Institute and
Séminaire de Mathématiques Supérieures on
Mathematical and Statistical Developments of Evolutionary Theory
Montréal, Canada
August 3–21, 1987

Library of Congress Cataloging in Publication Data

NATO Advanced Study Institute/Séminaire de Mathématiques Supérieures
on Mathematical and Statistical Developments of Evolutionary Theory
(1987 : Montréal, Québec)
  Mathematical and statistical developments of evolutionary theory :
proceedings of the NATO Advanced Study Institute/Séminaire de
Mathématiques Supérieures on Mathematical and Statistical
Developments of Evolutionary Theory, Montréal, Canada, August 3-21,
1987 / edited by Sabin Lessard.
     p.   cm. -- (NATO ASI series. Series C, Mathematical and
physical sciences ; v. 299)
   "Published in cooperation with NATO Scientific Affairs Division."
   ISBN-13:978-94-010-6717-1   e-ISBN-13:978-94-009-0513-9
   DOI: 10.1007/978-94-009-0513-9

   1. Evolution--Statistical methods--Congresses. 2. Evolution-
-Mathematics--Congresses.   I. Lessard, Sabin, 1951-   . II. North
Atlantic Treaty Organization. Scientific Affairs Division.
III. Title. IV. Series.
QH323.5N37 1987
575.01'0151--dc20   89-26702

ISBN-13:978-94-010-6717-1

Published by Kluwer Academic Publishers,
P.O. Box 17, 3300 AA Dordrecht, The Netherlands.

Kluwer Academic Publishers incorporates the publishing programmes of
D. Reidel, Martinus Nijhoff, Dr W. Junk and MTP Press.

Sold and distributed in the U.S.A. and Canada
by Kluwer Academic Publishers,
101 Philip Drive, Norwell, MA 02061, U.S.A.

In all other countries, sold and distributed
by Kluwer Academic Publishers Group,
P.O. Box 322, 3300 AH Dordrecht, The Netherlands.

Printed on acid-free paper

TABLE OF CONTENTS

PREFACE

Mathematical and statistical approaches to evolutionary theory are numerous. The NATO Advanced Study Institute (ASI) held at the Université de Montréal, Montréal, August 3-21, 1987, was an opportunity to review most of the classical approaches and to study the more recent developments. The participation of theoretical biologists and geneticists as well as applied mathematicians and statisticians made possible exchanges of ideas between students and scholars having different views on the subject.

These Proceedings contain the lecture notes of seven (7) of the eleven (11) series of lectures that were given.

ESS (Evolutionarily Stable Stragety) theory is considered from many perspectives, from a game-theoretic approach to understanding behavior and evolution (W.G.S. Hines), and a systematic classification of properties and patterns of ESS's (C. Cannings) to particular applications of the differential geometry of the Shahshahani metric (E. Akin). Extensions of ESS theory to sexual populations and finite populations, not to mention games between relatives, are presented (W.G.S. Hines). Special attention is given to the classical game called the War of Attrition but with n players and random rewards (C. Cannings). The Shahshahani metric is also used to show the occurrence of cycling in the two-locus, two-allele model (E. Akin).

Various inference problems in population genetics are adressed. Procedures to detect and measure selection components and polymorphism (in particular, the Wahlund effect) at one or several loci from mother-offspring combinations in natural populations are discussed at length (F.B. Christiansen).

The sampling theory of neutral alleles and the new retrospective theory in population genetics (as opposed to the classical prospective theory) are presented: the Ewens sampling formula and tests for neutrality, Kingman coalescent processes and procedures to determine the age of alleles in a sample or in the entire population (W. Ewens).

Topics in evolutionary ecology include coevolution of virulence and resistance in host-parasite and plant-herbivore systems and the effects of varying environments on dispersal and dormancy strategies. There is also a discussion of evolution as an optimization process (S. Levin and C. Castillo-Chavez).

Finally, evolution in sex-differentiated selection models is studied with regard to the population sex ratio. The validity of a product maximization principle in Mendelian populations with two sexes is discussed (S. Lessard).

The four series of lectures not reported in the Proceedings dealt with other major aspects of the subject: quantitative genetics and multifactorial inheritance, including mutation-selection balance (S. Karlin), molecular biology and DNA sequences, in particular the evolution of multigene families (T. Ohta-Harada), kin selection theory and the evolution of social behaviors (C. Matessi), and phylogenic trees and the differentiation of species (D. Sankoff).

Nine special lectures on related topics were also given by R. Chakrabarty, P. Donelly, J. Haigh, K. Holsinger, C. Lefèvre, D. Pierre-Loti-Viaud, Z. Taïb, E. Thompson, M. Uyenoyama.

My sincere thanks to all lecturers and all participants for their contribution. I also wish to thank the members of the Organizing Commitee, especially Aubert Daigneault, director of the ASI, Ghislaine David, secretary of the ASI, Gert Sabidussi, technical editor of the Proceedings, and Lise Perreault, who typed the manuscript. The credit for the success of the ASI must be given to all.

The ASI was made possible by financial support from NATO, the Natural Sciences and Engineering Research Council of Canada, the Ministère de l'éducation du Québec and the Université de Montréal. Special thanks for their efforts on behalf of the ASI are due to the NATO Scientific Affairs Division, in particular to Dr. Craig Sinclair, then Director of the ASI program.

Sabin Lessard
Scientific Director

LIST OF PARTICIPANTS

Frederick ADLER
Center for Applied Mathematics
Cornell University
312 Sage Hall
Ithaca, NY 14853
USA

Ethan AKIN
Department of Mathematics
The City College of CUNY
New York, NY 10031
USA

Dominique ANFOSSI
Department of Mathematics
    and Statistics
University of Guelph
Guelph, Ont. N1G 2W1
Canada

Adel ANTIPPA
Département de physique
Université du Québec à Trois-Rivières
C.P. 500
Trois-Rivières, Qué. G9A 5H7
Canada

Chris BASTEN
Department of Mathematics
Washington State University
Pullman, WA 99164-2930
USA

Marc BOURDEAU
Départment de mathématiques
    appliquées
École Polytechnique
C.P. 6079, Montréal, Qué. H3C 3A7
Canada

Lisa BROOKS
Biomed Department
Box G
Brown University
Providence, RI 02912
USA

David BROWN
Department of Zoology
University of Toronto
25 Harbord St.
Toronto, Ont. M5S 1A1
Canada

Robert BRUNET
Département de mathématiques
    et de statistique
Université de Montréal
C.P. 6128, Succ. A
Montréal, Qué. H3C 3J7
Canada

Chris CANNINGS
Department of Probability and Statistics
University of Sheffield
Sheffield S10 2TN
United Kingdom

Carlos CASTILLO-CHAVEZ
Section of Ecology and Systematics
Cornell University
Corson Hall
Ithaca, NY 14853-2701, USA

Ranajit CHAKRABORTY
Genetics Center
University of Texas
Health Science Center
P.O. Box 20334
Houston, TX 77225, USA

Freddy B. CHRISTIANSEN
Institute of Ecology and Genetics
University of Aarhus
Ny Munkegade, Bldg. 550
DK-8000 Aarhus C
Denmark

Mariana CONSTANTINESCU
Département de mathématiques
    et de physique
Université de Moncton
Moncton, New Brunswick E1A 3E9
Canada

Ross CRESSMAN
Department of Mathematics
Wilfrid Laurier University
Waterloo, Ont. N2L 3C5
Canada

Kalyan DAS
Département de mathématiques
    et de statistique
Université de Montréal
C.P. 6128, Succ. A
Montréal, Qué. H3C 3J7
Canada

Peter DONNELLY
Department of Statistical Science
University College
Gower Street
London WC1E 6BT
United Kingdom

Simplice DOSSOU-GBETE
Département de mathématiques
Université de Pau
Avenue de l'Université
64000 Pau
France

Guillemette DUCHATEAU
4, rue Casimir Perier
75007 Paris
France

Yves ESCOUFIER
Département d'informatique
Université des Sciences et
    Techniques du Languedoc
99, ave d'Occitanie
34075 Montpellier Cédex
France

Warren J. EWENS
Department of Mathematics
Monash University
Clayton, Victoria 3168
Australia

Vincent FERRETTI
6826, rue Boyer
Montréal, Qué. H2S 2V7
Canada

Carmela GUGLIELMINO
Istituto di Genetica, Biochimica
    ed Evoluzionistica, CNR
Università di Pavia
Via Abbiategrasso, 207
27100 Pavia
Italy

John HAIGH
Mathematics Division
University of Sussex
Mathematics & Physics Bldg.
Falmer, Brighton BN1 90H
United Kingdom

David HALL
Department of Zoology
Duke University
Durham, NC 27706, USA

André HARDY
Département de Mathématique
Fac. Universitaires de Namur
Rempart de la Vierge, 8
B-5000 Namur, Belgium

Rubén HERNANDEZ
Dpto de Estadistica
Univ. Nac. Autónoma de Mexico
Apdo. Postal 70-429
Ciudad Universitaria
04510 Coyoacán D.F.
Mexico

Gordon HINES
Department of Mathematics
    and Statistics
University of Guelph
Guelph, Ont. N1G 2W1
Canada

Howard HINES
Department of Mathematics
University of the West Indies
Mona, Kingston 7
Jamaica W.I.

Kenneth J. HOCHBERG
Department of Mathematics
    and Computer Sicence
Bar-Ilan University
52100 Ramat Gan, Israel

Kent E. HOLSINGER
Department of Ecology and
    Evolutionary Biology
Box U-43, Room TLS 314
University of Connecticut
Storrs, CT 06268, USA

Fred M. HOPPE
Department of Mathematics
    and Statistics
McMaster University
1280 Main Street West
Hamilton, Ont. L8S 4K1
Canada

Hélène JOLY
Division des Sciences Biologiques
Pièce 3121
Conseil National de Recherche
100, Promenade Sussex
Ottawa, Ont. K1A 0R6
Canada

Paul JOYCE
Department of Mathematics
University of Utah
233 Widtsoe Bldg.
Salt Lake City, UT 84112
USA

Mürüvvet KARA
Operations Research Unit
Marmara Scientific and
    Industrial Research Institute
P.O. Box 21
Gebze–Kocaeli
Turkey

Samuel KARLIN
Department of Mathematics
Stanford University
Stanford, CA 94305
USA

Eric KINDAHL
Section of Genetics and
    Development
Cornell University
Emerson Hall
Ithaca, NY 14853-1902
USA

Mokhtar KIRANE
4, rue des Sorbiers
92000 Nanterre
France

Michel KRIEBER
Department of Biology, LSB
Dalhousie University
Halifax, Nova Scotia B3H 3J5
Canada

Yolande LAVOIE
Département de démographie
Université de Montréal
C.P. 6128, Succ. A
Montréal, Qué. H3C 3J7
Canada

Claude LEFÈVRE
Avenue van Becelaere 24B
Boîte 12
B-1170 Bruxelles
Belgium

Sabin LESSARD
Département de mathématiques
    et de statistique
Université de Montréal
C.P. 6128, Succ. A
Montréal, Qué. H3C 3J7
Canada

Simon A. LEVIN
Section of Ecology and Systematics
Cornell University
Corson Hall
Ithaca, NY 14853-2701, USA

Carlo MATESSI
Istituto di Genetica, Biochimica
    ed Evoluzionistica, CNR
Università di Pavia
Via Abbiategrasso, 207
27100 Pavia
Italy

Francine MAYER
Centre de recherches caraïbes
Université de Montréal
C.P. 6128, Succ. A
Montréal, Qué. H3C 3J7
Canada

Marcel MONGEAU
Département de mathématiques
    et de statistique
Université de Montréal
C.P. 6128, Succ. A
Montréal, Qué. H3C 3J7
Canada

Michael MOODY
Department of Mathematics
Washington State University
Pullman, WA 99164-2930
USA

Annie MORIN
42, rue Vasselot
35100 Rennes
France

Tim MOUSSEAU
Department of Biology
McGill University
1205, Ave Docteur Penfield
Montréal, Qué. H3A 1B1
Canada

Olle NERMAN
Department of Mathematics
Chalmers University of Technology
S-412 96 Göteborg
Sweden

Tomoko OHTA
Department of Population Genetics
National Institute of Genetics
Yata 1, 111
Mishima, Shizuoka-Ken 411
Japan

Susan PAULSEN
Department of Zoology
Duke University
Durham, NC 27706
USA

Daniel PIERRE-LOTI-VIAUD
L.S.T.A.
3e étage, tour 45-55
Université Paris VI
4, Place Jussieu
75005 Paris Cédex 05
France

Gregory B. POLLOCK
W.K. Kellog Biological Station
Michigan State University
3700 Gull Lake Dr.
Hickory Corners, MI 49060-9516
USA

Rui M. RAMUSGA
Pr. José Maria da Silva No. 8
R/C Dtº 2900 Setubal
Portugal

Djamal SALAH
33, Allée Camille Flammarion
    app. 27
45100 Orléans-La-Source
France

David SANKOFF
Département de mathématiques
    et de statistique
Université de Montréal
C.P. 6128, Succ. A
Montréal, Qué. H3C 3J7
Canada

Yadavalli V.S. SARMA
Department of Mathematics
University of the West Indies
Mona, Kingston 7
Jamaica W.I.

Pedro J.N. SILVA
Department of Ecology and Evolution
State University of New York
Stony Brook, NY 11794-5245
USA

Hamish SPENCER
Museum of Comparative Zoology
Harvard University
Cambridge, MA 02138
USA

Steven STEWART
Department of Biology
McGill University
1205, Ave Docteur Penfield
Montréal, Qué. H3A 1B1
Canada

Ziyad TAÏB
Department of Mathematics
Chalmers University of Technology
S-142 96 Göteborg
Sweden

Andrée THÉRIAULT
529, rue Wellington sud
Sherbrooke, Qué. J1H 5E2
Canada

Elizabeth THOMPSON
Department of Statistics, GN-22
University of Washington
Seattle, WA 98185
USA

Marcy UYENOYAMA
Department of Zoology
Duke University
Durham, NC 27706
USA

Robert VAN HULST
Department of Biology
Bishop's University
Lennoxville, Qué. J1M 1Z7
Canada

Dumisani VUMA
Department of Mathematics
University of Zimbabwe
P.O. Box MP 167
Mount Pleasant, Harare
Zimbabwe

Dany WEVERBERGH
Département de Mathématique
Fac. Universitaires de Namur
8, Rempart de la Vierge
B-5000 Namur
Belgium

LIST OF CONTRIBUTORS

Ethan AKIN
Department of Mathematics
The City College of CUNY
New York, NY 10031
USA

Dominique ANFOSSI
Department of Mathematics
    and Statistics
University of Guelph
Guelph, Ont. N1G 2W1
Canada

Chris CANNINGS
Department of Probability
    and Statistics
University of Sheffield
Sheffield S10 2TN
United Kingdom

Carlos CASTILLO-CHAVEZ
Section of Ecology and Systematics
Cornell University
Corson Hall
Ithaca, NY 14853-2701
USA

Freddy B. CHRISTIANSEN
Institute of Ecology and Genetics
University of Aarhus
Ny Munkegade, Bldg. 550
DK-8000 Aarhus C
Denmark

Warren J. EWENS
Department of Mathematics
Monash University
Clayton, Victoria 3168
Australia

Gordon HINES
Department of Mathematics
    and Statistics
University of Guelph
Guelph, Ont. N1G 2W1
Canada

Sabin LESSARD
Département de mathématiques
    et de statistique
Université de Montréal
C.P. 6128, Succ. A
Montréal, Qué. H3C 3J7
Canada

Simon A. LEVIN
Section of Ecology and Systematics
Cornell University
Corson Hall
Ithaca, NY 14853-2701
USA

THE DIFFERENTIAL GEOMETRY OF POPULATION GENETICS AND EVOLUTIONARY GAMES

Ethan Akin
Mathematics Departement
The City College
137 Street and Convent Avenue
New York, NY 10031, USA

ABSTRACT. Via the Riemannian metric introduced by Shahshahani and Conley, some elementary ideas from differential geometry have proved broadly useful in mathematical biology. The mathematics is not very deep but it is unfamiliar to many, so we begin with a survey of the elements of linear algebra on Euclidean vector spaces and of calculus on Riemannian manifolds. We then apply the Shahshahani metric to population genetics, deriving the occurrence of cycling in the two locus, two allele model. Then we turn to evolutionary games and compare the ESS condition with other notions of stability.

These notes are intended to be introductory. While they require some familiarity with linear algebra and differential equations, I have tried to provide a review of the elements of those subjects most useful to us. My intentions are expository and so many results are described with proofs omitted or sketched. This heuristic mood changes in the long, last section on game dynamics. Partly because I am pulling together scattered results there, the development is more technical. Also, this section overlaps much more with the subjects of other speakers at the conference, notably Hines and Cannings. This multiple exposure allows a more detailed picture.

I would like to thank Prof. A. Daigneault and Prof. S. Lessard for their hospitality and my fellow conferees for their company. I hope the audience enjoyed listening to these talks as much as I enjoyed giving them.

Contents

S. Lessard (ed.), Mathematical and Statistical Developments of Evolutionary Theory, 1–93.
© 1990 by Kluwer Academic Publishers.

1. Linear algebra and Euclidean metrics

Most linear algebra texts begin with systems of equations and then move on to matrices. For many readers, even among those who use it actively, linear algebra remains a device for manipulating arrays of scalars. For a pure mathematician, oriented by the Bourbaki school, it rather concerns vector spaces and linear transformations. This is an example of the difference in philosophy, or style, which separates the mathematician from many prospective readers. With this initial review of linear algebra I intend partly to recall classic results and the associated notation, but partly to introduce, via familiar material, a possibly unfamiliar style.

A real vector space V is a set, whose elements are called vectors, together with a definition of addition of vectors and multiplication of vectors by real numbers (also called scalars). Addition and scalar multiplication are required to satisfy certain standard axioms including the two distributive laws relating the two operations:

$$t(\xi+\eta) = t\xi + t\eta \qquad \text{for } t,s \in \mathbb{R}$$

(1.1)

$$(t+s)\xi = t\xi + s\xi \qquad \xi,\eta \in V .$$

In particular, the question "What is a vector, really?" is inappropriate and the answer to it "A vector has magnitude and direction while a scalar has only magnitude," with which you were introduced to the subject, is misleading. Instead, the phrase describes a particular example: the system of arrows in the plane or in space with the parallelogram law of addition etc. From this original geometric system come all our intuitions about vector spaces, and within it we discover certain useful properties like (1.1). You get the concept of vector space by a process of –literal– abstraction. You throw away your original model after pulling out those nice properties which are then used as axioms to define a vector space in general. Afterwards you mean by a vector just a point in such an abstract vector space.

The most important example is $\mathbb{R}^n$, the set of ordered n-tuples with coordinatewise addition and multiplication:

$$(x_1,\ldots,x_n) + (y_1,\ldots,y_n) = (x_1+y_1,\ldots,x_n+y_n)$$

$$t(x_1,\ldots,x_n) = (tx_1,\ldots,tx_n) .$$

But let us, instead, write these equations as:

$$(x+y)_i = x_i + y_i \qquad x,y \in \mathbb{R}^n$$

(1.2)

$$(tx)_i = tx_i \qquad i = 1,\ldots,n .$$

This replacement illustrates the mathematician's apparently bizarre taste in notation. Why the preference for complication? Notice first that the latter equations emphasize that x and y are the points of the vector space $\mathbb{R}^n$ to which the various numbers x_i and y_i

are associated functionally. This, in turn, shows that the definition is a special case of a much more general one. If V is a vector space and I is any set, let V^I denote the set of all functions from I to V. So if $f \in V^I$ and $i \in I$, then $f(i)$ is the value of f at i, the point of V associated to i via f. V^I is a vector space by defining:

(1.3)
$$(f+g)(i) = f(i) + g(i) \qquad f,g \in V^I$$
$$(tf)(i) = tf(i) \qquad i \in I.$$

Notice that the formulae on the right are used to define new functions from old by expressing their value at each point i of I. In the special case when V is $\mathbb{R}$ and $I = \{1,2,...,n\}$, (1.3) reduces to (1.2). On the other hand, if I is an interval in $\mathbb{R}$, then we get the usual definition of addition of real- (or vector-) valued functions of a real variable.

Most of the examples we will meet will be subspaces of $\mathbb{R}^n$. A subset of a vector space is a subspace if it is closed under addition and under multiplication by all scalars. It is then a vector space in its own right. For any vector space, the entire space and the vector 0 alone are always subspaces, called the trivial subspaces. For Euclidean space the nontrivial subspaces are the lines and planes which contain the origin. When I is an interval the set of all continuous functions is a subspace of $\mathbb{R}^I$. This says that the sum of two continuous functions is continuous and the scalar multiple of a continuous function is continuous. Similarly the set of all differentiable functions forms a subspace.

Once some sort of structure is defined on a set, the functions which preserve the structure become important and are usually singled out by name. Continuous functions, for example, are those which preserve the concept of limit.

If V_1 and V_2 are vector spaces then a function between them $T : V_1 \to V_2$ is called a linear function or a linear map if it relates the vector space structures on V_1 and V_2 :

(1.4)
$$T(\xi+\eta) = T(\xi) + T(\eta) \qquad \xi,\eta \in V_1$$
$$T(t\xi) = tT(\xi) \qquad t \in \mathbb{R}.$$

Here the operations on the left are in V_1 and those on the right are in V_2. The demand that these linearity properties hold is a strong one. The false assumption of linearity underlies many mistakes in elementary algebra, e.g., $\sqrt{x+y} = \sqrt{x} + \sqrt{y}$.

The set of linear maps from V_1 to V_2 is a subspace of $V_2^{V_1}$ and so forms a vector space which we denote $L(V_1,V_2)$. The proof of closure is an excellent exercise. The steps in showing that the sum of linear maps is linear provides a good test in distinguishing between (1.1), (1.3) and (1.4) which all look alike.

If $A = (a_{ij})$ is an $m \times n$ matrix, then by regarding the elements of $\mathbb{R}^n$ and $\mathbb{R}^m$ as columns, i.e., as $n \times 1$ and $m \times 1$ matrices, we get a linear map from $\mathbb{R}^n$ to $\mathbb{R}^m$ via matrix multiplication $y = Ax$ with

(1.5)
$$y_i = \sum_{j=1}^{n} a_{ij} x_j \qquad i = 1,...,m .$$

Furthermore, these are the only such examples. Every linear map from $\mathbb{R}^n$ to $\mathbb{R}^m$ comes from a (unique) $m \times n$ matrix in this way.

The most important space of linear maps is the *dual space* of a vector space V . The dual space, denoted V^* , is $L(V,\mathbb{R})$ and so consists of the linear maps from V to the reals, also called *linear forms* on V . If $\xi \in V$ and $\omega \in V^*$, then $\omega(\xi)$, the value of ω at ξ , is also written as the *Kronecker product*

(1.6)
$$<\omega,\xi> = \omega(\xi) \qquad \omega \in V^* , \xi \in V .$$

The product $< , >$ is bilinear, that is, it is linear in each variable separately when the other is held fixed at any value.

Again the new notation has as its purpose a change in emphasis. By definition ω in V^* is a function on V transforming vectors in V to real numbers. But we can also regard each ξ in V as acting on V^* transforming ω to $\omega(\xi)$. It is this symmetry of viewpoint which is emphasized by the Kronecker product notation.

The linear operations allow us to construct new vectors from old. If $\xi^1,...,\xi^n$ is a list of vectors in V and $x_1,...,x_n$ is a list of scalars, then the vector

$$\xi = x_1\xi^1 + ... + x_n\xi^n = \sum_i x_i\xi^i$$

is called the linear combination of the vectors $\xi^1,...,\xi^n$ with coefficients $x_1,...,x_n$. The list of vectors is called linearly independent if we can equate coefficients, that is, if $\sum x_i\xi^i = \sum y_i\xi^i$ implies $x_i = y_i$ for $i = 1,...,n$. So we can translate a vector equation into a list of scalar equations. If $\xi^3 = \xi^1 + \xi^2$, for example, then $\{\xi^1,\xi^2,\xi^3\}$ is not linearly independent, and so is called linearly dependent, because $1 \cdot \xi^1 + 1 \cdot \xi^2 + (-1) \cdot \xi^3 = 0 \cdot \xi^1 + 0 \cdot \xi^2 + 0 \cdot \xi^3$. Sloppy use of language here can cause confusion. It is commonly, but incorrectly, said that $\xi^1,\xi^2,...,\xi^n$ are linearly independent vectors. But linear independence is not a property of individual vectors. Instead, it describes how the vectors of the list are related to one another and so is a property of the entire set of vectors. Thus, "The set is linearly independent (or not)" but never "The vectors are linearly independent".

If every vector in V is a linear combination of a linearly independent set $\{\xi^1,...,\xi^n\}$, and so the latter also spans V , then the list is called a basis for V . If such a finite set of vectors exists for V , then V is called finite dimensional. V will then have in general many different bases but the structure theorem of linear algebra says that any two bases have the same number of elements. This common number is called the dimension of V . It corresponds to the physicist's notion of the number of degrees of freedom in the system.

If $\{\xi^1,...,\xi^n\}$ is any list of vectors in V , then there is a linear map $T : \mathbb{R}^n \rightarrow V$ defined by:

$$(1.7) \qquad T(x_1,...,x_n) = \sum_{i=1}^{n} x_i \xi^i .$$

If $\{\xi^1,...,\xi^n\}$ is a basis then this map is onto, meaning that every vector in V is in the image of T, because the set spans. It is one-to-one, meaning that no two lists of coefficients hit the same vector in V, because the set is linearly independent. A one-to-one and onto linear map $T : V_1 \to V_2$ is called a *linear isomorphism*. If T is a linear isomorphism, then the inverse map $T^{-1} : V_2 \to V_1$ is defined and is automatically also linear. For the special case defined by (1.7) with $\{\xi^1,...,\xi^n\}$ a basis for V, this inverse map associates to every vector ξ in V the list of coefficients $(x_1,...,x_n)$ such that $\xi = \sum x_i \xi^i$. The scalars $(x_1,...,x_n)$ are then called the *coordinates* of ξ with respect to the basis. A different basis will in general give a different list of coordinates for the vector ξ. In general, if the vectors of a basis are indexed by a finite set I, then the above construction gives an isomorphism of $\mathbb{R}^I$ with V.

On $\mathbb{R}^I$ itself the *standard basis* $\{\delta^i : i \in I\}$ is defined by letting δ^i be zero at all indices j of I except $j = i$ at which it is 1. So δ^i_j is the Kronecker delta symbol defined by

$$(1.8) \qquad \delta^i_j = \begin{cases} 0 & i \neq j \\ 1 & i = j . \end{cases}$$

There is an ambiguity about the term "coordinates" which is resolved by using the standard basis. With $x \in \mathbb{R}^I$ we usually call the numbers x_i the coordinates of x. But we just defined the coordinates of a vector to be its coefficients when expanded using a basis. When the standard basis is used the coefficients for the vector x are the numbers x_i. That is, $x = \sum x_i \delta^i$ (check this). So the two notions then agree. Alternatively, this says that for the standard basis of $\mathbb{R}^I$ the map $T : \mathbb{R}^I \to \mathbb{R}^I$ given by (1.7) is the identity map.

These coordinate maps show that each finite dimensional vector space is isomorphic to $\mathbb{R}^n$, where n is the dimension of the space. Thus, by using a basis we can translate statements about vectors into statements about lists of numbers. For example, suppose $T : V_1 \to V_2$ is a linear map. If $\{\xi^1,...,\xi^n\}$ is a basis for V_1 and $\{\eta^1,...,\eta^m\}$ is a basis for V_2, then this choice of bases determines an $m \times n$ matrix $A = (a_{ij})$ so that the linear map T corresponds to matrix multiplication as in (1.5). Just let the j column of A consist of the coordinates of $T(\xi^j)$ with respect to $\{\eta^1,...,\eta^m\}$. If $\eta = T(\xi)$, then $y = Ax$, where x and y are the coordinate lists for ξ and η, respectively.

The compositon of two linear maps is again linear and the matrix of the composed map is the product of the matrices for the factors.

* * *

In order to speak of distance or angle between vectors we need more than the vector space structure. We define them by introducing a *Euclidean metric* or *inner product* on a vector space V. This is a function $(,) : V \times V \to \mathbb{R}$, a real-valued function of two vector variables. It is assumed bilinear (as was $< , >$ of (1.6)) and symmetric (meaning $(\xi,\eta) = (\eta,\xi)$). Furthermore, it satisfies the *positivity* condition:

(1.9) $(\xi,\xi) > 0$ if $\xi \in V$ and $\xi \neq 0$.

Notice that by bilinearity $(0,\xi) = (\xi,0) = 0$ for any vector ξ in V. A vector space with a given inner product is called a *Euclidean vector space*.

Positivity allows us to define the length, or norm, or absolute value, of a vector ξ in a Euclidean vector space by taking the square root:

(1.10) $\|\xi\| = (\xi,\xi)^{\frac{1}{2}}$.

By analogy with the real numbers we define the distance between ξ and η to be the absolute value of the difference $\|\xi-\eta\| = \|\eta-\xi\|$.

On $\mathbb{R}^I$ the *usual* or *standard inner product* $_0(,)$ is defined by

(1.11) $_0(\xi,\eta) = \sum \xi_i\eta_i$.

Then (1.10) reduces to the Pythagorean distance, the square root of the sum of the squares.

We define the angle θ between vectors ξ and η by:

(1.12) $(\xi,\eta) = \|\xi\|\|\eta\| \cos\theta$.

By using bilinearity to expand $\|\xi-\eta\|^2 = (\xi-\eta,\xi-\eta)$ we get the so-called polar identity:

(1.13)
$$\|\xi-\eta\|^2 = \|\xi\|^2 + \|\eta\|^2 - 2(\xi,\eta)$$
$$= \|\xi\|^2 + \|\eta\|^2 - 2\|\xi\|\|\eta\| \cos\theta.$$

In the original Euclidean plane example this is the law of cosines and yields (1.12) as a consequence. In an abstract vector space equipped with an inner product it suggests using (1.12) to define the angle θ. This approach requires that the ratio $(\xi,\eta)/\|\xi\|\|\eta\|$ have absolute value at most one so that regarding it as the cosine of an angle makes sense. This result, the Schwarz inequality, follows from the first form of (1.13) (Exercise with hint: replace η by $t\eta$ and recall that if a quadratic in t is always positive, then its discriminant is negative.). (1.13) is also important in showing that the inner product between different vectors (ξ,η) can be recovered from a knowledge of the lengths $\|\xi\|, \|\eta\|$, and $\|\xi-\eta\|$.

The angle between ξ and η is a right angle when the cosine is zero. So ξ and η are perpendicular, or orthogonal, if and only if $(\xi,\eta) = 0$. With respect to the usual inner product distinct members of the standard basis are orthogonal. Furthermore, the length of each basis vector is 1. In sum, with $\xi^i = \delta^i$ for $i \in I$ we have

(1.14) $(\xi^i, \xi^j) = \delta_{ij} = \begin{cases} 0 & i \neq j \\ 1 & i = j. \end{cases}$

In general, in a Euclidean vector space a basis which satisfies (1.14) is called *orthonormal*. For any finite dimensional Euclidean vector space orthonormal bases always exist. When a basis $\{\xi^1,...,\xi^n\}$ of V is orthonormal, then the coordinate isomorphism T of (1.7) is an isometry between $\mathbb{R}^n$ with the usual metric and V. A linear map $T : V_1 \to V_2$ between Euclidean vector spaces is called an *isometry* if it relates the inner products:

(1.15) $(T(\xi), T(\eta)) = (\xi, \eta)$ for all $\xi, \eta \in V_1$.

An isometry preserves length and distance (check that, conversely, $\|T(\xi)\| = \|\xi\|$ for all ξ implies (1.15); use the polar identity). So an isometry is one-to-one and is an isomorphism when it is onto. The inverse map is then also an isometry.

Every linear map $T : \mathbb{R} \to V$ can be naturally identified with a vector ξ in V, namely $\xi = T(1)$ because $T(t) = tT(1) = t\xi$. This gives a linear isomorphism between $L(\mathbb{R}, V)$ and the space V itself.

Using the inner product we can get a – quite different – isomorphism between V and the space of linear maps from V to $\mathbb{R}$, i.e., the dual space V^*. Every vector $\eta \in V$ defines a linear form $\eta^* : V \to \mathbb{R}$ via the inner product, namely $\eta^*(\xi) = (\eta, \xi)$. The association of η^* with η defines a linear map of V into V^* by bilinearity of (,). It is easily seen to be one-to-one because if $\eta^* = 0$, then $\eta^*(\eta) = (\eta, \eta) = 0$ and so $\eta = 0$. The *Riesz representation theorem* says that this map is onto and so defines a linear isomorphism between V and its dual:

1.1 Theorem. *Let* V *be a finite dimensional Euclidean space. For every linear form* $\omega : V \to \mathbb{R}$ *there exists a unique vector* $\eta \in V$ *such that* $\omega(\xi) = (\eta, \xi)$ *for all* $\xi \in V$.

Proof. Choose an orthornormal basis $\{\xi^i\}$ for V. With respect to this basis, and the number 1 chosen as a basis for $\mathbb{R}$, ω is represented by a $1 \times n$ matrix. These n numbers are the coordinates of η with respect to the ξ-basis. In more detail, if the matrix is (a_i), then $\omega(\xi) = \sum a_i x_i$, where $\{x_i\}$ are the coordinates of ξ with respect to $\{\xi^i\}$, i.e., $\xi = \sum x_j \xi^j$. Define $\eta = \sum a_i \xi^i$. Then by (1.14)

$$(\eta, \xi) = \sum a_i x_j (\xi^i, \xi^j) = \sum a_i x_j \delta_{ij} = \sum a_i x_i = \omega(\xi).$$

So this η works, i.e., $\eta^* = \omega$. It is the only one which does because the map $\eta \to \eta^*$ is one-to-one.   □

This simple result has many profound consequences. For the moment, we will use it to define the least squares approximation.

1.2 Theorem. *Let* V *be a finite dimensional Euclidean space and let* A *be a subspace of* V . *If* ξ *is a vector in* V *there is a unique vector* ξ_A *in* A *satisfying the following equivalent conditions:*

(i) $(\eta,\xi) = (\eta,\xi_A)$ *for all* η *in* A ;

(ii) $\xi - \xi_A$ *is orthogonal to every vector* η *in* A ;

(iii)   *for every vector* η *in* A *the Pythagorean identity holds:*

(1.16)
$$\|\xi-\eta\|^2 = \|\xi-\xi_A\|^2 + \|\xi_A-\eta\|^2 .$$

In particular, *if* $\eta = 0$, *then we have:*

(1.17)
$$\|\xi\|^2 = \|\xi-\xi_A\|^2 + \|\xi_A\|^2 .$$

Proof. ξ^* is the linear form on V defined by $\xi^*(\eta) = (\eta,\xi)$. Restricting to A we get a linear form on A and so by Theorem 1.1 there is a unique vector ξ_A in A such that $\xi^*(\eta) = \xi_A^*(\eta)$ for all η in A . This proves that a unique ξ_A satisfying (i) exists. Since $(\eta,\xi) = (\eta,\xi_A)$ if and only if $(\eta,\xi-\xi_A) = 0$, (i) is equivalent to (ii). (ii) implies (iii) by the law of cosines (1.13) applied to $\xi - \xi_A$, $\xi_A - \eta$ and their sum. Conversely, if (iii) holds we can replace η by the vector $\eta + \xi_A$ in (1.16) and get

$$\|(\xi-\xi_A)-\eta\|^2 = \|\xi-\xi_A\|^2 + \|\eta\|^2 .$$

Applying (1.13) to $\xi - \xi_A$, η and their difference we get $(\xi-\xi_A,\eta) = 0$. So (iii) implies (ii).   □

Equation (1.16) explains why we call ξ_A the best approximation of ξ by a vector in A , "best" in the least squares sense. ξ_A is also called the perpendicular projection of ξ into A . To make this precise define the map $P_A : V \to V$ by $P_A(\xi) = \xi_A$. By using (i) it is easy to check (do it!) that P_A is a linear map, and that $P_A(\xi) = \xi$ if ξ already lies in A . Of course, $P_A(\xi) = \xi_A$ lies in A for any ξ , and so the compositon $P_A \circ P_A = P_A$. P_A is called the orthogonal projection map of V onto A .

In general, a *projection* in V is a linear map $P : V \to V$ such that $P \circ P = P$, or equivalently, if $P(\xi) = \xi$ for ξ in the image of P , Im P = $\{\xi : P(\eta) = \xi$ for some $\eta\}$. Complementary to the image of P is the kernel, Ker P = $\{\xi : P(\xi) = 0\}$. For a projection map P every vector can be resolved uniquely as the sum of a vector in Im P and a vector in Ker P (write $\xi = P(\xi) + (\xi-P(\xi))$). A projection is called an orthogonal projection if every vector in Im P is orthogonal to every vector in Ker P . P_A is an orthogonal projection by (ii). It is a useful exercise to check that if P is a projection, then I - P is a projection, where I is the identity map, and Im P = Ker(I-P) , Ker (I-P) = Im P . In particular, I - P_A is the orthogonal projection $P_{A\perp}$ onto the subspace $A^\perp$ consisting of all vectors orthogonal to the vectors in A ,

(1.18)
$$A^\perp = \{\xi : (\xi,\eta) = 0 \text{ for all } \eta \in A\} .$$

When α is a unit vector in V, i.e., $\|\alpha\| = 1$, and A consists of all multiples of α, then writing P_α for P_A:

$$(1.19) \qquad\qquad P_\alpha(\xi) = (\xi,\alpha)\alpha$$

and the complementary projection onto the subspace of vectors perpendicular to α is thus given by:

$$(1.20) \qquad\qquad P_{\alpha\perp}(\xi) = \xi - (\xi,\alpha)\alpha .$$

* * *

So far, the only inner product we have introduced is the usual one. The covariance examples which we now take up allow us to apply our linear algebra machinery to probability problems.

With I a finite set let $\mathbb{R}_+^I$ consist of the non-negative vectors in $\mathbb{R}^I$:

$$(1.21) \qquad\qquad \mathbb{R}_+^I = \{x \in \mathbb{R}^I : x_i \geq 0 \text{ for all } i \in I\} .$$

(Check why this set is *not* a subspace). Within it define the probability simplex:

$$(1.22) \qquad\qquad \Delta = \{p \in \mathbb{R}_+^I : \textstyle\sum_i p_i = 1\} .$$

So each vector p in Δ can be regarded as a probability distribution on a finite set I.

Looking ahead towards applications, think of I as the set of aternative genotypes and p as the vector of relative frequencies of the different types. Every alternative does in fact occur in the population if p is an interior distribution, that is, if $p \in \mathring{\Delta}$ where

$$(1.23) \qquad\qquad \mathring{\Delta} = \{p \in \Delta : p_i > 0 \text{ for all } i \in I\} .$$

Any vector $\xi \in \mathbb{R}^I$ can be regarded as a random variable with respect to the distribution p. More concretely, think of it as the measurement of some character which depends on the genotype, having value ξ_i for genotype i in I. A character is *constant* if it has the same value for every i, i.e., if it is genotype independent. Such a vector is a multiple of u defined by

$$(1.24) \qquad\qquad u_i = 1 \qquad \text{for all } i \in I .$$

We now define for $p \in \mathring{\Delta}$ the *covariance inner product* $_p(,)$ on $\mathbb{R}^I$ by:

$$(1.25) \qquad\qquad _p(\xi,\eta) = \textstyle\sum_i p_i \xi_i \eta_i .$$

Bilinearity and symmetry are easy to check. It is positivity that requires p to be an interior distribution.

Observe that u has length one with respect to $_p(\ ,\ )$ and for any $\xi \in \mathbb{R}^I$:

(1.26)
$$_p(\xi,u) = \sum p_i\xi_i \equiv \bar{\xi}$$

is the mean or average value of ξ with respect to the distribution p . So (1.19) implies that $\bar{\xi}u$ is the projection of ξ onto the subspace of constant vectors. In this case (1.17) becomes:

$$\sum p_i(\xi_i-\bar{\xi})^2 = \sum p_i\xi_i^2 - (\bar{\xi})^2$$

the equivalence between the two ways of computing the variance of ξ .

To extend this, suppose that I is itself a product $I_1 \times ... \times I_L$. For example, if there are L loci and I_ℓ lists the alleles at locus ℓ , then a genotype $i \in I$ consists of a choice of allele at each of the L loci. $\pi_\ell : I \rightarrow I_\ell$ is the projection map which observes for genotype i the allele occurring at locus ℓ . With this projection we can push forward the probability distribution p on I to obtain the induced marginal distribution on I_ℓ . For example, with $\ell = 1$

(1.27)
$$p_{i_1} = \sum_{i_2,...,i_L} p_{i_1 i_2,...,i_L} .$$

So we obtain a map from Δ in $\mathbb{R}^I$ to the product $\Delta_1 \times \Delta_2 \times ... \times \Delta_L$ with Δ_ℓ the simplex in $\mathbb{R}^{I_\ell}$, associating to each probability distribution p on I the corresponding marginal distributions on $I_1, I_2,...,I_L$. For example, in the two-locus-two-allele (TLTA) case we have the diagram

(1.28)

	a	A	
B	p_{aB}	p_{AB}	p_B
b	p_{ab}	p_{Ab}	p_b
	p_a	p_A	1

Δ is 3-dimensional because the 4-vector $p = (p_{AB}, p_{Ab}, p_{aB}, p_{ab})$ is constrained to sum to 1. We obtain the two 1-dimensional marginal vectors (p_A, p_a) and (p_B, p_b) by summing vertically and horizontally. The remaining degree of freedom describes the amount of linkage disequilibrium or dependence between loci when the distribution is given by p .

A vector ξ is called additive between loci if it is the sum of functions each depending on a separate locus. To be precise, there must exist $\varphi^1 \in \mathbb{R}^{I_1},...,\varphi^L \in \mathbb{R}^{I_L}$ so that

(1.29)
$$\xi_{i_1 i_2 ... i_L} = \varphi^1_{i_1} + \varphi^2_{i_2} + ... + \varphi^L_{i_L} .$$

The set of all ξ additive between loci is a subspace $A^{(1)}$ of $\mathbb{R}^I$ which includes constants, i.e., multiples of u, but which is usually rather small when compared with all of $\mathbb{R}^I$. By throwing in functions which depend on pairs of loci we get a somewhat larger subspace $A^{(2)}$. Continuing we get a chain of subspaces

$$A^{(0)} \subset A^{(1)} \subset A^{(2)} \subset \ ... \ \subset A^{(L)} = \mathbb{R}^I,$$

where $A^{(0)}$ is defined to be the constants, multiples of u. Now given ξ we can project just to $A^{(0)}$, getting $\bar{\xi}u$ by (1.26). Project the remainder $\xi - \bar{\xi}u$ to $A^{(1)}$ to get a vector β^1 which has mean 0 and is additive between loci. Continuing we can write

$$(1.30) \qquad\qquad \xi = \bar{\xi}u + \beta^1 + \beta^2 + ... + \beta^L,$$

where the terms are mutually orthogonal and, for example, $\bar{\xi}u + \beta^1 + \beta^2$ represents the best approximation of ξ by a vector with only pairwise interactions at most. Here "best" means in the least squares sense weighted by p, in other words, the variance of the difference $\xi - \eta$ achieves its minimum among all vectors η in $A^{(2)}$ by the choice of $\eta = \bar{\xi}u + \beta^1 + \beta^2$

The explicit formulae for these projections can be quite nasty especially when there is dependence between loci. However, the projection description of these approximating vectors can often be used instead of a formula. In any case this description interprets in a nice way what these approximations mean and shows why, with its natural meaning from linear algebra, it is the least squares kind of approximation which is so convenient to use.

2.   Calculus and Riemannian manifolds

Calculus, properly interpreted, is the attempt to apply linear algebra methods to situations which are not linear. The fundamental technique, *linearization*, consists of two steps. First, you replace your given, bent function by a nice, flat one which approximates the original and you analyze the linear system. This is the easy step. The hard step is to show how much of your analysis applies to the nonlinar system you began with.

Understanding of this philosophy, which I adopt from Richard Palais' "Foundations of Nonlinear Analysis", is made difficult by the view of calculus we all began with: Given a function $y = f(x)$, the derivative at a point is a number $f'(x)$, and so by varying x we get a new function of the same sort as $f : y = f'(x)$. For example, the derivative of $\sin x$ is $\cos x$. While all correct, this is misleading, and the proper picture is only revealed when we look at vector functions of a vector variable.

Suppose U is an *open subset* of a Euclidean vector space V_1. This means that whenever x lies in U, then all points sufficiently close to x are also in U, i.e., there exists $\varepsilon > 0$, depending on x, such that $\|h\| < \varepsilon$ implies $x + h$ is in U. Thus, we can move from x in any direction a small distance and not leave U. Let f be a function defined on U with values in a Euclidean vector space $V_2, f : U \to V_2$. The derivative of f at a point $x \in U$ is a linear map written $d_x f : V_1 \to V_2$. It is the unique linear map such that the function $f(x) + d_x f(h)$ (with x fixed and h varying) gives the best approximation to f near x, that is, to $f(x+h)$. Here the notion of approximation has nothing to do with least squares. Instead we look at the error term $f(x+h) - f(x) - d_x f(h)$. Provided that f is continuous this expression will approach 0 in size when the length of h gets small (and so $x + h$ approaches x). In fact, this is true even without the linear map term $d_x f(h)$. So we demand that the size of the error tend to 0 faster than the length of h. The derivative is uniquely determined by the demand that the ratio between the error and $\|h\|$ tend to 0. We write

(2.1) $$f(x+h) \;=\; f(x) + d_x f(h) + o(h)$$

where the error term, denoted here $o(h)$, is defined for $\|h\|$ sufficiently small and satisfies:

(2.2) $$\text{limit } \frac{\|o(h)\|}{\|h\|} = 0 \quad \text{as } \|h\| \text{ tends to } 0.$$

Notice that lengths in the numerator and denominator are taken in V_2 and V_1, respectively.

The derivative of a function need not exist. For example, $f(x) = x^{\frac{1}{3}}$ from $\mathbb{R}$ to $\mathbb{R}$ is not differentiable at $x = 0$, nor is the absolute value function. But the functions we will encounter are well-behaved and we will assume that they are all differentiable unless otherwise mentioned. The real problem is: what does this linear map $d_x f$ mean? The following result interprets $d_x f(v)$ as the directional derivative of f in the direction v.

2.1 Lemma. *With $x \in U$ and $v \in V_1$ fixed let s vary through positive real values toward zero. Then*

(2.3) $$\lim_{s \to 0} \frac{f(x+sv)-f(x)}{s} = d_x f(v).$$

Proof. Both sides are zero when $v = 0$ so we will suppose v is not 0 and so has positive length. We substitute $h = sv$ and note that $\|h\| = s\|v\|$. Because v is fixed, $\|h\|$ tends to 0 precisely as s does. Because of (2.2) (actually multiplied by the constant $\|v\|$) we get

$$\lim_{s \to 0} \frac{f(x+sv)-f(x)}{s} = \lim_{s \to 0} \frac{d_x f(sv)}{s}.$$

But $d_x f$ is a linear map. So on the right the s pulls out and cancels, yielding $d_x f(v)$ for all s. $\qquad\square$

In the special case when $V_1 = \mathbb{R}$ and $V_2 = V$, f is a vector-valued function of a real variable t and $d_t f$ is a linear map from $\mathbb{R}$ to V. We saw in section 1 that such a linear map could be completely described by its value on $v = 1$ because $d_t f(s) = sd_t f(1)$ for all s in $\mathbb{R}$. So if we denote this vector by $f'(t)$, Lemma 2.1 says that, in this case,

$$(2.4) \qquad \underset{s \to 0}{\text{limit}} \ \frac{f(t+s)-f(t)}{s} = f'(t) .$$

If you imagine the variable t as standing for time, then the function f describes the motion of a particle along a path in V with $f(t)$ its position at time t. $f'(t)$ is the velocity vector at time t for this motion.

Now back in the general case of Lemma 2.1 we start at x and move along a line away from x in the v direction. The function f converts this to motion along a curved path in V_2 and the directional derivative $d_x f(v)$ is the initial velocity vector of this motion.

When $V_1 = \mathbb{R}^n$ and $V_2 = \mathbb{R}^m$ then f consists of m components

$$f(x) = (f_1(x),...,f_m(x))$$

with $x = (x_1,...,x_n)$. Varying along the standard basis vector δ^j in $\mathbb{R}^n$ (see (1.8)) amounts to fixing all the components x_i except for the jth and varying x_j alone. In other words, $d_x f(\delta^j)$ represents partial differentiation in the x_j direction. Any linear map like $d_x f : \mathbb{R}^n \to \mathbb{R}^m$ is given by an $m \times n$ matrix with respect to the standard bases (see (1.5)). This is the Jacobian matrix with jk entry given by $\partial f_k / \partial x_j$ evaluated at x. In particular, when $m = 1$ so that f is a real-valued function of a vector variable x in $\mathbb{R}^n$, the matrix is $n \times 1$ and we have the formula:

$$(2.5) \qquad d_x f(v_1,...,v_n) = \sum_{j=1}^{n} \frac{\partial f}{\partial x_j} v_j$$

with the partial derivatives evaluated at x.

Taking the derivative itself is a linear operation. If $f,g : U \to V_2$ and $t \in \mathbb{R}$, then

$$d_x(tf+g) = t(d_x f) + (d_x g) .$$

So in the standard case the Jacobian matrix of the sum of two functions is the sum of the correspondings Jacobian matrices. We will also need the *chain rule* which says that the derivative of a composite map is the composite of the derivatives. If $f : V_1 \to V_2$ and $g : V_2 \to V_3$ then the composite function is $g \circ f : V_2 \to V_3$ defined by $g \circ f(x) = g(f(x))$ for $x \in V_1$. Now for x in V_1 we can take the derivatives of f and $g \circ f$ at x and the derivative of g at $f(x)$. We get linear maps $d_x f : V_1 \to V_2$, $d_{f(x)} g : V_2 \to V_3$ and $d_x(g \circ f) : V_1 \to V_3$. The chain rule says:

$$(2.6) \qquad d_x(g \circ f) = (d_{f(x)} g) \circ (d_x f) .$$

In the standard case this implies that the Jacobian of the composite $g \circ f$ is the product of the Jacobians of g and of f.

In section 1 we showed that any linear map $T : \mathbb{R} \to V$ could be completely described by the single vector $T(1)$. We used this above to define the velocity vector $f'(t)$ when $V_1 = \mathbb{R}$ and $V_2 = V$.

On the other hand, when $V_1 = V$ and $V_2 = \mathbb{R}$, $d_x f$ is a linear form on V called the *differential* of f at x. If $f : U \to \mathbb{R}$ then the differential $df : U \to V^*$ associates to x the form $d_x f$. Now if we use the Euclidean metric on V, the Riesz representation theorem (Theorem 1.1) associates to $d_x f$ a vector in V. This is the gradient of f at x denoted $\mathrm{grad}_x f$. It is defined by:

$$(2.7) \qquad\qquad d_x f(v) = (\mathrm{grad}_x f, v) \,.$$

The gradient depends on the particular Euclidean metric on V. Up to now we have only needed the metric to make the limit statements like (2.2) make sense. But any Euclidean metric will give the same idea of limit, the same topology, on V. So the derivatives like $d_x f$ are independent of the choice of metric. This is not true of the gradient and we will later see different kinds of gradients.

Now suppose that the vector v has unit length $(\|v\| = 1)$ and use formula (1.12) to compute the directional derivative given by (2.7):

$$(2.8) \qquad\qquad d_x f(v) = \|\mathrm{grad}_x f\| \cos\theta \qquad (\text{when } \|v\| = 1) \,,$$

where θ is the angle between v and the gradient vector. Clearly this value is greatest when $\cos\theta = 1$, i.e., $\theta = 0$. So the gradient points in the direction of greatest increase of f.

In the special case when $V = \mathbb{R}^m$ and the inner product on V is the usual one given by (1.11), then formulae (2.7) and (2.5) say that the components of the gradient of f at x, which is a vector in $\mathbb{R}^m$, are given by the partial derivatives $\partial f / \partial x_j$ $(j = 1,...,m)$. This is the way the gradient is usually defined and it is important to notice that if assumes the usual inner product.

We can now be a bit more explicit about the idea of linearization with which we introduced this section. The derivative $d_x f$ as well as the higher derivatives computed at x (about which I am going to remain vague) constitute the *infinitesimal* information about f at x. The word is intended to indicate that such information is obtained from various limits as points approach x. In particular, if we change f far away from x it has no effect on these limits. Looked at the other way, this means that the infinitesimal information at x tells us nothing about how the funciton behaves far away from x. What we can hope to do is to get from the infinitesimal information a *local* description, indications of how f actually behaves in some small open set around x. This is what I called the hard, second step in the linearization program. The prototype and most important example of this kind of result is the *inverse function theorem*.

Suppose U_1 is open in V_1 and $f : U_1 \to V_2$. f is called *smooth* if it is differentiable and all those higher derivatives exist as well. A smooth map is called a *diffeomorphism* if it has a smooth inverse map, that is, if f maps U_1 one-to-one and onto an open subset U_2 in V_2, and the inverse map $f^{-1} : U_2 \to V_1$ is smooth as well.

When f is a diffeomorphism and $x \in U_1$, then the chain-rule (2.6) implies that the linear map $d_x f$ is a linear isomorphism and that its inverse is the derivative of f^{-1} taken at $f(x)$, symbolically:

$$(2.9) \qquad\qquad (d_x f)^{-1} = d_{f(x)}(f^{-1}) .$$

Thus, if f is invertible so it its derivative at each point. The inverse function theorem is the converse, at least locally. It says that if the derivative at a point   x   is a linear isomorphism, then f is a diffeomorphism on some little open set containing x . We state the result carefully but will omit the proof.

2.2 Theorem. *Let* $f : U \to V_2$ *be a smooth map with* U *open in* V_1 *and let* $x \in U$. *If the   derivative* $d_x f : V_1 \to V_2$ *is a linear isomorphism, then*   f   *is locally a diffeomorphism near* x , *i.e., there exists an open set* $U_1 \subset U$ *with* $x \in U_1$ *and f restricted to* U_1 *is a diffeomorphism.*   □

* * *

If the set I has n elements and k is a whole number with $k \le n$, then a k-dimensional manifold in the vector space $\mathbb{R}^I$ is a subset M of $\mathbb{R}^I$ which looks locally, near each point, like a curved piece of a k-dimensional subspace. There are two equivalent ways of making this precise.

First, we can define M near $x \in M$ explicitly by giving a coordinate system on M near x . This is a function $h : U \to M$, where U is open in $\mathbb{R}^k$ and h maps U one-to-one and onto all of the points of M near x (i.e., the intersection of M with some open set in $\mathbb{R}^I$). h is assumed to have rank k . This means that if we regard h as a function from U to $\mathbb{R}^I$ the derivative $d_u h : \mathbb{R}^k \to \mathbb{R}^I$ is one-to-one at every point u of U . This description is called *explicit* because it parametrizes the points of M near x by k real parameters For example, the piece of the circle of radius 1 in the interior of the first quadrant of $\mathbb{R}^2$ is the image of the function $f(t) = (\cos t, \sin t)$ with t varying in the open interval of $\mathbb{R}$ between 0 and $\pi/2$. Similar pieces can be constructed near any point of the circle. This example illustrates the typical fact that often no coordinate system can be found which works on the entire manifold. The manifold is obtained by gluing together many coordinate patches.

The *implicit* description of the manifold near x is as the level surface of a family of functions. This means we have a function $F : G \to \mathbb{R}^{n-k}$, with G some open subset of $\mathbb{R}^I$ containing x , such that the points of the manifolds in G are precisely the solutions of the equations $F(y) = \xi$ for some fixed vector ξ in $\mathbb{R}^{n-k}$, i.e., $M \cap G = F^{-1}(\{\xi\})$. F is assumed to have rank n - k at all points of $M \cap G$. This means that the derivative $d_y F : \mathbb{R}^I \to \mathbb{R}^{n-k}$ is onto for every point y in $M \cap G$. We can think of F as a list of n - k scalar functions and the equation $F(y) = \xi$ as a list of n - k constraints, which reduce the number of degrees of freedom (= dimension) from n to k . Frequently an implicit description can be given for the entire manifold.   For example, the (n-1)-dimensional sphere of radius r in $\mathbb{R}^I$ is given by the single scalar equation

$F(y) = r^2$ where $F(y) = \Sigma\, y_i^2$. The subset $\mathring{\Delta}$ of $\mathbb{R}^I$ is defined by the equation $F(y) = 1$ where $F(y) = \Sigma\, y_i$. Here the open set G consists of the set of vectors with positive coordinates.

Just as the derivative at a point of a function is a linear approximation to the function, there is at every point of a manifold a linear subspace which approximates the manifold.

A path through x in M is a function v from an open interval in $\mathbb{R}$ to M such that $v(t) = x$ for some t in the interval. Taking the derivative at t we get the vector $v'(t)$ which is called a tangent vector at x . The collection of all tangent vectors at x is a linear subspace of $\mathbb{R}^I$, called the *tangent space* of M at x and denoted T_xM . It is not clear from this definition that T_xM is a subspace, but T_xM can also be defined using the explicit or implicit description of M near x . If $h : U \rightarrow M$ with U open in $\mathbb{R}^k$ is a coordinate system near x , then every path in M through x can be described using these coordinates. It then follows from the chain rule (2.6) that T_xM is the image of the linear map d_xh . Since d_xh is one-to-one, T_xM is a k-dimensional subspace of $\mathbb{R}^I$. On the other hand, if $F : G \rightarrow \mathbb{R}^{n-k}$ with $M \cap G = F^{-1}(\{\xi\})$, then every path in M maps under F to a constant path in $\mathbb{R}^{n-k}$. Since constants have derivative 0 , T_xM is the kernel of the map $d_xF : \mathbb{R}^I \rightarrow \mathbb{R}^{n-k}$, i.e.,

$$T_xM = \{y \in \mathbb{R}^I : d_xF(y) = 0\}\ .$$

For example, if $F(y) = \Sigma_i\, y_i$, then $F : \mathbb{R}^I \rightarrow \mathbb{R}$ is already a linear mapping so for any x the best linear approximation d_xF is F itself, in other words, $d_xF(y) = \Sigma_i\, y_i$ independent of x . So for $p \in \mathring{\Delta}$ the tangent space $T_p\mathring{\Delta}$ is given by

(2.10) $$T_p\mathring{\Delta} = \mathbb{R}_0^I \equiv \{y \in \mathbb{R}^I : \Sigma_i\, y_i = 0\}\ .$$

On the other hand, if $F(y) = \Sigma_i\, y_i^2 = {}_o(y,y)$, then $d_xF(y) = 2\Sigma_i\, x_iy_i = 2\, {}_o(x,y)$ where ${}_o(\,,\,)$ is the usual inner product. So for the sphere of radius r , the tangent space at x consists of all vectors orthogonal to x . Notice that $T_p\mathring{\Delta}$ is the same subspace for all p (it is the hyperplane through the origin parallel to the one containing Δ), but the tangent space of the sphere at x changes as x does. In fact at x it consists of the subspace of all vectors perpendicular to x .

If M_1 is a manifold in $\mathbb{R}^{I_1}$, M_2 is a manifold in $\mathbb{R}^{I_2}$ and f is a function from M_1 to M_2 , we can extend the definition of f to a function from U to $\mathbb{R}^{I_2}$, where U is some open set in $\mathbb{R}^{I_1}$ containing M_1 . Then for x in M_1 we can define the derivative $d_xf : \mathbb{R}^{I_1} \rightarrow \mathbb{R}^{I_2}$. There are many different ways of extending f and d_xf will depend on which extension is used. However d_xf maps T_xM_1 into $T_{f(x)}M_2$ and this part of d_xf does not depend on the choice of extension, so we can define the linear map $d_xf : T_xM_1 \rightarrow T_{f(x)}M_2$ without ambiguity. The reason is that if v is a path in M_1 through x , then the composition $f \circ v$ is path in M_2 through $f(x)$, and $(f \circ v)'(t) = d_xf(v'(t))$ by the chain rule. But $f \circ v$ depends only on the original function f on M_1 . This allows us to do calculus on manifolds. For example, if d_xf is a linear isomorphism of T_xM_1 onto $T_{f(x)}M_2$, then one can extend the inverse function theorem to show that f is a diffeomorphism between some open set of M_1 containing x and some open set of M_2 containing f(x) .

In particular, if $\mathbb{R} = M_2 = \mathbb{R}^{I_2}$, then the differential, df , of the real-valued function $f : M_1 \rightarrow \mathbb{R}$ associates to each point x in M_1 a linear function $d_x f : T_x M_1 \rightarrow \mathbb{R}$, that is, $d_x f$ is a member of the dual space $(T_x M_1)^*$. So df is a function associating to each x a member of $(T_x M_1)^*$.

Dual to this view of the differential of a function is the idea of a vector field. A vector field X on an open set U of $\mathbb{R}^I$ is just a function $X : U \rightarrow \mathbb{R}^I$. A vector field on a manifold M in $\mathbb{R}^I$ is a function $X : M \rightarrow \mathbb{R}^I$ such that $X(x) \in T_x M$ for all $x \in M$, i.e. X is always tangent to M . Via the Kronecker product, see (1.6), we can associate to a function $f : M \rightarrow \mathbb{R}$ and a vector field X on M a new function <df,X> defined by:

(2.11) $<df,X> (x) = <d_x f, X(x)> = d_x f(X(x)) , \quad x \in M .$

With f fixed we can regard (2.11) as a way that functions operate on vector fields to get new functions, or with X fixed we can regard (2.11) as the way a vector field operates on functions.

Recall now from (1.8) the standard basis $\{\delta^i\}$ on $\mathbb{R}^I$. We can regard each δ^i as a constant vector field on $\mathbb{R}^I$. That is, associate to each point x of $\mathbb{R}^I$ the vector δ^i thought of as based at x . Recall that $d_x f(\delta^i)$, the directional derivative of f in the δ^i direction, is just the partial $\partial f / \partial x_i$ evaluated at x . In our new Kronecker product notation, (2.11), this says

(2.12) $<df, \delta^i> = \dfrac{\partial f}{\partial x_i} \quad \text{at each } x \in \mathbb{R}^I .$

Because $\{\delta^i\}$ is a basis every vector field X on M can be written as a linear combination $X = \sum X_i \partial_i$ where the coordinates X_i are real-valued functions of x in M . Notice that the δ^i 's themselves are usually not tangent to M , they do not lie in the subspace $T_x M$, and so not every choice of coordinate functions X_i will define a vector field on M . For example, when $M = \mathring{\Delta}$, a particularly easy case, the coordinate functions must satisfy $\sum_i X_i = 0$ to define a vector field on $\mathring{\Delta}$ (see (2.10)).

In any case, the Kronecker product is bilinear and so from (2.12) we get the formula

(2.13) $<df,X> = \sum_i X_i <df, \delta^i> = \sum_i \dfrac{\partial f}{\partial x^i} X_i .$

Notice that with x fixed this formula is (2.5). Note also that as the δ^i 's are not tangent to M , the partial derivatives $\partial f / \partial x_i$ will depend not merely on the original real-valued function f defined on M but on its extension to an open set containing M . However, because X is tangent to M , the sum <df,X> does not depend on this choice of extension.

A vector field on M is the manifold analogue of a differential equation. A solution path for the vector field is a path v(t) in M such that for all t :

(2.14)
$$\frac{dv}{dt} = v'(t) = X(v(t)) .$$

In local coordinates it is easy to check that this is just an ordinary differential equation in the k coordinates.

* * *

An important example of a vector field, a gradient field, requires the notion of a Riemannian metric.

Any inner product on $\mathbb{R}^I$ restricts to an inner product on each subspace and in particular on the tangent spaces of a manifold M . However, in many applications the inner product which arises naturally from the problem will be different at different points. A *Riemannian metric* on a manifold M is a smooth choice of inner product $(,)_x$ for each subspace T_xM . A manifold equipped with a Riemannian metric is called a Riemannian manifold.

On a Riemannian manifold M we define the gradient ∇f of a function $f : M \to \mathbb{R}$. d_xf is a linear form on T_xM and so by Theorem 1.1 there exists a unique vector $\nabla_xf \in T_xM$ such that:

(2.15) $(\nabla_xf, X)_x = <d_xf, X> = d_xf(X)$ $x \in M$, $X \in T_xM$.

The vector field ∇f thus depends not only on $f : M \to \mathbb{R}$ but also on the Riemannian metric. This is in contrast to df which depends only on f .

If M_1 is a submanifold of M , i.e., another manifold in $\mathbb{R}^I$ with $M_1 \subset M$, then for $x \in M_1$ the tangent space T_xM_1 is a subspace of T_xM since every path in M_1 lies in M . So a Riemannian metric on M restricts to define one on M_1 . If $f : M \to \mathbb{R}$, then f restricts to a function $f|M_1 : M_1 \to \mathbb{R}$. Now for $x \in M_1$ the two gradients ∇_xf and $\nabla_x(f|M_1)$ both satisfy (2.15) for vectors $X \in T_xM_1$. In addition, $\nabla_x(f|M_1)$ itself lies in T_xM_1 . So by Theorem 1.2, $\nabla_x(f|M_1)$ is the perpendicular projection of ∇_xf into T_xM_1 . For a connected submanifold, i.e., every pair of points in M_1 can be joined by a path in M_1 , this implies the following:

2.3 Proposition. *Let* M_1 *be a connected submanifold of a manifold* M *and let* f : $M \to \mathbb{R}$. *The following conditions are equivalent:*
  (i)   f *is constant on* M_1 ;
  (ii) $d_xf(X) = 0$ *for all* $X \in T_xM_1$, $x \in M_1$;
  (iii) $\nabla_x(f|M_1) = 0$ *for all* $x \in M_1$;
  (iv) ∇_xf *in* T_xM *is orthogonal to the subspace* T_xM_1 *for all* $x \in M_1$. $\square$

On a Riemannian manifold the arc length of a path is defined. Thinking of t as time the velocity vector at time t of the path v(t) in M is $v'(t) \in T_{v(t)}M$. The speed is the length of the velocity measured using the inner product at v(t) . So the speed is

$$\| v'(t) \| \; = \; (v'(t), \, v'(t))_{v(t)}^{\frac{1}{2}} \, .$$

Integrating the speed we get the length of the path. This enables us to define the distance between two points x_1 and x_2 of M as the greatest lower bound of the lengths of all paths in M connecting x_1 and x_2 . A path which achieves this length –the shortest distance between the two points– is called a *geodesic*. For example, on the sphere with the Riemannian metric obtained by restricting the usual inner product on $\mathbb{R}^I$, geodesics are pieces of great circles, i.e., the intersection of the sphere with two-dimensional subspaces (planes through zero).

A diffeomorphism g between two Riemannian manifolds M_1 and M_2 is called an *isometry* if $d_x g : T_x M_1 \to T_{g(x)} M_2$ is an isometry at every point x of M_1 . While diffeomorphisms between pieces of manifolds of the same dimension are quite common, isometries are quite rare. This is related to the rigidity of geometry as opposed to the flexibility of topology ("rubber-sheet geometry") . Isometries preserve the structures of Riemannian geometry. They map geodesics to geodesics and relate gradient vectorfields, i.e., $d_x g(\nabla_x(f \circ g)) = \nabla_{g(x)} f$ if g is an isometry but not, in general, if it is not.

3.   The Shahshahani metric

Suppose a population is subdivided into different types indexed by a finite set I . Let x_i denote the size of the subpopulation of type i so that the population state is described by a vector in $\mathbb{R}_+^I$, the set of nonnegative vectors in $\mathbb{R}^I$ (see (1.21)). The total population size, $|x|$, and the distribution vector, p , of relative frequencies are defined by:

(3.1)
$$|x| \; = \; \sum_i x_i$$

$$p_i \; = \; x_i/|x|$$

so that p lies in the simplex Δ of all probability distribution vectors. All types i occur in the population when all x_i's are strictly positive or equivalently when $p \in \mathring{\Delta}$ (see (1.22) and (1.23)).

We assume that the change of state of the population through time is given by a differential equation, defined by a vector field X on $\mathbb{R}_+^I$. That is, to each point x of $\mathbb{R}_+^I$ is attached a vector $X(x)$ and so each component X_i of X is a real-valued function of the position vector x . The corresponding differential equation, (2.14), written in component form is:

(3.2)
$$\frac{dx_i}{dt} \; = \; X_i \; = \; x_i \xi_i \qquad i \in I .$$

X_i is called the *absolute growth rate* of x_i. If x_i is measured in grams, then the units of X_i are grams per unit time. ξ_i, defined as the ratio X_i/x_i, is the *relative growth rate* with units inverse time. In many biological and economic contexts it is the relative, rather than the absolute, growth rates which are the logical focus of interest. The interest rate on money is a relative growth rate. Concepts such as the half-life of an isotope or the doubling time of a broth of bacteria assume a constant relative growth rate.

Notice that constant absolute rates, constant X_i's, lead to linear growth while constant relative rates, constant ξ_i's, lead to exponential growth. The latter because we can rewrite (3.2) as

$$(3.3) \qquad \frac{d \ln x_i}{dt} = \frac{1}{x_i} \frac{dx_i}{dt} = \xi_i,$$

where $\ln x_i$ is the natural log of x_i.

We will not require that these rates be constant, assuming instead that each rate is a function of the entire state vector x. However, there is still considerable simplification in these models. First, the use of continuous variables and a deterministic system requires that population sizes be large. Next, by assuming that the rates depend on the current state vector of the population, we ignore all other possible effects, e.g. environmental effects and the past history of the population. In particular, our equations are autonomous, time-independent. Finally, while we can always define the relative rates by dividing, when the x_i's are positive, using both forms of (3.2) all the way to the boundary assumes that when $x_i = 0$ its growth rate, $X_i = 0$. This means that the different types breed true and when type i is absent from the population initially, it never appears. Consequently, we anticipate difficulties handling processes like mutation and recombination which may introduce initially absent types.

By summing (3.2) on i we get the equation for the entire population:

$$(3.4) \qquad \frac{d|x|}{dt} = \sum_i X_i = |x|\bar{\xi},$$

where

$$\bar{\xi} = \sum_i p_i \xi_i.$$

To check this note that $x_i = |x|p_i$ for each i.

Hence, the relative growth rate of $|x|$, which is $d \ln|x|/dt$, is $\bar{\xi}$, the mean growth rate, i.e., the average weighted by the relative frequency of the different population types. Now observe that

$$\frac{d \ln p_i}{dt} = \frac{d \ln x_i}{dt} - \frac{d \ln|x|}{dt} = \xi_i - \bar{\xi}$$

and so the equations for the relative frequencies are

$$(3.5) \qquad \frac{dp_i}{dt} = p_i(\xi_i - \bar{\xi}) \ .$$

Our major example will be the dynamic for evolutionary games described by Taylor and Jonker. Let I list the different strategy types and a_{ij} be the payoff in fitness to an i player when he meets a j player. We assume that each individual's strategy type is a fixed characteristic which breeds true. The relative growth rate of the subpopulation of an i strategist is

$$(3.6) \qquad a_{ip} \equiv \sum_j a_{ij} p_j$$

because he meets a j player with probability p_j . Thus, the equations (3.2), (3.4) and (3.5) become:

$$(3.7) \qquad \frac{dx_i}{dt} = x_i a_{ip} \ , \quad \frac{d|x|}{dt} = |x| a_{pp}$$
$$\frac{dp_i}{dt} = p_i(a_{ip} - a_{pp})$$

where $a_{pp} \equiv \sum_i p_i a_{ip} = \sum_{i,j} p_i p_j a_{ij}$ is the mean fitness of the entire population. This is the so-called *pure strategy dynamic* because p represents a polymorphism, a distribution of different type players each of which plays a single pure strategy. We will consider mixed strategists later.

It is an odd coincidence that these equations are the same as Fisher's equation for natural selection in a diploid population. The set I then lists the different gamete genotypes so that p describes the relative frequencies of the different genotypes in the gametic gene pool. a_{ij} , usually written m_{ij} in this context, is the fitness of the diploid genotype ij . Here random mating and Hardy-Weinberg proportions are assumed, or I should say, are imposed upon the equations. Thus, a_{ip} is the average fitness of type i because it is paired with a j type with probability p_j . Similarly, a_{pp} is the mean fitness of the population. The equations derived for selection in continuous time with overlapping generations are exactly those of (3.7) provided the effects of recombination and mutation are omitted There is, however, an assumption in the genetic interpretation which is absent form the game versions which makes the former a special case of the latter. In a game the payoff to the i player, a_{ij} , is certainly not assumed to be the same as the payoff to the j player, a_{ji} . In the zero sum case they are assumed negatives of one another so that the payoff matrix is antisymmetric. But in the genetic case the zygote genotype is an unordered pair because we do not distinguish the parentage of the two gametes. Consequently, in the selection equations the fitness matrix is symmetric:

$$(3.8) \qquad a_{ij} = a_{ji} \qquad \text{for } i,j \in I \ .$$

We will soon see the big difference this makes.

* * *

On the interior set of position vectors:

$$(3.9) \qquad \mathbb{R}_+^I = \{x \in \mathbb{R}^I : x_i > 0 \text{ for all } i\},$$

we will study these systems of equations by using a particular Riemannian metric introduced by Shahshahani and Conley

$$(3.10) \qquad (Y^1, Y^2)_x = \sum_i \frac{1}{x_i} Y_i^1 Y_i^2.$$

Here Y^1 and Y^2 are vectors of $\mathbb{R}^I$, thought of as tangent vectors attached to the point x of $\mathbb{R}^I$. Remember that in contrast with a Euclidean metric on a vector space, a Riemannian metric on a manifold may vary from point to point. This dependence upon the attachment point x appears in formula on the right in (3.10) and we indicate this dependence by the subscript on the left. Formula (3.10) is related to an information measure of R.A. Fisher, and essentially the same inner product arose from the diffusion results of Antonelli. Shahshahani's interpretation of the formula as the definition of a Riemannian metric leads, as we will see, to a nice geometrical theory. It is distinctive enough that his name seems an appropriate label.

Notice that we can rewrite the formula in terms of relative rates. Thus, when $Y_i^1 = x_i \eta_i^1$ and $Y_i^2 = x_i \eta_i^2$ with the same x_i's because of the common attachment point, we have

$$(3.11) \qquad (Y^1, Y^2)_x = \sum \frac{1}{x_i} Y_i^1 Y_i^2 = \sum x_i \eta_i^1 \eta_i^2.$$

In the case where $x = p \in \mathring{\Delta}$ the latter formula is the covariance metric $_p(\eta^1, \eta^2)$ of (1.25). By way of contrast, recall the formula for the usual Euclidean metric $_o(\eta^1, \eta^2) = \sum \eta_i^1 \eta_i^2$, (1.11).

Notice that for $p \in \mathring{\Delta}$ we have

$$(3.12) \qquad (p, p)_p = \sum \frac{1}{p_i} (p_i)^2 = \sum p_i = 1$$

and so the position vector out to a distribution is a unit vector regarded as attached to itself. Similarly,

$$(3.13) \qquad (X, p)_p = \sum_i X_i.$$

So, in particular, a vector X at p is perpendicular to p if and only if $\sum_i X_i = 0$, i.e., $X \in \mathbb{R}_o$. Now remember that $\mathbb{R}_o$ is the space of vectors tangent to $\mathring{\Delta}$ at p for any p in $\mathring{\Delta}$ (cf. (2.10)). This was because as we move in $\mathring{\Delta}$ the total weight $\sum_i p_i$ remains constantly 1 and so the change in this total, $\sum X_i$, is 0. Thus, with respect to the

Shahshahani metric the simplex $\overset{\bullet}{\Delta}$ has the property that the vector out to a point of $\overset{\bullet}{\Delta}$ is the unit vector perpendicular to $\overset{\bullet}{\Delta}$ at that point. With respect to the usual metric $_o(\,,\,)$ it is the sphere which has this property while the vectors perpendicular to $\mathbb{R}_o$ are the multiples of the constant vector, u (see (1.24)). This analogy between simplex and sphere is no accident. If we define the map

$$(3.14) \qquad\qquad F(x)_i = 2\sqrt{x_i} \quad , \quad i \in I$$

it is not hard to show that F is an isometry of $\overset{\bullet}{\mathbb{R}}{}_+^I$ equipped with the Shahshahani metric onto $\overset{\bullet}{\mathbb{R}}{}_+^I$ with the usual metric $_o(\,,\,)$. This maps pops the simplex $\overset{\bullet}{\Delta}$ out to the $\overset{\bullet}{\mathbb{R}}{}_+^I$ piece of the sphere with radius 2 . Using this sphere trick it is possible to change coordinates so as to avoid the Shahshahani metric. I will now abandon this approach because it is my intention to show you how natural the Shahshahani geometry itself is. I will merely assert that for problems, the change of variables given by (3.14) makes things harder.

Given X any vector $\mathbb{R}^I$ attached to $p \in \overset{\bullet}{\Delta}$ we can use fomula (1.20) to define the perpendicular projection of X onto $\mathbb{R}_o^I$:

$$P_p : \mathbb{R}^I \to \mathbb{R}_o^I$$

(3.15)

$$P_p(x) = X - (X,p)_p p \ .$$

The projection depends on the attachment point because the inner product does. In coordinates the formula is:

$$P_p(X)_i = X_i - (\Sigma_j X_j)p_i$$

(3.16)

$$= p_i(\xi_i - \bar{\xi}) \ ,$$

where $X_i = p_i\xi_i$ at p and so $\Sigma X_i = \Sigma p_i\xi_i \equiv \bar{\xi}$. If one used the usual metric, then the projection to $\mathbb{R}_o^I$ would have coordinates $X_i - (\Sigma_j X_j)n^{-1}$, where n is the number of elements in I .

The coincidence between formulae (3.16) and (3.5), arrived at in completely different ways, is our first indication of the naturalness of the Shahshahani metric. For another we turn to the subject of gradients.

$$* \quad * \quad *$$

Let U be a smooth real-valued function on $\mathbb{R}_+^n$. Recall the formula for the directional derivative of U in the X direction where X is a vector attached at x . Equivalently, this is the differential of U applied to X (see (2.5) and (2.13)):

(3.17)
$$d_x U(X) = \sum_i \frac{\partial U}{\partial x_i} X_i .$$

In particular, if $x(t)$ is a path in $\mathbb{R}^I_+$ with velocity vector $x'(t) = dx/dt$, then the derivative of U along the path (the ordinary derivative of the real-valued function $U(x(t))$) is given by the chain rule formula:

(3.18)
$$\frac{dU}{dt} = \sum_i \frac{\partial U}{\partial x_i} \frac{dx_i}{dt} = d_{x(t)} U(\frac{dx}{dt}) .$$

Now with respect to the usual inner product $_0(\ ,)$ on $\mathbb{R}^I$, the gradient vector field of U, denoted grad U, is given by:

(3.19)
$$(grad_x U)_i = \frac{\partial U}{\partial x_i} \quad \text{evaluated at } x .$$

For then we get from (3.17) the formula which truly defines the gradient (see (2.15) and also (2.7)):

(3.20)
$$_0(grad_x U, X) = d_x U(X) \quad \text{for all } X \text{ in } \mathbb{R}^I .$$

Our original dynamical system (3.2) is called a *gradient dynamical system* with *potential function* U if the vector field X is the gradient, grad U. So in vector notation (3.2) or (2.14) becomes in that case:

$$\frac{dx}{dt} = grad_x U , \quad \text{i.e.,}$$

(3.21)

$$\frac{dx_i}{dt} = \frac{\partial U}{\partial x_i} .$$

We can compute now dU/dt, the rate of change of U as we move along a solution path for the differential equation. We apply (3.18) and substitute from (3.21):

$$\frac{dU}{dt} = {}_0(grad_x U, \frac{dx}{dt}) = {}_0(grad_x U, grad_x U)$$

(3.22)

$$= {}_0\| grad_x U \|^2 \geq 0 \quad \text{at } x = x(t) ,$$

with equality only when the gradient vanishes at x, i.e., when x is an equilibrium point for (3.21). Thus, U is always strictly increasing along the solution path of (3.21), excluding only the equilibrium point case when U remains constant.

Because the concept of gradient depends upon inner product we can do all this, at least on interior points of $\mathbb{R}_+^I$, using the Shahshahani metric instead. The analogue of (3.19) defining the Shahshahani gradient ∇U , is:

(3.23) $$(\nabla_x U)_i = x_i \frac{\partial U}{\partial x_i} \quad \text{evaluated at } x .$$

For we then have for $X \in \mathbb{R}^I$:

(3.24) $$(\nabla_x U, X)_x = \sum_i (x_i)^{-1}(x_i \frac{\partial U}{\partial x_i})X_i = d_x U(X)$$

(cf. (3.18) again).

(3.2) is a gradient dynamical system – potential function = U – with respect to the Shahshahani metric if $X = \nabla U$, i.e., if (3.21) is replaced by

$$\frac{dx}{dt} = \nabla_x U , \quad \text{i.e.,}$$

(3.24)

$$\frac{dx_i}{dt} = x_i \frac{\partial U}{\partial x_i} ,$$

The analogue of (3.22) is:

$$\frac{dU}{dt} = (\nabla_x U, \frac{dx}{dt})_x = \| \nabla_x U \|_x^2$$

(3.25)

$$= \sum_i x_i (\frac{\partial U}{\partial x_i})^2 \geq 0$$

with equality only at equilibrium. So U is increasing on non-equilibrium solution paths of (3.24) as well.

As the formal manipulations are the same the question arises: why –or better, when– should this fancy differential geometry be introduced to replace the usual Euclidean geometry and the ordinary notion of gradient familiar to us all? The answer comes from the component versions of (3.21) and (3.24). Our original system is an ordinary gradient of U if $X_i = \partial U/\partial x_i$ at all points, that is, if the absolute growth rates are the partials of U . But comparing (3.2) with (3.24) we see that our system is a Shahshahani gradient of U if $\xi_i = \partial U/\partial x_i$ at all points, that is, if the relative growth rates are the partials of U . From this arises my central claim about all this machinery: *Where the natural focus of a problem is upon absolute growth rates, then the usual Euclidean geometry is appropriate. Where, instead, we focus upon relative growth rates, then it is Shahshahani geometry which naturally applies.* Recall the ubiquity of relative growth rates mentioned earlier and you will see that this assertion, while vague, is quite strong. In its defense I will cite the example which originally inspired Shahshahani and Conley.

As a preliminary we have to extend these concepts to systems defined on the simplex Δ instead of on all of $\mathbb{R}_+^I$.

Now $\mathring{\Delta}$ is a submanifold of $\mathbb{R}_+^I$ on which the Shahshahani metric is defined. If we restrict U to the simplex we get a function on $\mathring{\Delta}$ whose gradient *with respect to* $\mathring{\Delta}$ must be tangent to $\mathring{\Delta}$, that is, it must lie in $\mathbb{R}_o^I$, the subspace of vectors which sum to 0. This restricted gradient $\bar{\nabla} U$ must satisfy:

$$\bar{\nabla}_p U \in \mathbb{R}_o^I$$

(3.26)

$$(\bar{\nabla}_p U, X)_p = d_p U(X) \qquad \text{for all } X \in \mathbb{R}_o^I$$

for $p \in \mathring{\Delta}$. As discussed before Proposition 2.3 this means that $\bar{\nabla}_p U$ is just the perpendicular projection of $\nabla_p U$ onto the subspace $\mathbb{R}_o^I$ (perpendicular with respect to the inner product $(\ ,\ )_p$). This is a good exercise (compare (3.24) and (3.26) with (i) of Theorem 1.2).

So by applying the projection formula (3.16) to the definition (3.23) we get:

$$(\bar{\nabla}_p U)_i = p_i \left(\frac{\partial U}{\partial x_i} - \frac{\overline{\partial U}}{\partial x} \right), \quad \text{where}$$

(3.27)

$$\frac{\overline{\partial U}}{\partial x} = \sum_i p_i \frac{\partial U}{\partial x_i}.$$

Notice that the partial derivatives make no sense when you restrict yourself to the simplex because you can't there vary x_i leaving the remaining x_j's all fixed. This is because on the simplex $\sum x_i = 1$. In practice we will often begin with a function U defined only on $\mathring{\Delta}$. To apply a formula like (3.27) (or (3.17) for that matter) we must extend U to define a smooth function on $\mathbb{R}_+^I$ (or technically on an open subset of $\mathbb{R}^I$ containing $\mathring{\Delta}$). There are always many different ways to do this. We will usually do it by replacing p_i's by x_i's in various formulae. The partial derivatives will depend upon the choice of extension. A different extension for U will lead to different formulae for the partials but the final result in (3.27) (or (3.17) provided $X \in \mathbb{R}_o^I$) will be independent of the choice. The differences arising from an alternative choice cancel in subtracting $\overline{\partial U / \partial x}$.

* * *

When R.A. Fisher introduced the symmetric version of system (3.7), he proved his Fundamental Theorem of Natural Selection. This says that, under assumption (3.8), the mean fitness function, a_{pp}, is constantly increasing on nonequilibrium solution paths of (3.7) Furthermore, Kimura's Maximum Principle says, in addition, that at each point p

of $\mathring{\Delta}$ the vector field defining (3.7) points in the direction of greatest increase of mean fitness. This latter phrase is essentially the definition of the gradient direction. But when one computes the gradient of mean fitness according to formula (3.19) the wrong vector field is produced even accounting for the need to project to $\mathbb{R}_o^I$.

The solution of the puzzle can be found in Kimura's discussion of his maximum principle. If you go back to the discussion around equation (2.8) you will discover that the left side, apparently independent of the inner product, and the right side, with its metric concepts of gradient, length and angle, are tied together by the notion of a unit vector, a vector of length 1. Kimura introduced what seemed at the time to be a rather puzzling concept of unit vector, which he called unit variance. As the reader has no doubt guessed, this turns out to be the unit vector concept using the Shahshahani metric rather than the usual one. The variance language comes from equation (3.11).

Letting $\bar{a}$ denote the mean fitness function $\bar{a}(p) = a_{pp}$ on $\mathring{\Delta}$, we extend $\bar{a}$ to $\mathbb{R}^I$ by $\bar{a}(x) = a_{xx} = \sum x_i x_j a_{ij}$ and compute:

$$(3.28) \qquad \frac{\partial \bar{a}}{\partial x_i} = a_{ip} + a_{pi} = 2a_{ip} .$$

The first equation comes by differentiating a_{xx} and the second then follows from the symmetry assumption (3.8). Hence, from (3.27):

$$(3.29) \qquad (\bar{\nabla}_p \tfrac{1}{2} \bar{a})_i = p_i(a_{ip} - a_{pp}) ,$$

and so in the symmetric case the system (3.7) is a gradient system but only when the Shahshahani geometry is used. In particular, as in (3.22) and (3.25):

$$(3.30) \qquad \frac{d\bar{a}}{dt} = \tfrac{1}{2} \| \bar{\nabla}_p \bar{a} \|_p^2 = 2 \sum p_i(a_{ip} - a_{pp})^2 \geq 0 .$$

The latter expression is twice the additive variance and this is Fisher's equation leading to the fundamental theorem.

It is a good exercise here for the reader to check that if $\bar{a}$ is extended to $\mathring{\mathbb{R}}_+^I$ by using $\bar{a}(x) = \sum p_i p_j a_{ij}$ with $p_i = x_i / \sum_k x_k$, then the partial derivatives will not be given by (3.28) but the formula (3.29) still results.

* * *

In applying the Shahshahani metric there are two especially useful classes of functions on Δ. The first are the linear functions:

$$E^a(p) = \sum_i p_i a_i \qquad \text{for } a \in \mathbb{R}^I$$

$$(3.31)$$

$$(\bar{\nabla}_p E^a)_i = p_i(a_i - E^a(p)) .$$

The gradient formula follows easily from (3.27), as does the formula for the log linear function on $\overset{\circ}{\Delta}$:

(3.32)

$$L^b(p) = \sum_i p_i \ln p_i \quad \text{for} \quad b \in \mathbb{R}^I$$

$$(\overline{\nabla}_p L^b)_i = b_i - p_i(\sum_k b_k) .$$

Notice that the second term vanishes if $b \in \mathbb{R}_o^I$. In that case, we derive:

(3.33)

$$(\overline{\nabla}_p E^a, \overline{\nabla}_p L^b)_p = \sum_i a_i b_i = {}_o(a,b) ,$$

provided $b \in \mathbb{R}_o^I$.

This simple formula is very important because the right side is independent of p . It is the key to the Product Theorem which we now look at.

Let A be a subspace of $\mathbb{R}^I$ containing the constant vectors, the multiples of u , and let B be the perpendicular complement of A with respect to the usual metric ${}_o(\,,\,)$. Thus, every vector in $\mathbb{R}^I$ can be decomposed uniquely as the sum of a vector in A and one in B . Furthermore, for all $a \in A$ and $b \in B$, ${}_o(a,b) = 0$. In particular, since $u \in A$, ${}_o(u,b) = \sum_i b_i = 0$ for $b \in B$ and so $B \subset \mathbb{R}_o^I$.

Now choose $\{u, a^1, ..., a^{k-1}\}$ and $\{b^1, ..., b^{n-k}\}$ bases for A and B , respectively. On $\overset{\circ}{\Delta}$ define the maps $E^A : \overset{\circ}{\Delta} \to \mathbb{R}^{k-1}$ and $L^B : \overset{\circ}{\Delta} \to \mathbb{R}^{n-k}$ by

(3.34)

$$E^A(p) = (E^{a^1}(p), ..., E^{a^{k-1}}(p)) ,$$

$$L^B(p, = (L^{b^1}(p), ..., L^{b^{n-k}}(p)) .$$

3.1 Theorem. E^A *is a linear map whose image on* $\overset{\circ}{\Delta}$ *is an open, convex subset of* $\mathbb{R}^{k-1}$ *which we will denote* O_A . *The product map*

$$E^A \times L^B : \overset{\circ}{\Delta} \to O_A \times \mathbb{R}^{n-k}$$

is a diffeomorphism, that is, it is one-to-one and onto with a smooth inverse.

Sketch of Proof. Using independence of the basis vectors and (3.33) it is not hard to show that the differential

$$d_p(E^A \times L^B) : \mathbb{R}_o^I \to \mathbb{R}^{k-1} \times \mathbb{R}^{n-k}$$

is a linear isomorphism. It then follows from the inverse function theorem (our Theorem 2.2) that $E^A \times L^B$ is locally a diffeomorphism. The global results, that every point of $O_A \times \mathbb{R}^{n-k}$ is hit and that no two distinct points of $\overset{\circ}{\Delta}$ are mapped to the same place, require a further argument using some topology.   □

The product theorem says that we can use the $n - 1$ coordinate functions $E^A \times L^B$ as a curvilinear coordinate system on $\mathring{\Delta}$. More important are the two mutually orthogonal *foliations* of $\mathring{\Delta}$ the map induces.

If we fixed $\alpha \in O_A$, then the set of points p in $\mathring{\Delta}$ which map to α form a convex cell in $\mathring{\Delta}$ of dimension $n - k$. In fact, $E^A(p) = E^A(q)$ if and only if $E^a(p) = E^a(q)$ for all a in A (general E^a's are linear combinations of the components of E^A) . We can write this:

$$0 = E^a(p) - E^a(q) = \Sigma_i a_i(p_i - q_i) = {}_0(a, p-q) .$$

Because B is the perpendicular complement of A this says exactly that the difference vector $p - q$ lies in B . Thus, the preimage of α under E^A is just the $\mathring{\Delta}$ part of a translation of the subspace B :

$$(E^A)^{-1}(\alpha) = \mathring{\Delta} \cap \{q+b : b \in B\} ,$$

where q is any fixed member of $\mathring{\Delta}$ such that $E^A(q) = \alpha$. If we choose a different α in O_A we get the intersection with a different parallel translate of B . Thus, $\mathring{\Delta}$ is sliced up – "foliated" is the technical term– into disjoint, parallel slices. Following the picture in Figure 1 we will call this the *vertical foliation*. The separate slices, the cells in this case, are called the leaves of the foliation.

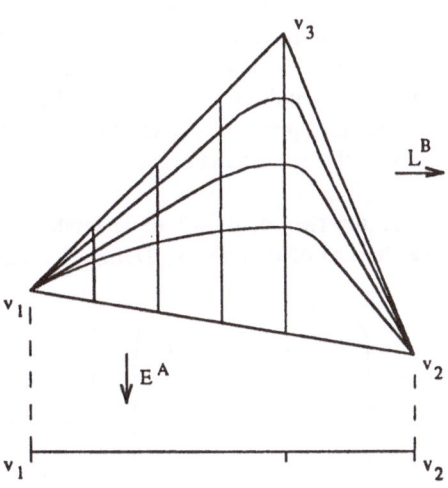

Figure 1

If we instead fix $\lambda \in \mathbb{R}^{n-k}$ and look at the preimage $(L^B)^{-1}(\lambda)$ we get the *horizontal foliation*. Again $\overset{\circ}{\Delta}$ is sliced up into disjoint sheets, now of dimension $k - 1$. But this time the leaves of the foliation are not flat convex cells. Because the functions L^b are log-linear rather than linear, the constraints imposed by the $n - k$ esquations $L^{bi}(p) = \lambda_i$ yield curved submanifolds.

The two foliations are related in that each leaf of the vertical foliation intersects each leaf of the horizontal foliation in exactly one point p of $\overset{\circ}{\Delta}$, and at that point the tangent spaces of the leaves are mutually perpendicular with respect to the Shahshahani metric $(\ ,\ )_p$. The unique point intersection property is just a restatement of the product theorem. For the other result let us go slowly.

For $p \in \overset{\circ}{\Delta}$ the horizontal leaf through p is a manifold defined implicitly by $L^B = \lambda$. This uses the language of section 2 where we then remarked that the tangent space of the leaf at p consists of exactly those vectors X in $\mathbb{R}_o^I$ such that $d_p L^B(X) = 0$. Thus, X is horizontal, tangent to the leaf of the horizontal foliation, precisely when:

$$(3.35) \qquad\qquad 0 = d_p L^b(X) = (\overline{\nabla}_p L^b, X)_p$$

for all $b \in B$. This is really a restatement of part of Proposition 2.3: because L^b is constant along the horizontal leaf its gradient is perpendicular to the leaf. But A is the $_o(\ ,\ )$ orthogonal complement of B. So (3.33) says that all the vectors X of the form $\overline{\nabla}_p E^a$ satisfy (3.35). So $\overline{\nabla}_p E^a$ is tangent to the horizontal leaf at p. Reversing the argument we get that the $\overline{\nabla}_p E^a$'s are perpendicular and the $\nabla_p L^b$'s are tangent, respectively, to the vertical leaf at p.

Thus, if we solve the gradient dynamical system associated with $\overline{\nabla} E^a$ we move everywhere along a fixed horizontal leaf (the L^b's remain constant) and always cutting the vertical leaves perpendicularly. Dually, moving along $\overline{\nabla} L^b$ we stay on a fixed vertical leaf, with the E^a's constants of the motion, and cut each horizontal leaf perpendicularly. That the curves in Figure 1 do not appear to cut orthogonally should serve as a reminder that the appropriate geometry is the Shahshahani one.

As an example, let us use the two-locus-two-allele model with the usual labelling and numbering conventions (see the end of section 1). Define the matrix:

$$(3.36) \qquad\qquad J = \left. \begin{pmatrix} 1 & 1 & 1 & 1 \\ 1 & 1 & -1 & -1 \\ 1 & -1 & 1 & -1 \\ 1 & -1 & -1 & 1 \end{pmatrix} \right\} \begin{matrix} A \\[2pt] - B \end{matrix}$$

This is a symmetric 4×4 matrix whose rows are mutually perpendicular with respect to the usual dot product. As indicated on the right we let A be the 3-dimensional subspace spanned by the first three rows (note that the first row is u). Its perpendicular complement B in $\mathbb{R}^4$ consists of multiples of the last row. So E^A has two real linear components which we denote by x and y, and L^B is one log-linear fucntion which we denote by L:

$$x = p_1 + p_2 - p_3 - p_4 = p_A - p_a = 2p_A - 1$$

$$(3.37) \quad y = p_1 - p_2 + p_3 - p_4 = p_B - p_b = 2p_B - 1$$

$$L = \ln p_1 - \ln p_2 - \ln p_3 + \ln p_4 = \ln p_1 p_4 - \ln p_2 p_3 = \ln(p_1 p_4 / p_2 p_3) .$$

The first two coordinates indicate the marginal gene frequencies at the first and second locus, respectively. Over each point of the xy square lies a vertical line segment upon which the frequencies at each locus are constant. For example, over the origin is the segment of points p such that $p_A = p_a = p_B = p_b = 1/2$. L is the log of the ratio measure of linkage disequilibrium. In particular, the horizontal leaf where $L = 0$ is the manifold of linkage equilibrium, independence between loci, or (following Freddy Christiansen) Robbins proportions.

In multi-locus models this picture can be generalized by using as E^A the marginal projection $\mathring{\Delta} \to \mathring{\Delta}^1 \times \ldots \times \mathring{\Delta}^L$ discussed in section 1. The horizontal leaves then consist of manifolds upon which all of the ratio measures of linkage disequilibrium remain constant. For the dynamical system (3.2) it can be proved that at each point p the relative growth rate vector $\xi_i(p)$ is additive between loci (here p is fixed and we are varying i) if and only if the horizontal leaves are invariant manifolds for the system. In particular, Fisher's selection equations show no epistasis in the additive sense precisely when the ratio measures of linkage disequilibrium remain unchanged in time.

4. Equilibrium analysis

Instead of writing a dynamical system in component form, let us return to the vector field formulation:

$$(4.1) \qquad\qquad \frac{dx}{dt} = X(x) .$$

Here x is assumed to vary in an open subset U of a finite dimensional vector space V. X is a vector field on U, that is, a smooth function $X : U \to V$ associating to each point $x \in U$ a vector $X(x)$ regarded as attached to x (we are thinking of our original arrows-in-the-plane example as a motivating picture). A solution path for (4.1) is a path in U, that is, a function x(t) where t is a real time variable, whose velocity vector at time t, $x'(t)$, is exactly the vector which X attaches to the position x(t). The Existence-and-Uniqueness Theorem for differential equations says that the initial value problem consisting of the equation of motion, (4.1), and the specification of a position at a single time, $x(t_0) = x_0$, together determine a unique solution path. The result, whose proof we omit, requires smoothness of X . In our example, the path functions will usually be defined for all positive and negative time but this need not be true in general. Instead the solution may in finite time "blow up", approaching infinity, or, alternatively, approach the boundary of U .

A point $x = e$ is called an equilibrium if X is the zero vector at e , $X(e) = 0$. If a solution is at an equilibrium e at any time, then its position does not change. It remains at the rest point e . You should not accept this argument too quickly. As an exercise, resolve the paradox that $x = t^3$ is a solution of the real differential equation $x' = 3x^{2/3}$ which passes, at time $t = 0$, through the equilibrium $x = 0$ without remaining there. This pathological behavior does not arise among practical examples because the uniqueness theorem quoted above says that the constant, resting solution is the only solution through an equilibrium point.

The stability analysis of the solutions near equilibrium is perhaps the best known application of linearization. The first step is to replace the original function X by its linearization at the equilibrium point, the derivative $d_e X$, a linear map from $V \to V$. Recall from section 2 that it is the linear map T such that $T(h)$ is the best approximation to the function $X(e+h)$. When $V = \mathbb{R}^I$, $T = d_e X$ is represented by the Jacobian matrix $(\partial X_i / \partial x_j)$ evaluated at e .

Instead of the original system, (4.1), we analyze the linear system:

$$(4.2) \qquad\qquad \frac{dh}{dt} = T(h)$$

with equilibrium thus translated to $h = 0$.

At least as far as stability is concerned, the behavior of the linear system is described by the real parts of the eigenvalues of the map T . Let us review why this is true.

First, if a is a real eigenvalue with corresponding eigenvector v , so that $T(v) = av$, then the line through v is invariant for the system. By solving the real differential equation:

$$(4.3) \qquad\qquad \frac{dx}{dt} = ax \Rightarrow x = x_0 e^{at}$$

we get the solution $h(t) = x_0 e^{at} v$ which begins at $x_0 v$ and remains on the line of multiples through v (check that $h(t)$ is a solution of (4.2)). As t tends to infinity this solution tends to zero or blows up according to whether a is negative or positive. In the boundary case where a is zero the entire line through v consists of equilibria. See the left side of Figure 2.

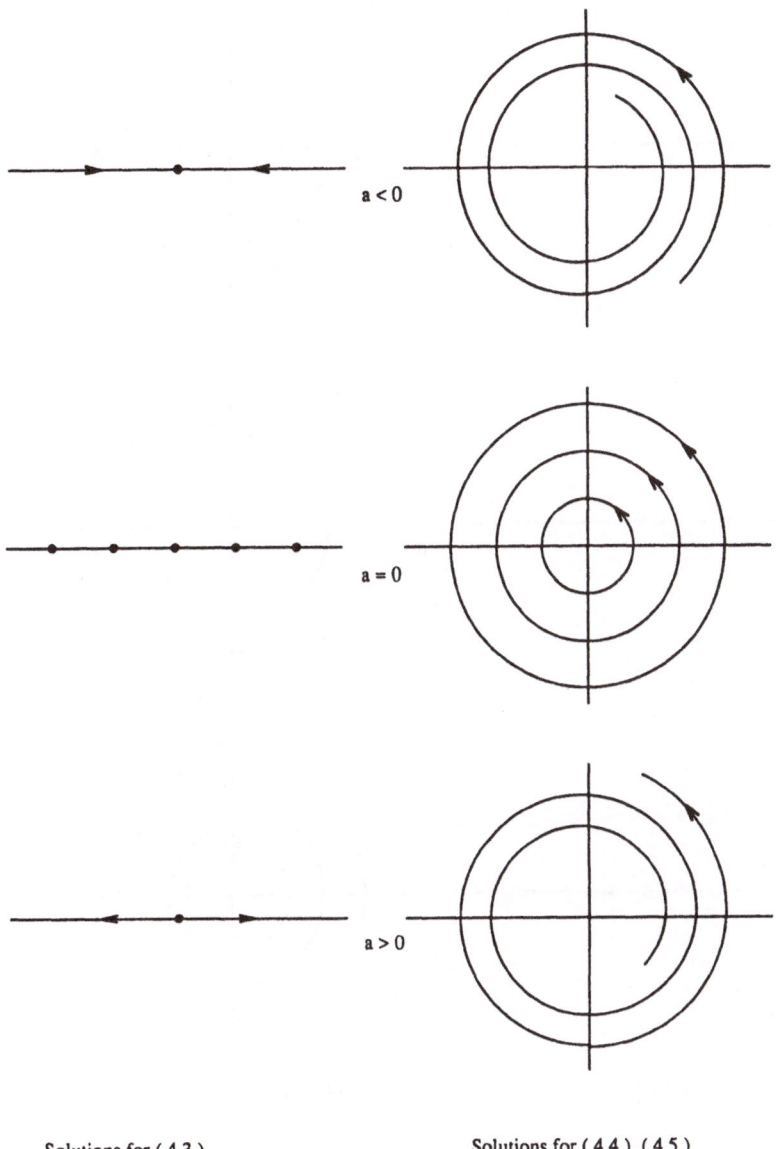

Solutions for (4.3)   Solutions for (4.4), (4.5)

Figure 2

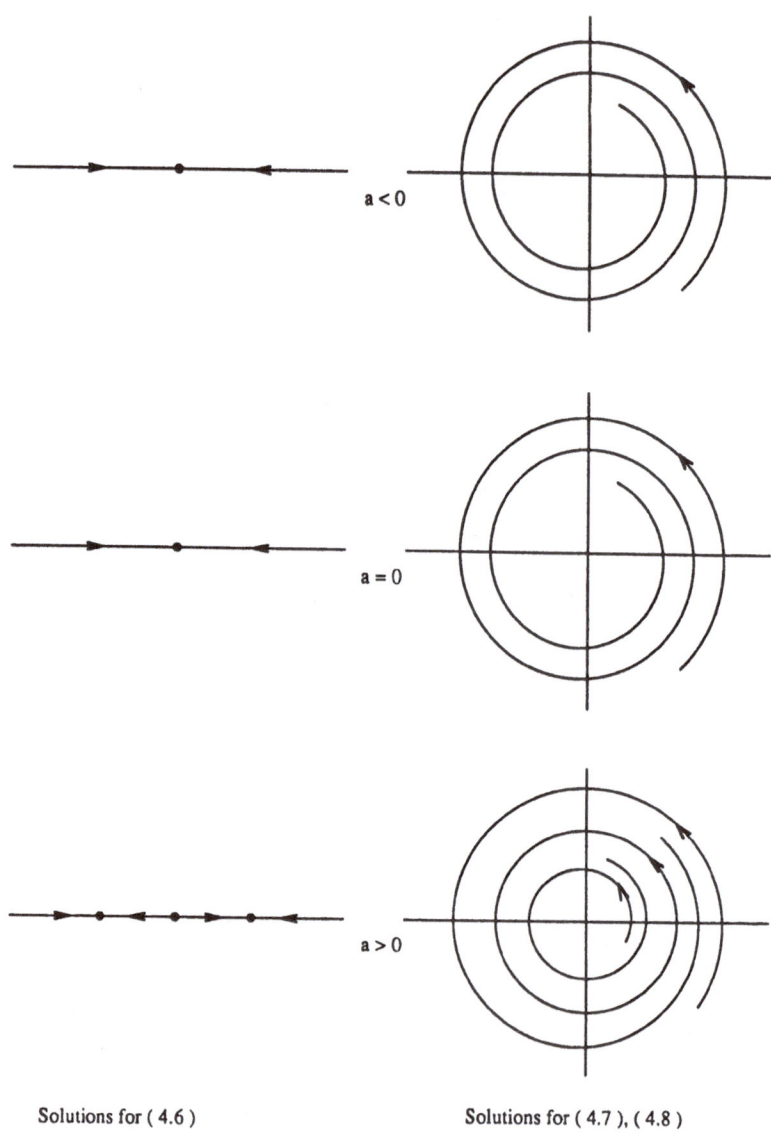

a < 0

a = 0

a > 0

Solutions for (4.6)   Solutions for (4.7), (4.8)

Figure 3

If an eigenvalue has a nonzero imaginary part, then the complex conjugate is an eigenvalue as well and corresponding to such a conjugate pair of eigenvalues $a \pm ib$ there is an invariant plane upon which coordinates can be chosen so that the system (4.2) restricts there to

$$\frac{dx}{dt} = ax - by$$

(4.4)

$$\frac{dy}{dt} = bx + ay$$

which is solved by introducing polar coordinates: $r \cos \theta = x$, $r \sin \theta = y$ or $r^2 = x^2 + y^2$, $\tan \theta = y/x$. This coordinate change yields:

(4.5)
$$\frac{dr}{dt} = ar \qquad r = r_0 e^{at}$$
$$\Rightarrow$$
$$\frac{d\theta}{dt} = b \qquad \theta = \theta_0 + bt .$$

Notice that the radial equation is essentially (4.3) and so the solution tends to zero or infinity according to whether a is negative or positive. Meanwhile the angular variable θ increases at a constant rate b (or decreases if b is negative). Thus, while the sign of the real part a determines stability as before, the imaginary part b introduces spiraling motion about the origin. See the right side of Figure 2. Notice also the difference between the two sides for the special case $a = 0$. This time the origin remains the only equilibrium point. The plane is filled with invariant circles centered at the origin about which the solutions all cycle with frequency b .

These two cases show why stability is determined by the real part of the eigenvalues. After choosing a basis so that the matrix of T is in Jordan canonical form the general situation of (4.2) adds only such complications as are illustrated by the system:

$$\frac{dx}{dt} = ax$$

$$\frac{dy}{dt} = x + ay .$$

You should check by solving this system that new terms of the form te^{at} appear. However, because the exponential factor dominates the polynomial, the limit of such a term is determined by the sign of a , just as with e^{at} (provided $a \neq 0$) .

The hard part of the linearization program is the use of our analysis of the linear system (4.2) to describe the original system (4.1). For our first result we define the equilibrium e to be *nondegenerate* if the linearization at e , $T = d_e X$, is nonsingular, i.e., is a linear isomorphism. This is the same as saying that 0 is not an eigenvalue for T . The inverse function theorem, quoted as Theorem 2.2, then implies that the vector field function X is a diffeomorphism near e . This means there is a little open subset U_0 of U which contains e and on which X maps one-to-one and onto a little open set about $0 = X(e)$ in V . In particular, e is the only point in U_0 which maps to 0 . So we see that e is an isolated equilibrium for (4.1), meaning there are no other equilibria arbitrarily close to it, just as 0 was for (4.2). Nondegeneracy includes the possibility that an eigenvalue be purely imaginary, complex with real part 0 . We call e a *hyperbolic* equilibrium if no eigenvalue has real part 0 . It is a theorem of Hartman that for a hyperbolic equilibrium e the behavior of the nonlinear system near e resembles the linearized picture. To see what this means we look at nonlinear systems related to (4.3) and (4.4). For

$$(4.6) \qquad\qquad \frac{dx}{dt} = ax - x^3$$

with equilibrium $e = 0$, the linearization at e is (4.3). If a is negative then the derivative dx/dt is negative when x is positive and vice-versa. Starting from any nonzero position the solution path approaches 0 , and the picture is the same as that for the linear case. Compare the top left of Figures 2 and 3. In particular, e is the only equilibrium. Now if a is positive there are, in addition to 0 , two other equilibria, namely $\pm\sqrt{a}$. Despite the hyperbolicity of the equilibrium and Hartman's theorem, the picture in the nonlinear case is now more complicated than the linear one, see the bottom left of Figures 2 and 3. But this serves to illustrate our warning in section 2 about the usefulness of the derivative. Even with a strong hypothesis like hyperbolicity the best we can hope to get from the infinitesimal description given by the derivative is local information. In this case the two pictures do look alike provided you restrict your gaze to a little interval about 0 .

Observe as the eigenvalue a changes sign, passing through the degenerate value 0 , a characteristic pattern, called bifurcation. The equilibrium $e = 0$ which was stable for $a < 0$ becomes unstable when a becomes positive. Meanwhile a pair of stable equilibria appear, flanking 0 , whose distance apart grows with a . This particular bifurcation is called the *pitchfork bifurcation* because of the picture, the so-called bifurcation diagram shown in Figure 4 below. The horizontal axis represents the bifurcation parameter a and the graph of x against a marks the positions of equilibrium. The $x = 0$ position changes from stable (-) to unstable (+) and around the unstable half is the parabola $x^2 = a$ of new stable equilibria.

Notice that, as shown in Figure 3, the picture for $a = 0$ looks similar to that for $a < 0$. In complete contrast with its linearization, the degenerate system has a unique equilibrium and it is stable. It is called a *vague attractor* because its attracting character cannot be observed from the linearization. The rate of attraction towards 0 is rather different though (solve (4.6) explicitly with $a \leq 0$ to see this). Also, if we add the x^3 term in (4.6) instead of subtracting it, the pitchfork curls around the other way. Then with $a = 0$, the equilibrium at 0 becomes a vague repellor and for $a \geq 0$ the picture is similar to the linear case. This time the stable equilibrium for $a < 0$ is flanked by a pair of unstable ones.

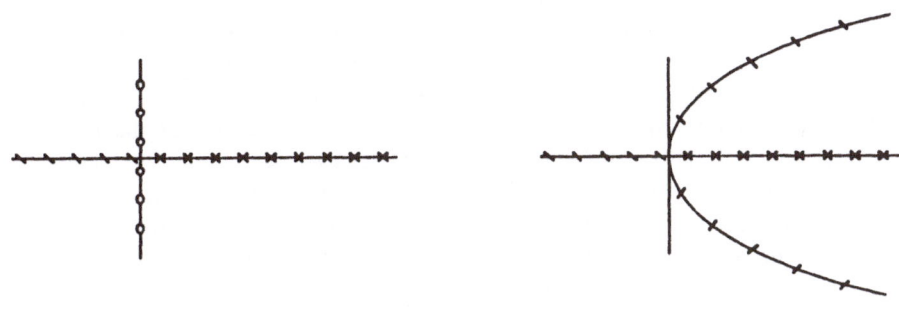

The linear case   The nonlinear case

Figure 4
Bifurcation diagram

With all this talk of changing a , I want to emphasize that it is a parameter. In any particular run of the dynamical system it is fixed and constant. We are comparing the results of setting it at different values.

For the complex conjugate pair of eigenvalues we compare (4.4) with

(4.7)

$$\frac{dx}{dt} = ax - by - (x^2+y^2)x$$

$$\frac{dy}{dt} = bx + ay - (x^2+y^2)y$$

or in polar coordinates:

(4.8)

$$\frac{dr}{dt} = ar - r^3$$

$$\frac{d\theta}{dt} = b$$

The radial part is just (4.6). So we known that for $a \leq 0$, r tends towards 0 . But now the angular motion means that the approach to 0 is via a counterclockwise spiral about the origin. Also, when a is positive we know that for $r = \sqrt{a}$ there is no radial motion. But this time such points are not at rest. Instead, there are periodic solutions which cyle about the origin on the circle of radius $\sqrt{a}$.

Again, observe that, as described by Hartman's theorem, the picture of the linear system is similar to that of the nonlinear system for $a \neq 0$, provided you restrict to a little open set containing the origin. For the nonhyperbolic (but still nondegenerate) value a = 0 the pictures do not look alike. In this example the origin is again a vague attractor when a = 0 .

The change in behavior as a changes sign in this case is called a *Hopf bifurcation*. In general, suppose we are given a system of differential equations depending upon a parameter and as this parameter changes the real part of a complex conjugate pair of eigenvalues at an equilibrium changes sign. Meanwhile, the remaining eigenvalues have real parts nonzero throughout. Then a Hopf bifurcation occurs and the system has nonequilibrium periodic solutions for some parameter values at or near the bifurcation value. System (4.7) is an example of the supercritical case with stable, attracting limit cycles surrounding the equilibrium when it changes from stable to unstable, observable for positive values of a . In the reverse, subcritical case unstable orbits surround the stable equilibrium for negative values of a shrinking to it at the value when the equilibrium becomes unstable. As with the pitchfork, this occurs when the cubic terms are added in (4.7) and (4.8) instead of subtracted.

To decide analytically whether the bifurcation is supercritical, subcritical or of some degenerate sort like the linear case shown in Figure 2, one must go beyond the linearization to compute higher order terms at the bifurcation value of the parameter. In this case what one needs is essentially the sign of the coefficient of the cubic part. Alternatively, one can observe that if the equilibrium is a vague attractor at the bifurcation value, then the case is supercritical with limit cycle solutions.

We should remark upon the odd mood which has entered our discussion. Earlier we described linear systems as the good ones, because they are easy to analyze. The repeated story of linearization was the move away from bad nonlinear systems to nice linear ones with the hope that the return trip would be as short as possible. Now, suddenly, in bifurcation theory it is the linear systems we call degenerate or critical and it is the nonlinear systems with their limit cycles whose picture we hope our simulations will draw. Let us see why this is so.

Think of the difference between (4.4) and (4.7) as analogous to the difference between classical and relativistic physics. In the former any change in positon of a particle is instantly felt by masses arbitrarily far away because the associated gravitational field has changed. But in general relativity the speed of light not only bounds the speed of motion of objects and radiation but also the rate at which any kind of information can be propagated. For both (4.4) and (4.7) with a < 0 solutions beginning at any point in the plane spiral in toward the attracting equilibrium at the origin. Now when a changes to a positive valve in (4.4) the resulting change of the equilibrium from attractor to repellor is felt everywhere in the plane and all solutions now spiral out. The periodic, cycling behavior occurs only at the bifurcation value a = 0 . On the other hand, when the origin becomes repelling in (4.7) it is still true (Hartman's theorem) that orbits begining near the origin spiral out. But now news of the change has not reached points far from the origin. From such points the orbits continue to spiral in as in the old regime. The limit cycle separates the inward spiraling orbits from the outward ones. The cycle grows as the parameter a increases, enlarging the region governed by the new repelling regime at the origin. The cycles occur for all values of a in an interval on one side of the bifurcation value. This is why the nonlinear system is good. It is *structurally stable* meaning that once a is positive a little change in the parameter leaves the picture with its cycle essentially unaltered. In the linear case any change of a away from 0 causes all the cycles to disappear.

The synthesis à la Hegel of these dissonant views comes in the concept of normal form about which we can say only a little here. From this perspective linearization is a special case of throwing away higher order terms when the behavior of the simpler, truncated system is similar to the original. When the equilibirum is hyperbolic the first order, linear terms do suffice to describe the system near equilibrium. However, when the equilibrium is not hyperbolic many fundamentally different systems have the same linearization. The best we can then hope for is that certain quadratic or cubic terms be nonzero, for then we can throw away the terms beyond these.

* * *

You will have noticed that the word "stable" has a number of different uses. We are calling an equilibrium stable if it is attracting. Solution paths beginning near enough to such an equilibrium move back toward it, approaching it in the limit as time tends to infinity. Technically, this is called asymptotic stability.

We have seen that 0 is a stable equilibrium for a linear system like (4.2) if and only if all of the eigenvalues have negative real parts. In that case, we call the linear map T or the associated system, (4.2), stable as well.

For stability of an equilibrium e of (4.1) stability of the linearized system with $T = d_e X$ is sufficient. As such an equilibrium is hyperbolic, this result follows from Hartman's theorem but is actually easier and is better proved directly. An equilibrium with a stable linearization is called *linearly stable*. The vague attractors discussed above are stable but not linearly stable.

In a Euclidean vector space we can use the inner product to devise a sufficient condition for linear stability. This strong stability condition does not always hold; it is not a necessary condition. But testing for it is usually easier than computing the eigenvalues, and when it does hold it yields a strong, explicit description of the stability.

First, we define a linear map $T : V \rightarrow V$, where V is a Euclidean vector space, to be symmetric if:

$$(4.9) \qquad (T(v),w) = (T(w),v) \qquad \text{for all } v,w \in V .$$

In the case where $V = \mathbb{R}^I$ with the usual inner product $_0(,)$ of (1.11), and T is represented by the matrix $A = (a_{ij})$, then T is symmetric precisely when the matrix is symmetric:

$$(4.10) \qquad a_{ij} = a_{ji} \qquad i, j \in I .$$

This is because when v and w are regarded as column vectors, then the inner product is given by the matrix formula:

$$(4.11) \qquad _0(T(v),w) = w^T A v .$$

Recall from section 1 that for any finite dimensional Euclidean space we can choose an orthonormal basis (cf. (1.14)). Then the coordinate map of (1.7) is an isometry with $\mathbb{R}^I$ and the usual metric. This means that for the purposes of proofs we can always go to the standard case and use matrices.

The following lists the results about symmetric maps which are important for us.

4.1   Proposition. *If* T *is a symmetric linear mapping of a Euclidean vector space, then all of the eigenvalues of* T *are real. Thus,* T *is stable when all of its eigenvalues are negative. A necessary and sufficient condition for stability is that* T *be negative definite:*

$$(4.12) \qquad\qquad (T(v),v) < 0 \qquad \textit{for all } v \neq 0 \textit{ in } V .$$

In the case when T *is not symmetric, condition* (4.12) *is sufficient, though not necessary, for stability.* □

Notice that (4.9) and (4.12) together say exactly that the formula $- (T(v),w)$ defines a new inner product on V . (4.12) is just the reverse of the positivity condition (1.9). Now if you think back to the polar identity (1.13) you can see how from just the diagonal, the values where v = w , we can recover all of the information about this new inner product and so about the symmetric map T . When T is not symmetric we lose information when we known only the diagonal values $(T(v),v)$. In the matrix formula (4.11) you get the same number $v^T A v$ if you replace A by its transposed matrix A^T . In matrix language the second part of Propostion 4.1 says that if the "symmetrized matrix" $\frac{1}{2}(A+A^T)$ is stable, then the original matrix A is, but the converse is not necessarily so.

In the language of symmetric matrices the first part of Proposition 4.1 appears in all linear algebra books. Both because it is less common and because the method is important we will prove the second part.

4.2   Theorem. *Let* X *be a vector field on an open set in a Euclidean vector space and let* e *be an equilibrium for the associated dynamical system* (4.1). *We call* e *strongly stable if for all nonzero* v *in* V :

$$(4.13) \qquad\qquad (d_e X(v),v) < 0 .$$

Define the real-valued function N^e *on* V *to be the square of the distance from* e :

$$(4.14) \qquad\qquad N^e(x) = \| x-e \|^2 = (x-e,x-e) .$$

If e *is a strongly stable equilibrium, then on some little open set* U_0 *containing* e , N^e *decreases monotonically on solution paths of* (4.1):

$$(4.15) \qquad\qquad \frac{dN^e}{dt} \leq 0$$

at points of U_0 *with equality only at the equilibrium* e .

Proof. We will assume that the vector space is $\mathbb{R}^I$ with the usual inner product. As we mentioned above, we can always translate into this case. For notational convenience we will assume e is 0 . Then

$$N(x) = \|x\|^2 = \sum x_i^2 ,$$

and the vector field is described by its component functions X_i .

(4.16) $$\frac{dN}{dt} = d_x N(\frac{dx}{dt}) = d_x N(X) = 2 \sum x_i X_i ,$$

where x is the point $x(t)$ at which the derivative $d_x N$ is computed and at which the vector field functions X_i are evaluated (see 2.5) and (2.13)). Now let us define by the right side (times 1/2) a real-valued function of x :

(4.17) $$\dot{N}(x) = \sum x_i X_i(x) .$$

This function of x vanishes at x = 0 . We will show by using strong stability that at x = 0 it has a strict local maximum. It will then follow that on some neighborhood of 0 , $\dot{N} \leq 0$ vanishing only at 0 . Then (4.16) implies (4.15).

There are two conditions we must check:

$$\frac{\partial \dot{N}}{\partial x_j} = 0 \qquad \text{at } x = 0 \qquad \text{for all } j$$

and

$$(\frac{\partial \dot{N}}{\partial x_j \partial x_k}) \qquad \text{at } x = 0 \qquad \text{is negative definite matrix.}$$

We compute:

$$\frac{\partial \dot{N}}{\partial x_j} = X_j + \sum x_i \frac{\partial X_i}{\partial x_j} .$$

This vanishes at 0 because 0 is an equilibrium.

$$\frac{\partial \dot{N}}{\partial x_j \partial x_k} = \frac{\partial X_j}{\partial x_k} + \frac{\partial X_k}{\partial x_j} + \sum x_i \frac{\partial X_i}{\partial x_j \partial x_k} .$$

At the origin the summation terms vanish and we are left with $A + A^T$ where A is the Jacobian matrix of X at 0. Strong stability, (4.13), says that $v^T A v < 0$ for all $v \neq 0$. This says exactly that $A + A^T$ is negative definite. $\square$

4.3 Corollary. *Strong stability implies linear stability.*

Proof. Apply the theorem not to the original system but to the linearized system (4.2) with $T = d_e X$. $\square$

It is (4.15) which describes what is strong about strong stability. Not only do orbits near e move back generally toward e but the distance from e decreases monotonically. Furthermore, by using the function $\overset{\circ}{N}$ of (4.17) we can estimate how fast the return to equilibrium is.

On the other hand, it is a theorem of Lyapunov that every linearly stable equilibrium is strongly stable with respect to some choice of inner product. Thus, strong stability as a separate condition arises in problems where there is a natural inner product.

It is emphatically not true that every linear map is symmetric with respect to some inner product. What is really special about the symmetric case is that eigenvalues are real. Remember it is the nonzero imaginary parts which induce circular motion around an equilibrium.

$$* \quad * \quad *$$

We now begin to apply all this to dynamical systems on the open simplex $\overset{\circ}{\Delta}$ in $\mathbb{R}^I$, the set $\{p \in \mathbb{R}^I : p_i > 0 \text{ for all } i \text{ and } \Sigma_i p_i = 1\}$. We then write (4.1) in component form:

$$(4.18) \qquad\qquad \frac{dp_i}{dt} = X_i = p_i \xi_i$$

with the absolute rates X_i and the relative rates ξ_i functions of the position p. Of course, we assume that

$$\Sigma_i X_i = \Sigma p_i \xi_i = \overline{\xi} = 0$$

which says that at each point p the vector $X(p)$ lies in $\mathbb{R}_o^I$, the tangent space of $\overset{\circ}{\Delta}$. Equivalently, this says that because the total mass Σp_i remains constant at 1, its change ΣX_i is 0. Thus, the vector field X is a function from $\overset{\circ}{\Delta}$ to $\mathbb{R}_o^I$ and at each point p of $\overset{\circ}{\Delta}$ its derivative is a linear mapping:

$$d_p X : \mathbb{R}_o^I \to \mathbb{R}_o^I .$$

Instead of studying this linear map directly, it is often easier to deal with the associated bilinear real-valued map called the *Hessian* of X. This function, denoted by $H_p(X)$, or simply H_p when the vector field is understood, is defined to be:

(4.19) $$H_p(Y^1,Y^2) = (d_pX(Y^1),Y^2)_p, \quad Y^1, Y^2 \in \mathbb{R}_o^I.$$

Notice that the point p enters in two different ways. First we are taking the derivative at p and then we are using the Shahshahani metric at p as inner product. Recalling the definition, (3.10), of the latter we compute:

(4.20)
$$H_p(Y^1,Y^2) = \sum_{ij} p_i^{-1} \frac{\partial X_i}{\partial x_j} Y_j^1 Y_i^2$$

$$= \sum_i p_i^{-1}\xi_i Y_i^1 Y_i^2 + \sum_{ij} \frac{\partial \xi_i}{\partial x_j} Y_j^1 Y_i^2 \qquad Y^1,Y^2 \in \mathbb{R}_o^I.$$

Remember what these partial derivative formulae mean. We first extend the component functions X_i or ξ_i from $\overset{\circ}{\Delta}$ to get smooth functions on $\mathbb{R}^I$. Only then does the partial with respect to x_j make sense: vary x_j keeping all other x_i's fixed. The partial derivatives you get will depend on how the extension was defined but the formula for the Hessian will not because we only use Y^1 and Y^2 in $\mathbb{R}_o^I$.

For example, in the game-dynamic case (3.7) with $X_i = p_i(a_{ip}-a_{pp})$, the formula is:

(4.21) $$H_p(Y^1,Y^2) = \sum p_i^{-1}(a_{ip}-a_{pp})Y_i^1 Y_i^2 + \sum a_{ij}Y_j^1 Y_i^2$$

for Y^1, Y^2 in $\mathbb{R}_o^I$. In particular, check that the terms obtained by differentiating $-a_{pp}$ with respct to x_j drop out because $\sum Y_i^2 = 0$.

Note that of the two sums in each of (4.20) and (4.21) the first is symmetric in Y^1 and Y^2 and also vanishes if $p = e$ is an equilibrium point. So using the Shahshahani metric to define the idea, we call an equilibrium e of (4.18) strongly stable if

(4.22) $$H_e(Y,Y) = \sum \frac{\partial \xi_i}{\partial x_j} Y_i Y_j < 0 \qquad \text{for } 0 \neq Y \in \mathbb{R}_o^I,$$

where the partials are evaluated at e. In the game dynamic case this says precisely that e is an ESS.

The use of this version of strong stability and its relationship with the ESS idea will be picked up again in section 6. We turn instead to the role of symmetry of the Hessian and its relationship with the gradient concept.

In section 3 we saw that a vector field X on $\mathbb{R}^I$ is the usual gradient of some potential function U precisely when the components X_i are partials $\partial U/\partial x_i$ (see (3.19)). How can you recognize whether X is a gradient or not? Notice that if we take partials again we get:

(4.23) $$\frac{\partial X_i}{\partial x_j} = \frac{\partial^2 U}{\partial x_i \partial x_j}.$$

The order in which you take repeated partials does not matter. Thus, the expression on the right in (4.23) is always symmetric. So it is a necessary condition for X to be a gradient field, with respect to the usual metric, that the matrix $(\partial X_i/\partial x_j)$ be symmetric at every point. The Poincaré lemma says that, except for some topological difficulties which don't apply here, this condition is sufficient as well. The Poincaré lemma appears in disguise, in every first differential equations course, Writing a first order equation as $Mdx + Ndy = 0$ the exactness test is: $\partial M/\partial y = \partial N/\partial x$? Providing the answer is yes, you apply a little integration routine to find U such that $\partial U/\partial x = M$ and $\partial U/\partial y = N$.

The analogous recognition test for the Shahshahani metric uses the Hessian:

4.4 Propositon. *Let X be a smooth vector field on the open simplex $\mathring{\Delta}$ in $\mathbb{R}^I$. X is a gradient field with respect to the Shahshahani metric, i.e. there exists a smooth real-valued function U on $\mathring{\Delta}$ such that $X = \overline{\nabla} U$, if and only if the Hessian is symmetric at every point p of $\mathring{\Delta}$, that is, if for all p in $\mathring{\Delta}$:*

$$(4.24) \qquad H_p(Y^1, Y^2) = H_p(Y^2, Y^1) \qquad Y^1, Y^2 \in \mathbb{R}_o^I.$$

Sketch of Proof. From the formula (3.27) for the Shahshahani gradient we see that X is the gradient of U precisely when $\xi_i = \partial U/\partial x_i - \overline{\partial U/\partial x}$. Then the second sum in (4.20) becomes

$$\sum \frac{\partial^2 U}{\partial x_i \partial x_j} Y_j^1 Y_i^2$$

which is symmetric in Y^1 and Y^2 because the second partials are symmetric in i and j. Notice that $\partial \xi_i/\partial x_j$ itself is not symmetric in i and j but that the peculiar terms $\partial(-\overline{\partial U}/\partial x)/\partial x_j$ drop out in the sum because $\sum Y_i^2 = 0$. Thus, (4.24) is necessary. After some preliminary spadework sufficiency is proved by using the Poincaré lemma itself. $\square$

For example, recall that if the payoff matrix a_{ij} is symmetric, then the game dynamic is the gradient of $U = \frac{1}{2} \bar{a}$ where $\bar{a}$ is the mean payoff a_{pp} at p. In that case, of course, the Hessian given by (4.21) is symmetric. Using that formula we describe an important result.

4.5 Theorem. *Let X be a smooth vector field on $\mathring{\Delta}$ which is not a gradient with respect to the Shahshahani metric. For a symmetric matrix (a_{ij}) define the vector field*

$$X^a = \overline{\nabla}(\tfrac{1}{2} \bar{a}) + X$$

so that $X_i^a = p_i(a_{ip} - a_{pp}) + X_i$. There exists a symmetric matrix (a_{ij}) so that the dynamical system associated with X^a has cyclical solutions, i.e., non-equilibrium periodic motions.

Proof. We will show that in the family of vector field X^a Hopf bifurcations occur.

Because X is not a gradient, Propositon 4.4 says that a some point of $\overset{\circ}{\Delta}$ the Hessian of X is not symmetric. Fix such a point and label it e. At e we can write the Hessian $H_e(X)$ as the sum of two pieces H^+ and H^- defined by:

(4.25)
$$H^+(Y^1,Y^2) = \tfrac{1}{2}(H_e(Y^1,Y^2) + H_e(Y^2,Y^1))$$

$$H^-(Y^1,Y^2) = \tfrac{1}{2}(H_e(Y^1,Y^2) - H_e(Y^2,Y^1)).$$

If we use an $(\,,\,)_e$ orthonormal basis to translate into matrix language this amounts to writing the matrix associated with $H_e(X)$, or equivalently with the linear map $d_e X$, as the sum of a symmetric and an antisymmetric matrix. By choice of e, the antisymmetric part H^- is not zero.

Now if (a_{ij}) is a symmetric matrix define the symmetric Hessian:

(4.26)
$$H^a(Y^1,Y^2) = \sum a_{ij} Y_j^1 Y_i^2 .$$

By the right choice of matrix (a_{ij}) we can get any symmetric term we want. In fact, there is extra slack which we will need. If we write

(4.27)
$$a_{ij} = \alpha_{ij} + \beta_i + \beta_j + \gamma$$

then you can check that H^a is independent of the choice of the β_i's or γ. We will construct in pieces a family of matrices depending on a real parameter ε. The variation is because of a special family of symmetric bilinear forms labelled H^ε which I will describe at the end.

So first choose α_{ij} depending on ε so that:

(4.28)
$$H^a = H^\varepsilon - H^+ .$$

Now we are still free to choose the β's. We choose them so that the point e is an equilibrium for $X^a = \overline{\nabla}(\tfrac{1}{2}\overline{a}) + X$, i.e., so that

$$e_i(\alpha_{ie} + \beta_i - \beta_e - \alpha_{ee}) + X_i(e) = 0 .$$

These equations determine the β_i's up to a common constant γ. They depend upon ε through α_{ij}. We can choose that constant γ to give $\overline{a}$ any value we want at e. Select it to be 0. In other words, the symmetric matrix a_{ij} is uniquely determined by three things: (1) the choice of $H^a \cdot$ (2) the demand that e be an equilibrium and (3) the value of $\overline{a}$ at e.

For the combined system X^a it follows that e is an equilibrium and at e :

$$H_e(X^a) = H^{\varepsilon} + H^- .$$

The H^+ terms have cancelled.

H^- is represented by a nonzero antisymmetric matrix and so its nonzero eigenvalues are pure imaginaries. Pick out a conjugate pair of eigenvalues and the associated invariant plane. On the vectors perpendicular to this plane in $\mathbb{R}_o$ (using $(\,,\,)_e$) use as matrix for H^{ε} the identity times -1. But on the plane itself use the identity times ε.

For the sum $H^{\varepsilon} + H^-$ all of the eigenvalues have real part -1 except for a single conjugate pair for which the real part is ε. As ε changes sign the equilibrium e for X^a undergoes a Hopf bifurcation and so cyclic solutions occur for some a_{ij} corresponding to values of ε at or near 0. $\square$

This result has wide application because many vector fields which arise naturally are not gradients. The vector field modeling the effects of mutation is usually not a gradient and the one for recombination never is. Thus, once the dynamic for selection is adjusted to include these effects cycling is possible.

Left open by the proof is the question of what kind of Hopf bifurcations occur. Do supercritical bifurcations leading to stable limit cycles occur? At the other extreme do only critical bifurcations occur so that the cycles disappear when the selection parameters are perturbed at all? The answers depend upon X. In the next section we investigate the two-locus-two-allele model to answer these questions for the vector field modeling recombination.

5.   The two locus, two allele model

Following the standard notation for these TLTA models we let A, a and B, b denote the alleles at the first and second loci, respectively. We number the four gamete types $1 = AB$, $2 = Ab$, $3 = aB$ and $4 = ab$, so that the dynamical systems lie on the unit simplex Δ in $\mathbb{R}^4$. The p_i's are constrained to sum to 1. So the system is three dimensional. As often happens in applied problems the choice of appropriate coordinates is an important decision. Following the picture in Figure 5 we define coordinates by using the matrix J introduced in (3.36) and (3.37). Thus, we let

(5.1)
$$x = p_1 + p_2 - p_3 - p_4 = p_A - p_a = 2p_A - 1$$

$$y = p_1 - p_2 + p_3 - p_4 = p_B - p_b = 2p_B - 1 .$$

These two coordinates are indicators of the marginal gene frequencies at each of the two separate loci. They are preferable to p_A and p_B in that A, a and B, b play symmetrical roles in their definition. As shown by the figure the simplex maps down to the square with x and y varying between -1 and 1. The points which map to the origin form the segment where

(5.2) $x = y = 0$ $\Leftrightarrow$ $p_1 = p_4$ and $p_2 = p_3$ $\Leftrightarrow$ $p_A = p_a = p_B = p_b = \frac{1}{2}$.

We will see that this segment has special properties worth singling out.

For the third coordinate we use the fourth row of J which is important enough that we will label it $\varepsilon = (1,-1,-1,1)$, i.e., $\varepsilon_1 = \varepsilon_4 = 1$ and $\varepsilon_2 = \varepsilon_3 = -1$. There are three alternate possibilities for a fourth coordinate:

$$z = p_1 - p_2 - p_3 + p_4$$

(5.3) $$L = \ln p_1 - \ln p_2 - \ln p_3 + \ln p_4 = \ln p_1 p_4 - \ln p_2 p_3 = \ln(p_1 p_4 / p_2 p_3)$$

$$d = p_1 p_4 - p_2 p_3$$

z has the advantage that it, like x and y , is linear in the original p_i's . On the other hand, it does not have a nice genetic interpretation. When we fix x and y we are restricted to a segment in the simplex lying over a point in the square below. As you move along this segment, the marginal frequencies are fixed at each locus. What changes is the extent to which the two loci are independent. At exactly one point, the point of so-called *linkage equilibrium,* the gamete distribution is the product of the marginals. It is d which most naturally measures the deviation from this point because as you can check by substituting and multiplying out:

$$p_1 = p_A p_B + d$$

$$p_2 = p_A p_b - d$$

(5.4)

$$p_3 = p_a p_B - d$$

$$p_4 = p_a p_b + d .$$

Thus, exactly when $d = 0$ the probability of AB is the product of the probabilities of A and B separately etc. d and z are related by the formula:

(5.5) $4d = z - xy$

and so the surface of linkage equilibrium is the hyperboloid $z = xy$. Also this gives another interpretation of d . If we denote by ξ^1 and ξ^2 , respectively, the second and third rows of the matrix J of (3.36), then we can think of ξ^1 and ξ^2 as scores for the first and second locus respectively, for ξ^1 is 1 when the first locus has allele A ($i = 1,2$) , and -1 when it has allele a ($i = 3,4$) . Similarly, ξ^2 is ± 1 according to whether the second locus shows a B or a b . By definition x is the mean of ξ^1 and, similarly, y is the mean of ξ^2 . Furthermore, $\varepsilon_i = \xi_i^1 \xi_i^2$ ($i = 1,2,3,4$), and so the mean of the product $\xi^1 \xi^2$ is z . Thus, $z - xy = d/4$ is the covariance of ξ^1 and ξ^2 .

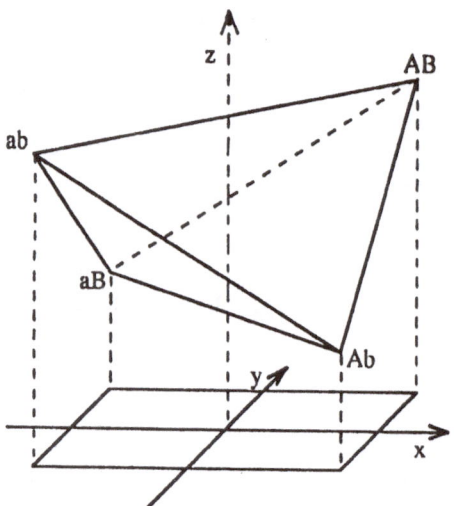

Figure 5

The equivalent expressions for L provide different interpretations of this useful function. The first form shows it is the log linear analogue of z, while the third shows it is the log of the so-called ratio measure of linkage disequilibrium, where d is the difference measure. The second form shows it has the same sign as d. In fact, we can define:

(5.6)
$$Q = \frac{d}{L} = \begin{cases} \dfrac{p_1p_4 - p_2p_3}{\ln p_1p_4 - \ln p_2p_3} & L \neq 0 \\[2em] p_1p_4 = p_2p_3 & L = 0. \end{cases}$$

5.1 Lemma. *On* $\mathring{\Delta}$, *Q is a smooth, positive function with value at* p *between* p_1p_4 *and* p_2p_3.

Proof. The ratio $R = (b-a)/(\ln b - \ln a)$ is sometimes called the logarithmic average of the positive numbers a and b. Since the log is increasing $\ln b > \ln a$ if and only if $b > a$. So the ratio is positive for $b \neq a$. Its reciprocal is the difference quotient for the function $y = \ln x$ and by the mean value theorem of first year calculus, this difference quotient equals the derivative at some intermediate point. So $1/R = d \ln x/dx$ evaluated at some point θ between a and b, i.e., $= 1/\theta$. Thus, $R = \theta$. As b and a approach a common value so does R.   □

In particular, it follows that $L = 0$ exactly when $d = 0$, the hyperboloid of linkage equilibrium. However, on the sheet defined by L equals some constant c, d is not constant unless $c = 0$. This is exactly why we use L in addition to d. Recall that by the Product Theorem, 3.1, the surfaces obtained by fixing L at different constant values provide a foliation of $\mathring{\Delta}$, the horizontal foliation, each leaf of which is mapped one-to-one and onto the open square below in Figure 5. As we discussed at the end of section 3, the horizontal foliation is everywhere perpendicular (in the Shahshahani sense) to the vertical foliation by the line-segments obtained by fixing x and y. In particular, the gradient of L points vertically along these segments. Applying (3.32) with $b = \varepsilon$ we see that

$$(5.7) \qquad\qquad (\overline{\nabla}_p L)_i = \varepsilon_i$$

at every point p of $\mathring{\Delta}$.

The dynamical system for selection is given by the game dynamic equations (3.7) with a symmetric payoff matrix (a_{ij}). As we saw in section 3, symmetry implies that the associated vector field is the gradient of $\frac{1}{2}\bar{a}$ where $\bar{a}$ is the mean fitness function $= a_{pp}$ at p. In the TLTA case we can represent the entries of the symmetric 4×4 matrix by using the following 3×3 table:

(5.8)

	aa	Aa	AA
BB	a_{33}	a_{13}	a_{11}
Bb	a_{34}	$a_{14}=a_{23}$	a_{12}
bb	a_{44}	a_{24}	a_{22}

where the remaining entries like a_{31}, a_{21}, etc are obtained by symmetry. To interpret the table observe that if the first locus is heterozygous and the second is homozygous for B, then the two gametes of the zygote must be 1 and 3 with corresponding fitness value $a_{13} = a_{31}$. The only ambiguity is for the double heterozygotes which can be obtained by pairing 1 and 4 or 2 and 3 (so-called "cis" versus "trans" pairings) and we assume that the two fitnesses are the same. This means we are ignoring the possibility of position effects and assume that the fitness depends on the genes present at the two loci independent of how they are paired up on the chromosomes.

While the mean fitness value describes the rate of growth of the total population size, the dynamical system on $\mathring{\Delta}$, which represents competiton between gamete types, is unaffected by the addition of a common constant to all of the payoff values. The constant is added to the a_{ip}'s and then subtracted away again with a_{pp}. Consequently, we can normalize the equations by adding such a constant to make the common central value 0. That is, we will assume, in addition to symmetry of (a_{ij}) that:

$$(5.9) \qquad\qquad a_{14} = a_{23} = 0.$$

Thus, the family of selection matrices for the TLTA model with no position effects is an 8 parameter family represented by the remaining 8 entries in table (5.8). We should remark upon an oddity that the reader may have noticed in the table. Putting AA on the right instead of the left is deliberate because it makes the pattern of this square table coincide with motion in the xy square. As we move up, varying y from -1 to +1 we move from b at the second locus to B. As we move right, varying x from -1 to +1, we move from a to A at the first locus. We will see later that this apparently meaningless harmony will prove useful.

Unless the two loci are completely linked, recombination perturbs the time path of the gene pool away from the solution of the system described by selection alone. To correct for its effect we add the recombination vector field first described by Kimura with components:

$$(5.10) \qquad\qquad R_i = -rd\varepsilon_i .$$

Thus, R_1 and R_4 are $-rp_1p_4 + rp_2p_3$ because a crossover in the 14 zygote causes a loss of gametes of types 1 and 4, while a crossover in the 23 zygote produces new gametes of these types. The reverse is true for 2 and 3.

Notice that the constant r represents the number of crossovers per unit time. It is thus the probability of crossover per birth times the birth rate (assumed the same for 14 and 23 types). In particular, r is not restricted to lie between 0 and $\frac{1}{2}$ but can take on any nonnegative value. Thus, r increases with looser linkage but increases with higher birth rate as well. For most biological species population size is roughly steady implying a mean fitness near 0. This reflects a balance between birth and death rates, either both are low or both are high. In the latter case, the high birth rate implies that the constant r is large unless linkage is very tight.

If we apply (5.7) we can rewrite the recombination vector field as:

$$(5.11) \qquad\qquad R = -rd\overline{\nabla}L = -rQ\overline{\nabla}_{\frac{1}{2}}L^2 .$$

The latter formula comes from the chain rule computation: $\overline{\nabla}L^2 = 2L\overline{\nabla}L$ and the definition of $Q : d = QL$.

The second formula shows that R is a negative function times the gradient of $\frac{1}{2}L^2$. In particular, in the selectively neutral case, with only recombination acting, the solution curve moves along the vertical foliation, with L decreasing in absolute value monotonically and approaching 0. In short, recombination moves a population toward linkage equilibrium leaving the marginal frequencies at each locus unaffected.

If Q were only a constant, and thus d only a constant multiple of L, then the combined vector field of selection plus recombination would be the gradient of

$$\frac{1}{2}(\overline{a}-rQL^2) = \frac{1}{2}(\overline{a}-rdL) .$$

This would mean that the combined field while not maximizing mean fitness, $\bar{a}$, as selection alone does, would be acting to maximize a fitness function consisting of mean fitness and an adjustment term for recombination. A version of this argument would still work if d were some complicated nonlinear function of L. For R would still be the gradient of some function of L. But we mentioned above that d is not constant on the leaves where L is constant with the exception of the zero leaf of linkage equilibrium. We can show that R is not the gradient of anything by using the test given by Proposition 4.4.

5.2 Propositon. *At p in $\dot{\Delta}$ the Hessian of the recombination field R is given by the formula:*

$$\text{(5.12)} \qquad H_p(R)(Y^1, Y^2) = -r(\bar{\nabla}_p d, Y^1)_p (\bar{\nabla}_p L, Y^2)_p.$$

It is symmetric in $Y^1, Y^2 \in \mathbb{R}_o^I$ if and only if either $L = d = 0$ at p (p is in linkage equilibrium) or $x = y = 0$ at p (p is in the origin segment of (5.2)). In particular, the vector field R is not a gradient with respect to the Shahshahani metric.

Proof. Apply the first formula in (4.20) observing that the ε_i's are constants:

$$\text{(5.13)} \qquad H_p(Y^1, Y^2) = -r \left(\sum_j \frac{\partial d}{\partial x_j} Y_j^1 \right) \left(\sum_i \frac{\varepsilon_i}{p_i} Y_i^2 \right).$$

Here we are using the Σ-notation result that the double sum on i and j of a product $\alpha_i \beta_j$ is the product of $\Sigma_j \beta_j$ with $\Sigma_i \alpha_i$. The first sum is the differential of d at p applied to Y^1, $(\bar{\nabla}_p d, Y^1)_p$ and the second, by (5.7), is $(\bar{\nabla}_p L, Y^2)_p$.

For the symmetry result we write $d = x_1 x_4 - x_2 x_3$ and compute partial derivatives. We can write the result as:

$$\frac{\partial d}{\partial x_i} = \frac{\varepsilon_i}{p_i} \gamma_i \quad \text{with} \quad \gamma_1 = \gamma_4 = p_1 p_4 \quad \text{and} \quad \gamma_2 = \gamma_3 = p_2 p_3.$$

Thus, we get symmetry of $H_p(R)$ if γ_i is constant in i, $p_1 p_4 = p_2 p_3$ and so p is in linkage equilibrium. More generally, because $Y^1 \in \mathbb{R}_o^I$ we have symmetry (at p fixed) if and only if:

$$\text{(5.14)} \qquad \gamma_i = A(p_i/\varepsilon_i) + B$$

for some A and B independent of i. If $A = 0$ this is the previous case. If $A \neq 0$ we substract the equations for 4 and 3 from those for 1 and 2 to get $A(p_1 - p_4) = 0$ and $A(p_2 - p_3) = 0$. So $p_1 = p_4$ and $p_2 = p_3$, i.e., p lies on the origin segment. Conversely, we can solve (5.14) for $i = 1,2$ to define A and B. If $p_1 = p_4$ and $p_2 = p_3$ this choice of A and B satisfies (5.14) for $i = 3,4$ as well.

If R were a gradient field, then its Hessian would be everywhere symmetric by Proposition 4.4.   □

This result dooms our hope for some sort of generalized fitness function which is being maximized by the combined effect of selection and recombination. In fact Theorem 4.5 now implies that for some choice of selection numbers the combined field of selection and recombination admits non-equilibrium periodic solutions. Of ocurse, as mentioned after that theorem we still don't know if attracting solutions, limit cycles occur. Because of noise in real systems it is only these which are observable. Furthermore, the proof required us to range among all symmetric matrices. It could be that cycling examples only occur for systems which show the position effects excluded by the assumption $a_{14} = a_{23}$. To resolve these issues and to generate examples we can look at, we study symmetries of the two locus model.

* * *

We have used the word "symmetric" for matrices satisfying (3.8). Interchanging the indices, or equivalently transposing rows and columns, leaves the matrix unchanged. See also condition (4.24). In general, the symmetries of an object or mathematical structure are the motions or transformations which leave it unchanged. For example the group of symmetries of a square (which we think of centered at the origin in the xy-plane) consists of 8 motions. In addition to the identity, these consist of rotations through 90° , 180° and 270° and reflections across the diagonals and across the two axes. A rectangle is less symmetric. Its group has only half as many motions: identity, 180° rotation and two axis reflections.

The symmetries of the simplex Δ in $\mathbb{R}^3$ are obtained by permuting the 4 vertices. There are factorial 4 = 24 such permutations. This is in fact the group of the underlying structure for 1 locus, 4 allele models. But not all of these symmetries map the picture shown in Figure 5 to itself. In fact, the group of symmetries of the TLTA picture is the square group, discussed above, of size 8.   There are several ways to see these eight possibilities. First of all, from Figure 5, only the motions of the simplex are allowed which map the xy-square to itself and there are 8 of those. We can also think of these eight possibilities as relabeling choices. We can relabel by interchanging the two loci or not, interchanging A with a or not and interchanging B with b or not, for a total of $2 \cdot 2 \cdot 2 = 8$ possiblilities. Similarly, in d and L , p_1 is paired with p_4 and p_2 with p_3 . The 8 possibilities consist of interchanging the two pairs or not and within each pair interchanging the coordinates or not. The three most important elements in the group are described in the following table which illustrates the different ways of thinking of the motions:

$$\pi_o \quad \begin{pmatrix} p_4 \\ p_3 \\ p_2 \\ p_1 \end{pmatrix} \quad \begin{pmatrix} -x \\ -y \\ z \end{pmatrix} \qquad \begin{matrix} a \leftrightarrow A \\ b \leftrightarrow B \end{matrix} \qquad \begin{matrix} x = 0 \text{ and } y = 0 \\[4pt] p_A = p_B = \tfrac{1}{2} \end{matrix}$$

$$(5.15) \quad \pi_y \quad \begin{pmatrix} p_3 \\ p_4 \\ p_1 \\ p_2 \end{pmatrix} \quad \begin{pmatrix} -x \\ y \\ -z \end{pmatrix} \qquad \begin{matrix} a \leftrightarrow A \\ b \quad B \end{matrix} \qquad \begin{matrix} x = 0 \text{ and } z = 0 \\[4pt] p_1 = p_3 \text{ and } p_2 = p_4 \end{matrix}$$

$$\pi_+ \quad \begin{pmatrix} p_1 \\ p_3 \\ p_2 \\ p_4 \end{pmatrix} \quad \begin{pmatrix} y \\ x \\ z \end{pmatrix} \qquad \begin{matrix} a \quad A \\ \updownarrow \quad \updownarrow \\ b \quad B \end{matrix} \qquad \begin{matrix} x = y \\[4pt] p_2 = p_3 \end{matrix}$$

Next to the name π for each map, is listed the point $\pi(p)$ and the same point in xyz coordinates. In the fourth column is the relabeling corresponding to π and in the fifth is the fixed point set, $\{p : \pi(p) = p\}$. In terms of their action on the square, π_o is the origin symmetry, the 180° rotation, while π_y and π_+ are the reflections about the y-axis and the $x = y$ diagonal, respectively.

Now we want to consider the systems which are invariant with respect to one of these symmetries. What does this mean?

Suppose X is a smooth vector field on a manifold like $\overset{\circ}{\Delta}$ and suppose h is a smooth map from this manifold to itself. $X(p)$ is the vector based at p which comes from the vector field X. Recall that the linear mapping, the derivative, $d_p h$ takes tangent vectors based at p to tangent vectors based at $h(p)$. So at $h(p)$ we see two vectors. There is the image $d_p h(X(p))$ obtained by moving $X(p)$ to the point $h(p)$ by using the derivative, and then there is the vector originally attached to $h(p)$ by X, $X(h(p))$. We call the vector field X h-*invariant* if these two vectors are the same for every point p :

$$(5.16) \qquad\qquad d_p h(X(p)) = X(h(p)) .$$

The main usefulness of h-invariance comes from:

5.3   Proposition. *Assume* X *is a smooth vector field invariant with respect to a smooth map* h *. If* x(t) *is a solution path for the associated dynamical system,* x'(t) = X(x(t)) , *then the image* h(x(t)) *is also a solution path for the system.*

Proof. Letting y(t) = h(x(t)) we have to show that

$$y'(t) = X(y(t)) .$$

By the chain rule (2.6) applied to the composed function of t : y = h ∘ x :

$$y'(t) = d_{x(t)}h(x'(t)) = d_{x(t)}h(X(x(t))) ,$$

where the second equation is true because x(t) solves the system. Now use (5.16) to continue the chain of equations and complete the proof:

$$= X(h(x(t))) = X(y(t)) .$$ □

5.4   Corollary. *If, in addition, the solution path* x(t) *lies in the fixed point set of* h *:*

$$(5.17) \qquad\qquad Fix(h) = \{p : h(p) = p\}$$

at some time t_0 , *then it remains in* Fix(h) *for all* t . *In other words, if* h(x(t)) = x(t) *at* t_0 , *then it holds for all* t .

Proof. y(t) = h(x(t)) and x(t) are both solutions and if $x(t_0)$ lies in Fix(h) then $y(t_0) = x(t_0)$. The uniqueness theorem says that there is only one solution passing through this common point at time t_0 . So y(t) and x(t) are the same for all t .   □

Remember that if the map h is linear, then it is its own best linear approximation, $d_p h = h$. In particular, the symmetries of (5.15) are all linear and so the invariance condition (5.16) becomes in these cases:

$$(5.18) \qquad\qquad \pi(X(p)) = X(\pi(p)) .$$

As an exercise, check that the recombination vector field R given by (5.10) is invariant with respect to all three symmetries. Note that d at π(p) equals ±d at p with the sign change only for π_y . Also, π(ε) = ε for π = π_0 and π_+ and -ε for π = π_y .

The selection vector field is usually not invariant with respect to these symmetries. In fact the design I insisted upon for table (5.8) was precisely so that we can easily recognize the invariant systems by using the following result, which I will merely assert.

5.5   Proposition. *The selection vector field* $\bar{\nabla}(\frac{1}{2}\bar{a})$ *is invariant with respect to a symmetry of the two locus, two allele model if and only if the selection numbers displayed as the* 3 × 3 *square of table (5.8) are invariant with respect to the corresponding motion of the square.*   □

Thus, the selection field is π_0, π_y or π_+ invariant if the numerical table (5.8) is left unchanged by a 180° rotation, y-axis reflection or x = y diagonal reflection, respectively.

Under the name "the symmetric viability model" the π_0 invariant examples have been widely studied. This is because the fixed point set of π_0 is the origin segment described by (5.2). Corollary 5.4 says that for a π_0 invariant system this segment is an invariant manifold. Solutions which start in it remain in it. So we can restrict the combined selection-plus-recombination field, and get a one-dimensional system within the three-dimensional one. As one-dimensional systems are quite easy to analyze this was a good place to begin to see how the combined field behaves. Karlin and Feldman completed this early work by studying π_0 symmetric systems without restricting themselves to the fixed point set.

For π_y the fixed point set is the segment by x = 0 and z = 0 or equivalently p_1 = p_3 and p_2 = p_4. Observe that along this segment d = 0. Thus, it is a straight segment in the hyperboloid of linkage equilibrium. In Figure 5 it maps to the y-axis. Because the recombination field vanishes at linkage equilibrium, restricting to Fix(π_y) is not very interesting, and π_y invariant systems have not, to my knowledge, been studied.

Now with the diagonal symmetry π_+ all this changes. The purpose of this digression into group theory is to get some help for the construction of Hopf bifurcation examples. The proof of Theorem 4.5 showed that such bifurcations come from points at which the Hessian of the recombination field is not symmetric. As we showed in Proposition 5.2 it is symmetric when d = 0 or x = y = 0 and these two sets include the fixed point sets of π_0 and π_y. But they do not include Fix(π_+). Furthermore, this set consists of the entire two-dimensional set of points in $\overset{\circ}{\Delta}$ which project down to the x = y diagonal in the square. Hopf bifurcation is a plane phenomenon. So we may hope that we can find Hopf bifurcations within the family of π_+ invariant examples with the cycles in the x = y plane. As we will see, such two-dimensional examples can be constructed and so can be drawn in the plane.

Before proceeding to the construction I should remark on what happened to the rest of the symmetries of the system. Why look only at the three listed in (5.15)? First, if a system is invariant with respect to the 90° or 270° rotation, then it remains so when the motion is applied twice to get the 180° rotation π_0. So, invariance with respect to 90° or (equivalently) the 270° rotation is a special case of π_0 invariance. On the other hand, the other two reflections, the x-axis reflection π_x and x = -y diagonal reflection π_-, are conjugate in the group theory sense to π_y and π_+, respectively. In practice this means that a π_x or π_- invariant example is mapped by the 90° rotation to become a π_y or π_+ invariant example, respectively. So the two diagonal reflections and the two axis reflections are essentially the same.

* * *

We are now going to focus on systems invariant with respect to the reflection $\pi_+(p_1,p_2,p_3,p_4) = (p_1,p_3,p_2,p_4)$ whose fixed point set is the intersection of $\overset{\circ}{\Delta}$ with the plane defined by the equation $p_2 = p_3$. Our first task is to find good coordinates. We replace our original choice xyz , appropriate for the general system, with a special choice which exploits the symmetry π_+ . First define the geometric average of p_2 and p_3:

(5.19)
$$p_+ = \sqrt{p_2 p_3}$$

so that on the fixed point set p_+ is the common value $p_2 = p_3$. Now on $\mathring{\Delta}$ we define:

$$u = p_1/p_+$$

(5.20)
$$v = p_4/p_+$$

$$w = \tfrac{1}{2} (\ln p_2 - \ln p_3) = \ln \sqrt{p_2/p_3} \ .$$

This coordinate system maps $\mathring{\Delta}$ onto the open subset of $\mathbb{R}^3$ $\{(u,v,w): u,v > 0\}$. The coordinate w is chosen to measure deviation from the fixed point set Fix(π_+) . The log-linear version turns out to be preferable to the linear alternative $p_2 - p_3$. Similarly, u and v turn out to work better than p_1 and p_4 . Some motivation may come from the observations:

$$e^w = p_2/p_+ \ , \ e^{-w} = p_3/p_+$$

(5.21)
$$u + e^w + e^{-w} + v = 1/p_+ \ .$$

Thus, to get back to the original point p in $\mathring{\Delta}$ we normalize (u, e^w, e^{-w}, v) by dividing by the sum of the components.

Also, it is easy to check that

$$d = (uv-1)p_+^2$$

(5.22)
$$d = 0 \Leftrightarrow uv = 1 \ .$$

In these new coordinates the formulae for π_0 and π_+ are:

$$\pi_0(w,u,v) = (-w,v,u)$$

(5.23)
$$\pi_+(w,u,v) = (-w,u,v) \ .$$

Thus, as was built into the design, the set Fix(π_+) is defined by w = 0 . Fix(π_0) is the line u = v in the w = 0 plane. Recall that the latter is the origin segment. So in looking for points from which Hopf bifurcation occurs in the w = 0 plane, we will have to avoid the hyperbola uv = 1 , where d = 0 and the origin segment u = v .

To get a π_+ invariant system we must restrict the selection numbers so that table (5.8) is diagonal symmetric:

(5.24) $$a_{13} = a_{12} , a_{33} = a_{22} , a_{34} = a_{24} .$$

With the normalization (5.9): $a_{14} = a_{23} = 0$ this reduces the number of selection parameters from eight to five: a_{11} , a_{33} , a_{44} , a_{13} , a_{34} . The rest of the matrix (a_{ij}) is then determined by (5.24), (5.9) and symmetry (3.8).

Beginning with a π_+ invariant system we change variables to (w,u,v) and compute the formulae for (w',u',v') . There is a common factor p_+ which occurs in all three of these formulae. It is helpful to eliminate it by defining the new time variable τ :

(5.25) $$\frac{d\tau}{dt} = p_+ .$$

Suppose that we have solved the equation using the new time variable. We get a path function $p(\tau)$ on which p_+ is a real-valued function of τ . Integrating (5.25) we get the true time t as a function of τ , and vice-versa. We can then substitute to get the path function $p(\tau(t))$ in the original time variable t . It is important to notice that the clocks converting t time to τ time run at different speeds at different points. That is what (5.25) says. Consequently, the time function relating t and τ will be different along different solution paths. On the other hand, this complexity is irrelevant for the pictures we draw. The path functions in terms of the two times run along the same set of points, they just proceed at different speeds. So for our purposes the choice of time variable makes little difference. It only becomes important at the distant moment when we want to return to reality ("Tap your heels three times and say 'There's no place like home'.") to answer such questions as "What is the approximate period of the cycles we have discovered?"

In the equation for dw/dt , w is also a common factor. We expect this from Corollary 5.4, because $Fix(\pi_+)$ defined by $w = 0$ is supposed to be an invariant manifold. If $w = 0$ to begin with it remains 0 . We will restrict to the plane where $w = 0$, and there write the equations for the dynamical system in the coordinates u and v . It is helpful to replace the five selection parameters by certain simple combinations which arise in these equations:

(5.26)
$$\alpha = a_{11} - a_{13} \qquad \beta = -a_{34} \qquad \gamma = -a_{13}$$
$$\delta = a_{44} - a_{34} \qquad \varepsilon = 2a_{13} + 2a_{34} - a_{33} .$$

It is easy to check that given α through ε we can recover the values of a_{11} , a_{13} , etc.

On the uv-plane where $w = 0$, using the time change (5.25) the system becomes:

(5.27)
$$\frac{du}{d\tau} = [r+(\varepsilon+2\beta+r)u+\alpha u^2] + [u(\beta-r-ru)]v \equiv F(u,v) ,$$

$$\frac{dv}{d\tau} = [r+(\varepsilon+2\gamma+r)v+\delta v^2] + [v(\gamma-r-rv)]u \equiv G(u,v) .$$

At first glance these equations may not strike you as the elegant outcome of an especially crafty choice of coordinate system. To generate some enthusiasm for them observe that F is linear in the variable v and G is linear in u . In particular, we can rewrite the equation F(u,v) = 0 by solving for v = f(u) and for any particular choice of parameters α through ε and r graphing this function is an easy curve sketching exercise. Similarly, solve G(u,v) = 0 to get u = g(v) and sketch. The intersections of these two graphs –usually several points– are the equilibria. So the equilibria are quite easy to find graphically. This is unusually simple for a system with multiple equilibria. Furthermore, the sketching problems are easy enough that one can observe by hand the result of varying a single parameter like r .

For the cycles we use a special case of the parametrization method which describes in the general TLTA model all of the locations (= choice of selection numbers and associated equilibria) where Hopf bifurcation occurs. The parametrization method is the reverse of the above procedure of starting with the equation and finding the equilibria. Instead we start with an equilibrium point plus some conditions imposed upon the equilibrium and then determine the selection numbers.

5.6 Theorem. *Given* $u^*,v^* > 0$ *there exist parameter values* $(\alpha,\beta,\gamma,\delta,\varepsilon,r)$ *unique up to a positive multiple so that at* (u^*,v^*) *the following equations hold (for some real* λ*):*

$$(5.28) \qquad (F,G) = (0,0) \quad and \quad \begin{pmatrix} \dfrac{\partial F}{\partial u} & \dfrac{\partial F}{\partial v} \\[2mm] \dfrac{\partial G}{\partial u} & \dfrac{\partial G}{\partial v} \end{pmatrix} = \begin{pmatrix} 0 & \lambda \\ -\lambda & 0 \end{pmatrix}.$$

The ratios α/r , β/r , γ/r , δ/r , ε/r *and* λ/r *are rational functions of* u^* *and* v^* *(the ratio of polynomials).* $\lambda/r = 0$ *if and only if* $u^* = v^*$ *or* $u^*v^* = 1$.

Sketch of Proof. While dignified by the name of a theorem because of its consequences the proof is rather direct. Divide by r or equivalently set r = 1 so that α , β etc. really stand for the ratios α/r , β/r , etc. Then with (u*,v*) fixed, the system (5.28) consists of six equations in six unknowns α,β,...,λ . The complexity of the formulae lies entirely with u* and v* , they are all first order in the parameters. It is fairly easy to solve by bare hands methods the six linear equations in six unknowns. The observation about λ follows because its final formula is a positive multiple of (u*-v*)(1-u*v*) .   □

The equations of (5.28), determine selection parameters so that (u*,v*) is an equilibrium and at that equilibrium the eigenvalues are the conjugate imaginary pair ±iλ . Notice that the points where λ vanishes are exactly those where we knew we could not get Hopf bifurcation.

In the original three-dimensional system (in τ time) the three eigenvalues consist of ±iλ and the coefficient of w in dw/dτ at the point (0,u*,v*) . Calling this coefficient

μ, it can be shown that the ratio μ/r is also a simple rational function of (u^*,v^*) and has the same sign as $1 - u^*v^*$.

Finally, remember from section 4 that the Hopf bifurcation through these parameter values is supercritical with limit cycles resulting, if the equilibrium (u^*,v^*) is a vague attractor. Furthermore, this is determined by the coefficient of a cubic term. Marsden and McCracken produced a formula for computing this coefficient which I will denote MARMC. The limit cyle case is MARMC < 0. Because of the simple form of the linearization at (u^*,v^*) given by (5.28) it is possible to apply their computational procedure to get an explicit formula for the ratio MARMC/r as well. It is negative when $u^*v^* < 1$. This is too bad because in this region $\mu > 0$. Thus, while these cases lead to limit cycles which are stable in the plane, if the initial point is not exactly on the plane the solution path moves away from the plane and a fortiori away from the cycle. Happily, as shown in Figure 6 there are two other regions where MARMC is negative. These cases yield examples of limit cycles which attract all nearby points and not just those in the plane.

The question arises why this unexpected behavior was not discovered by random simulations in the parameter space of TLTA models. The selection numbers in the examples I have computed do not look outlandish.

The trouble – or good news depending on your viewpoint – seems to be that the cycles occur for only a thin range of parameter values. Furthermore, they remain rather small. Before they grow in size they appear to be eliminated by a so-called saddle crossing bifurcation.

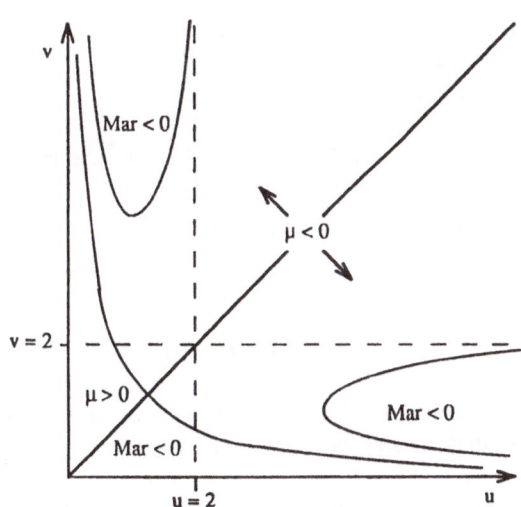

Figure 6

6.   Game dynamics

We consider again a general dynamical system on the unit simplex Δ of $\mathbb{R}^I$:

$$(6.1) \qquad \frac{dp_i}{dt} = X_i = p_i \xi_i$$

with X a smooth vector field on Δ so that

$$(6.2) \qquad 0 = \sum X_i = \sum p_i \xi_i \equiv \bar{\xi}$$

at every point p of Δ.

An interior equilibrium e is a point of $\overset{\circ}{\Delta}$ at which all of the absolute rates X_i vanish, or equivalently, since all e_i's are positive, the relative rates ξ_i vanish. Recall that an equilibrium is linearly stable if all of the eigenvalues of the linearization $d_e X : \mathbb{R}_o^I \to \mathbb{R}_o^I$ have negative real part. Given an inner product we saw that strong stability is a sufficient condition for linear stability, see Corollary 4.3. We begin by investigating the meaning of strong stability defined using the Shahshahani metric at the equilibrium point:

$$(6.3) \qquad (d_e X(Y), Y)_e < 0 \qquad \text{for } 0 \neq Y \in \mathbb{R}_o^I$$

or equivalently using the Hessian of X defined in section 4, see (4.19):

$$(6.4) \qquad H_e(X)(Y, Y) < 0 \qquad \text{for } 0 \neq Y \in \mathbb{R}_o^I.$$

Our first step is a theorem analogous to Theorem 4.2 but using a measure of deviation from equilibrium different from N^e of (4.14). S. Kullback defined what he called a measure of information discrimination:

$$(6.5) \qquad L^e(p) = -\sum e_i \ln(p_i/e_i) = -\sum e_i \ln p_i + \sum e_i \ln e_i.$$

For e in $\overset{\circ}{\Delta}$ this function is a smooth function of p in $\overset{\circ}{\Delta}$ approaching infinity as p approaches the boundary, i.e., as some p_i approaches 0. In general, if p is in Δ we define the support of p:

$$(6.6) \qquad \text{supp}(p) = \{i : p_i > 0\}.$$

If e is any point of Δ we can define L^e on the open set of p such that $\text{supp}(p)$ contains $\text{supp}(e)$. The sum must then be interpreted as extending only of those indices i which are in the support of e.

The first useful property about L^e is that, like N^e, it provides a measure of deviation from e. That is,

$$(6.7) \qquad L^e(p) \geq 0$$

with equality only at $p = e$. (6.7) follows from Jensen's inequality. This result begins with $f(x)$ the natural log function. Suppose we have some distribution on the x values. It says that f at the average value is greater than the average of the f's:

(6.8) $$f(\bar{x}) \geq \overline{f(x)}$$

with equality only for the trivial case where the distribution is concentrated at the single point $\bar{x}$. As is shown at the top of Figure 7, the geometry is that the graph arches over the chord between any two distinct points. Feller's proof of Jensen's inequality is so elegant it is worth digressing to look at it. Instead of the chord below look at the tangent line above, see the bottom of Figure 7. Take the tangent line at the point $(\bar{x}, f(\bar{x}))$. Because the graph lies below the tangent line, we have, for $x \neq \bar{x}$:

$$f(x) < f(\bar{x}) + f'(\bar{x})(x - \bar{x}) .$$

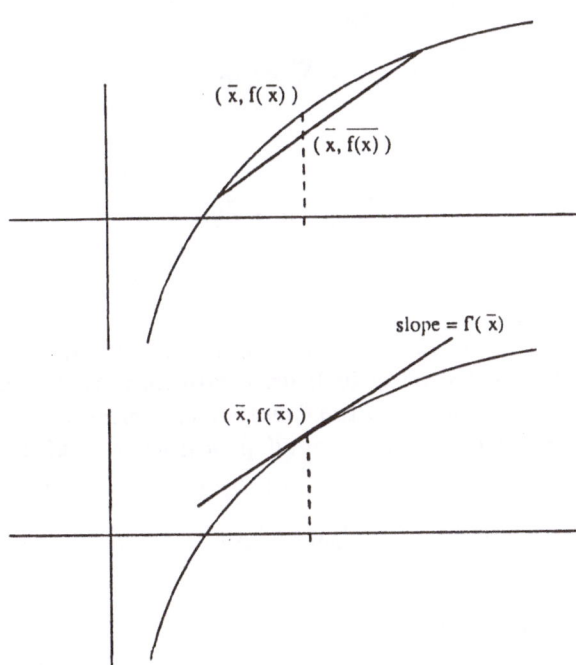

Figure 7

Here $\bar{x}$, $f(\bar{x})$ and $f'(\bar{x})$ are constants and so the left side is the original function $f(x)$ while the right is the tangent line function. Now average over the variable x . That is, take expected values. The constants all pull out and we get

$$\overline{f(x)} < f(\bar{x}) + f'(\bar{x})\,(\bar{x}\text{-}\bar{x}) \;=\; f(\bar{x}) \;,$$

which is (6.8).

In particular in (6.5):

$$L^e(p) \;=\; -\sum_i e_i \ln(p_i/e_i) \geq -\ln \sum_i e_i(p_i/e_i) \geq -\ln 1 \;=\; 0 \;.$$

The first inequality is Jensen's inequality and the second comes from $\sum e_i(p_i/e_i) \leq 1$. Of course, when $p,e \in \mathring{\Delta}$ this sum equals 1 but for e in the boundary the sum is only over those p_i's with i in the support of e . This may be less than 1. Furthermore, at least one of the inequalities is strict unless $p = e$ (in the boundary case this requires a bit of thought. Exercise!).

Now in preparation for the analogue of Theorem 4.2 we introduce some convenient notation: For $p \in \mathring{\Delta}$ and $X \in \mathbb{R}^I$ we define

(6.9) $$\xi_X(p) \;=\; \sum_i X_i \xi_i(p) \;.$$

So, in particular, when $X = p$

(6.10) $$\xi_p(p) \;=\; \sum_i p_i \xi_i(p) \;=\; \bar{\xi}(p) \;=\; 0$$

by (6.2).

6.1 Theorem. *Suppose* e *is an interior equilibrium for the dynamical system* (6.1). *If* e *is strongly stable in the sense of* (6.3) *or, equivalently,* (6.4), *then the following conditions hold. Furthermore these two conditions are equivalent.*

  (a) *For* p *close enough to* e *(that is, for all* p *in some open subset of* $\mathring{\Delta}$ *containing* e):

(6.11) $$0 \;=\; \xi_p(p) \leq \xi_e(p)$$

with equality only at $p = e$.

  (b) *For* p *close enough to* e

(6.12) $$\frac{dL^e}{dt} \leq 0$$

with equality only at $p = e$.

Proof. First, observe from (6.5) that L^e differs from the function denoted L^b (with $b = -e$) in section 3, only by addition of the constant $-\sum e_i \ln e_i$ (the entropy of e). Consequently the gradient of L^e is given by (3.32):

$$(6.13) \qquad\qquad (\bar{\nabla}_p L^e)_i = p_i - e_i .$$

From (3.18) and (3.26) we compute:

$$\frac{dL^e}{dt} = d_p L^e(X(p)) = (\bar{\nabla}_p L^e, X(p))_p$$

and by using the Shahshahani metric at p and (6.13):

$$(6.14) \qquad \frac{dL^e}{dt} = \sum (p_k - e_k)\xi_k(p) = \xi_p(p) - \xi_e(p) \equiv \dot{L}^e.$$

Thus, we see that (6.11) is equivalent to (6.12), i.e. (a) is equivalent to (b). Furthermore, (a) says exactly that the difference function we have denoted $\dot{L}^e$, which vanishes at e, has a strict local maximum at e.

Now the analogy with the proof of Theorem 4.2 is apparent. We will use strong stability to check that $\dot{L}^e$ has a strict local maximum at e by using the second derivative test. First, extend as usual so that we can take partial derivatives. In particular, use $x_k - e_k$ to extend $p_k - e_k$ and compute:

$$\frac{\partial \dot{L}^e}{\partial x_i} = \xi_i + \sum_k (p_k - e_k)\frac{\partial \xi_k}{\partial x_i} .$$

At $p = e$ this vanishes and so $\dot{L}^e$ has a critical point at e. For the second derivative:

$$(6.15) \qquad \frac{\partial^2 \dot{L}^e}{\partial x_i \partial x_j} = \frac{\partial \xi_i}{\partial x_j} + \frac{\partial \xi_j}{\partial x_i} + \sum_k (p_k - e_k)\frac{\partial^2 \xi_k}{\partial x_i \partial x_j} .$$

But with respect to motions in Δ, that is vectors Y in $\mathbb{R}_o^I$, this is negative definite at e because:

$$\sum \frac{\partial^2 \dot{L}^e}{\partial x_i \partial x_j} Y_i Y_j = 2\sum \frac{\partial \xi_i}{\partial x_j} Y_i Y_j = 2H_p(Y,Y) < 0$$

by formula (4.20) for the Hessian. Note that the sum in (6.15) and the first sum in (4.20) vanish at the equilibrium point $p = e$. □

Notice that if $\overset{\bullet}{L}{}^{e}$ has a local maximum at e, then while the Hessian may not be negative definite it is at least negative semidefinite. Thus, we get a weak converse of the theorem: conditions (a) and (b) imply (6.3) and (6.4) but with the strict inequality replaced by $\leq$.

Instead of building an analogy with Theorem 4.2, we could have applied that result directly. In this case, it says strong stability implies for p close enough to e

$$(6.16) \qquad\qquad \frac{dN^e}{dt} \leq 0$$

with equality only at $p = e$, where N^e is defined by

$$(6.17) \qquad\qquad N^e(p) = \sum \frac{1}{e_i} (p_i - e_i)^2 .$$

That L^e is really more natural and useful is shown by the simple and suggestive formula for the derivative along the vector field, $\overset{\bullet}{L}{}^{e}$, given by (6.14). You can check that the formula for $\overset{\bullet}{N}{}^{e}$ is a mess.

* * *

We are now ready to look at game models in some detail. They are built from a finite list of strategies I and a matrix whose entry a_{ij} is the payoff to the i player when he meets a j opponent. We introduced in section 3 the Taylor-Jonker dynamic of the relative frequencies of the different strategies in a population:

$$(6.18) \qquad\qquad \frac{dp_i}{dt} = p_i(a_{ip} - a_{pp}) \qquad p \in \Delta ,$$

where $a_{ip} = \sum a_{ij}p_j$ is the average payoff to the i player because he meets a j opponent with probability p_j, and $a_{pp} = \sum p_i a_{ip} = \sum p_i p_j a_{ij}$ is the average payoff for the population.

Notice that the strategy type i is a fixed character of the individual directly passed on to its offspring. Genetics are ignored or are assumed to be haploid. Because a special case of these equations models natural selection some confusion arises here. To see the difference notice that in the game model a pairing ij is an encounter between two strategists who retain their individuality and receive separate payoffs. In the selection model ij is the zygotic union of two gamete types and it is this which induces the symmetry condition $a_{ij} = a_{ji}$ for the selection interpretation.

Thus, the Taylor-Jonker equations describe the *pure-strategy* dynamic. The distribution vector p in Δ describes a population polymorphism. The relative frequency of the i strategy cohort is p_i and its growth rate is a_{ip}.

In classical game theory there is an alternative interpretation of the distribution p as a mixed strategy. A p player randomizes. At each encounter he makes an independent selection of a pure strategy using strategy i with probability p_i . Thus, the expected payoff he receives against a j player is a_{pj} or against a q player is

$$(6.19) \qquad a_{pq} = \sum p_i q_j a_{ij}$$

where q is another mixture in Δ .

There are dynamic models in classical game theory. They model designs I might use to improve my play by changing the relative weighting in the mixed strategy I use, e.g., I discover that strategy i seems to do well. I adjust the probabilities to put more weight on i . This element of reflective choice by a rational player is completely absent from evolutionary game models. I emphasize again that for such models the strategy type is a fixed characteristic of the player. Selection can only change the relative frequencies of a population polymorphism. If strategy i does well, then the cohort of i strategists will increase in size relative to the others and the population distribution will show larger weight at i . If, instead, the population were monomorphic, all playing a single mixed strategy, then because this strategy is inherited as a fixed character the population state remains unchanged. No matter how inefficient the state, selection cannot act without variation in the population. Of course, mutation can introduce such variation but it is not included in the models we are looking at.

While we will use the language of classical game theory and mixed strategies for motivation we will be examining the pure strategy dynamic almost exclusively. At the end of the section we will discuss the complications introduced by allowing individuals to randomize.

A point e of Δ is an equilibrium for (6.18) when for all i

$$(6.20) \qquad \text{either } a_{ie} = a_{ee} \text{ or } e_i = 0 .$$

In game theory language, the strategies i in the support of e , that is, where $e_i > 0$, are called the *active* strategies because they are used with positive probability. Using this language we can restate (6.20) as the Bishop-Cannings condition for equilibrium:

6.2 Proposition. $e \in \Delta$ *is an equilibrium if and only if all of the active strategies have the same payoff against* e .

Proof. (6.20) says that for $e_i > 0$ the payoffs a_{ie} are all a_{ee} . On the other hand, if these payoffs are all a common constant C , multiply the equation $a_{ie} = C$ by e_i for $i \in \text{supp}(e)$ and sum on i . Hence, $a_{ee} = C$, and we get (6.20). $\square$

e is called an *interior equilibrium* if $e \in \mathring{\Delta}$, i.e., if all strategies are active. (6.20) says that $e \in \mathring{\Delta}$ is an equilibrium if and only if $a_{ie} = a_{ee}$ for all i . If we let u be the barycentre of Δ :

$$(6.21) \qquad u_i = 1/n$$

where n is the number of strategies in I, then we can average the equation $a_{ie} = a_{ee}$ over u to get $a_{ue} = a_{ee}$. Thus, $e \in \overset{\circ}{\Delta}$ is an equilibrium if and only if $a_{ie} = a_{ue}$ for all i. Finally, this says that an interior equilibrium is precisely a positive solution of the system of $n + 1$ equations in n unknowns:

(6.22)

$$\sum_j (a_{ij}-a_{uj})e_j = 0$$

$$\sum_j e_j = 1.$$

Observe that the sum of the first n of these equations is identically 0. This is at least one redundancy among these $n + 1$ equations. Thus, finding the interior equilibrium consists of solving the linear system (6.22) (use Gaussian elimination or let UNIVAC do it), and then picking out the positive solutions.

To get all equilibria, observe that an equilibrium e with supp(e) equal the subset J of I is just an interior equilibrium for the game dynamic restricted to $\mathbb{R}^J$. This consists of forgetting about the existence of the $I - J$ strategies becaue they do not appear in the population.

In particular, with J consisting of the single strategy i, we get the vertex strategy δ^i with

(6.23)
$$\delta^i_j = \begin{cases} 0 & i \neq j \\ 1 & i = j. \end{cases}$$

Each vertex is always an equilibrium. This amounts to saying again that a monomorphic population remains unaffected by selection.

Since interior equilibria play a special role it is worth looking at the nice game theory characterization of when such equlibria do not exist.

A useful idea is the concept of *domination*. Strategy i_1 dominates i_2 if $a_{i_1 j} \geq a_{i_2 j}$ for all j with some inequalities strict. This means that i_1 does at least as well as i_2 against all opponents and better against some. In a population where all strategies are present it seems intuitive that the i_2 players are competitively eliminated by the i_1's. As we will now see, this is true.

In general, for $q_1, q_2 \in \Delta$ we say q_1 *dominates* q_2 if

(6.24)
$$a_{q_1 j} \geq a_{q_2 j}$$

for all j in I with strict inequality for some j. This is just the mixed strategy analogue of the idea.

Notice that these imply that for p in the interior $\overset{\circ}{\Delta}$:

(6.25)
$$a_{q_1 p} > a_{q_2 p}.$$

6.3 Theorem. *An interior equilibrium e for (6.18) does not exist if and only if there exist q_1 and q_2 in Δ such that q_1 dominates q_2. Furthermore, q_1 and q_2 can then be chosen with disjoint supports.*

Partial Proof. We will prove the easy direction. If q_1 dominates q_2, then with $b = q_1 - q_2$ we have $b \in \overset{I}{\mathbb{R}_o}$ and so with $L^b(p) = \sum b_i \ln p_i$ on $\overset{\circ}{\Delta}$:

$$(6.26) \qquad \frac{dL^b}{dt} = (\overline{\nabla}_p L^b, X)_p = \sum b_i(a_{ip} - a_{pp}) = a_{q_1 p} - a_{q_2 p} > 0$$

Thus, the function L^b is constantly increasing in the interior $\overset{\circ}{\Delta}$. In particular, there is no equilibrium there.

The converse applies convex set arguments when (6.22) has no positive solution to find $b \in \overset{I}{\mathbb{R}_o}$ such that $a_{bp} > 0$ for p in $\overset{\circ}{\Delta}$.

$$(q_1)_i = \begin{cases} b_i, & b_i > 0 \\ 0, & b_i \le 0 \end{cases} \qquad\qquad (q_2)_i = \begin{cases} -b_i, & b_i < 0 \\ 0, & b_i \ge 0. \end{cases}$$

After multiplying by a common positive constant we get q_1, q_2 in Δ with disjoint supports and $b = q_1 - q_2$ up to a constant multiple. We then recover (6.25) and so (6.24). $\qquad\qquad\qquad\qquad\qquad\qquad\qquad\qquad\qquad\qquad\qquad\qquad\qquad\qquad\qquad\square$

When $b = q_1 - q_2$, L^b is the log of the ratios:

$$(6.27) \qquad\qquad\qquad L^b(p) = \ln(p^{q_1}/p^{q_2}).$$

Here we are denoting by p^{q_1} the weighted geometric average:

$$p^q = \prod (p_i)^{q_i}.$$

When q_1 dominates q_2 it can be shown from (6.26) that as time it tends to infinity;

$$(6.28) \qquad\qquad\qquad \text{limit } p(t)^{q_2} = 0,$$

for a solution path $p(t)$ in $\overset{\circ}{\Delta}$. It is in this sense that the q_2 family of strategies is competitively eliminated. You should be careful to avoid the error which Hofbauer showed me I had made. (6.28) says that for the p_i's with i in the support of q_2 the product tends to 0. This does not imply that any one of them tends to 0 alone!

For game dynamics we can sharpen the strong stability results of Theorem 6.1.

6.4 Theorem. *An interior equilibrium e of the Taylor-Jonker dynamic, (6.18), is called strongly stable if it is strongly stable with respect to the Shahshahani metric at e . Each of the following conditions is equivalent to strong stability.*

(a) *For all nonzero* Y *in* $\mathbb{R}_o^I$:

(6.29)
$$a_{YY} = \sum a_{ij}Y_iY_j < 0.$$

(b) *For* p *in* Δ *with* $p \neq e$

(6.30)
$$a_{pp} < a_{ep}.$$

(c) *For the function* L^e *defining by (6.5) on* $\mathring{\Delta}$

(6.31)
$$\frac{dL^e}{dt} \leq 0$$

at every point p *of* $\mathring{\Delta}$ *with equality only at* p = e.

In fact, if (6.30) or (6.31) holds for p *close enough to* e, *then* e *is strongly stable and so they hold for all* p *in* $\mathring{\Delta}$.

Proof. Use the Hessian computation (4.21) for the game dynamic case. Since e is an equilibrium, (6.29) is just a restatement of the Hessian condition (6.4) strong stability.

Now rewrite (6.30) as:

(6.32)
$$a_{p-e\,p} < 0$$

for $p \neq e$. Then, because e is an interior equilibrium, $a_{ie} = a_{ee}$ for all i and so averaging using p we get $a_{pe} = a_{ee}$ or

(6.33)
$$a_{p-e\,e} = 0.$$

Subtracting the equilibrium condition (6.33) we get that (6.30) is equivalent to

(6.34)
$$a_{p-e\,p-e} < 0$$

for p in Δ with $p \neq e$. But Y = p - e lies in $\mathbb{R}_o^I$, and so (6.29) implies (6.34). Conversely, if (6.34) holds for p in some small $\mathring{\Delta}$ neighborhood of e and $Y \in \mathbb{R}_o^I$, then $p = e + \varepsilon Y$ lies in the neighborhood for $\varepsilon > 0$ small enough. So (6.34) implies

$$\varepsilon^2 a_{YY} < 0$$

and hence (6.29) is true.

Finally, in the proof of Theorem 6.1 we computed that at any point p of $\mathring{\Delta}$:

$$\frac{dL^e}{dt} = a_{pp} - a_{ep}$$

and so (6.31) is equivalent to (6.30). $\square$

These results show that for game models a strongly stable interior equilibrium is not merely stable, with points near e attracted to e , but globally stable. Along every solution path in the interior $\mathring{\Delta}$, L^e decreases monotonically toward 0 and so the solution path approaches e . By contrast, Zeeman constructed a payoff matrix example with three strategies, admitting a linearly stable interior equilibrium which is not globally stable.

Furthermore, strong stability is much easier to test for than linear stability. First, notice that the condition (6.29) does not even depend on the equilibrium e . Another advantage is that we can avoid choosing coordinates on the simplex Δ . This statement sounds paradoxical so let us pause here for a moment. In the last section I discussed how important the right choice of coordinates can be. In fact, it is so important that if a nice choice of coordinates does not present itself, you are better off trying to operate without coordinates. This seems beside the point here because the p_i's are coordinates on Δ , aren't they? Actually, they are not because they are not independent. They are constrained to sum to one. You can get a coordinate system by omitting one of the p_i's , but this singles out one of the p_i's , breaking the natural symmetry among them. These aesthetic issues can have important consequences. I mentioned that Antonelli and Strobeek discovered Shahshahani's metric before he did. But they were operating in exactly this, all but p_n , coordinate system and so the resulting formulae were rather ugly. As a result most of the beautiful geometry associated with the metric remained hidden.

We now use these preliminaries about interior equilibria to study equilibria in general.

Let e be an equilibrium with support J , a subset of I . From (6.20) this means $a_{ie} - a_{ee}$ is 0 for all indices i in the support set J of e . We use the terms the *restricted system* or the *restriction to* J to refer to system (6.18) applied to indices i in J and to p's only in the simplex Δ_J of $\mathbb{R}^J$. This means we consider the dynamic of the game restricted to use only strategies in J or, equivalently, only J strategies appear in the population.

We regard X in $\mathbb{R}^J$ as a member of $\mathbb{R}^I$ by defining $X_i = 0$ for $i \notin J$. This includes $\mathbb{R}^J$ of a subspace of $\mathbb{R}^I$ and so the simplex Δ_J as a subset, a face, of Δ . Δ_J is an invariant subset for the original system. This is why we can define the system restricted to J . It amounts to saying again that if $p_i = 0$ to begin with for $i \notin J$, then it remains 0 . New strategies are not introduced by the model (6.18).

Thus, an equilibrium of support J is exactly an interior equilibrium for the restricted system. This allows us to break the question of stability into two parts. First, we call e *internally stable* if it is a stable interior equilibrium for the system restricted to J = supp(e) . This means that if we perturb e to a new point p near e but with supp(p) = J , then the solution path is attracted back to e . The word "internal" emphasizes that we are looking only at little changes in the relative frequencies of the existing strategies. No new strategies are introduced. Similarly, we will use the regrettably cumbersome phrases *internally linearly stable* and *internally strongly stable* to denote linear and strong stability, respectively, of e in the restricted system.

For an internally stable strategy the next question is how it fares against the introduction of new strategies from I - J . By the Bishop-Cannings equilibrium conditions, $a_{ie} = a_{ee}$ for all i in J . All strategies active in e do equally well. But the remaining differences $a_{ie} - a_{ee}$ need not be 0. Intuitively, if $a_{ie} < a_{ee}$, then i is competitively inferior against e and should be eliminated if it is introduced in small numbers. The precise result which justifies this intuition is:

6.5 Proposition. *The eigenvalues of the linearization of (6.18) at an equilibrium e with support J consist of the eigenvalues of the restriction to J and the numbers $a_{ie} - a_{ee}$ for $i \in I - J$.*

Proof. Let $L : \mathbb{R}_o^I \to \mathbb{R}_o^I$ denote the linearization at e. Pick a basis for $\mathbb{R}_o^J$, and then, with j a fixed member of J, choose $\delta^i - \delta^j$ for $i \notin J$. $\delta^i - \delta^j$ is the vector which is $+1$ in the ith place, -1 in the jth, and 0 elsewhere. We get a basis for $\mathbb{R}_o^I$.

With respect to this basis the matrix for L has the bloc triangular form:

$$L = \begin{pmatrix} A & B \\ O & D \end{pmatrix},$$

where A is the linearization of the restricted system on $\mathbb{R}_o^J$. The O bloc below A comes from the invariance of Δ_J which implies that L maps the subspace $\mathbb{R}_o^J$ into itself.

It follows that the eigenvalues of L are those of A together with those of D, with B irrelevant. (Hint: the determinant of $L - \lambda I$ is the product of $\det(A - \lambda I)$ and $\det(D - \lambda I)$).

Finally, D is a diagonal matrix with entries $a_{ie} - a_{ee}$ down the diagonal for all $i \in I - J$. The reason for this is that $p_i = 0$ at e for such i's. So if we take the partial derivative $\partial/\partial x_j$ of $x_i(a_{ix} - a_{xx})$ at e we get 0 except for $j = i$, when we get $a_{ie} - a_{ee}$.
□

6.6 Corollary. e *is linearly stable if and only if it is internally linearly stable and, in addition,* $a_{ie} < a_{ee}$ *for all* i *not in the support of* e. □

These results lead naturally to an equilibirum concept from classical game theory:

6.7 Proposition. *For the dynamical system (6.18) a point* e *of* Δ *is called a* Nash equilibrium *if it satisfies the following equivalent conditions:*
(a) *For all* i *in* I

(6.35) $a_{ie} \le a_{ee}$.

(b) *For all* p *in* Δ

(6.36) $a_{pe} \le a_{ee}$.

(c) e *is an equilibrium and for* i *not in the support of* e *inequality (6.35) holds.*
e *is called a* regular Nash equilibrium *if (c) is replaced by*
(c') e *is an equilibrium and for* i *not in the support of* e:

(6.37) $$a_{ie} < a_{ee} .$$

An interior equilibrium $(e \in \overset{\circ}{\Delta})$ is a regular Nash equilibrium. Every system admits at least one Nash equilibrium though not necessarily a regular one.

Proof. (6.35) is the special case of (6.36) where p is the vertex δ^i. Conversely, if (6.35) holds, then

$$a_{pe} - a_{ee} = \sum_i p_i(a_{ie} - a_{ee}) \leq 0 .$$

It follows that (a) and (b) are equivalent. Applying this with $p = e$ we get that 0 is the sum of nonpositive terms $e_i(a_{ie} - a_{ee})$, and so all of these are 0. Thus, (6.35) is equivalent to the equilibrium condition $a_{ie} - a_{ee} = 0$ for i in the support of e and the inequality for the remaining i's. This is what (c) says.

For an interior equilibrium the support is all of I and so the additional condition (6.37) is vacuous.

The existence proof for Nash equilibria is interesting in that it uses a dynamical system which allows new strategies to appear in the population. This is in contrast with both (6.18) and its discrete time analogue (see below).

Define for $p \in \Delta$:

(6.38) $$k(p)_i = \max(a_{ip} - a_{pp}, 0) .$$

So $k(p)_i \geq 0$ and equals 0 for all i if and only if p satisfies (6.35), that is, p is a Nash equilibrium. Now define a continuous map K from Δ to itself.

(6.39) $$K(p)_i = (p_i + k(p)_i)/(1 + \sum_j k(p)_j) .$$

By the Brouwer fixed point theorem K has some fixed point. That is, for some $e \in \Delta$, $K(e) = e$ which we can rewrite as

$$k(e)_i = Ce_i$$

for all i with $C = \sum_j k(e)_j$. We show that C is 0 implying that e is Nash.

If C were not 0, then for i in the support of e, $k(e)_i$ is not 0 and so equals $a_{ie} - a_{ee}$. Hence,

$$C\sum_i e_i^2 = \sum_i e_i k(e)_i = \sum_i e_i(a_{ie} - a_{ee}) = 0$$

because each of these sums is the same whether i varies over all of I or just the support of e. Dividing by $\sum_i e_i^2$ we get that $C = 0$ after all. $\qquad \square$

The Nash idea is the natural equilibrium concept from the viewpoint of classical game theory. It says that there is no strategy i which improves the payoff of the current mixture e . So there is no incentive for a rational player to move his mixture from e . There is also no reason to restrict the options of such a decision maker only to strategies having positive weight in the current mixture.

On the other hand, for evolutionary game models, where the dynamics does not introduce new strategies, the Nash condition is an odd hybrid. For i in the support of e it is exactly the equilibrium condition. It says nothing about stability. This is why any interior equilibrium is Nash. However, for the indices not in the support, the Nash condition describes a kind of partial stability. By Proposition 6.5 it says that the eigenvalues beyond those of the restriction are all negative (in the regular Nash case) or at least nonpositive. We can thus restate Corollary 6.6 as:

6.8 Corollary. *An equilibrium e is linearly stable for (6.18) if and only if it is internally, linearly stable and is a regular Nash equilibrium.*   □

The Nash condition is an intermediate one between mere equilibrium and various kinds of stable equilibrium conditons. The most interesting version of the latter is Maynard Smith's ESS condition. The initials originally stood for "evolutionarily stable strategy" but as we want to apply it to a population polymorphism rather then to a single strategy (even a mixed one) it is perhaps better to replace the third word by "state" which begins, happily, with the same letter.

6.9 Proposition. e *is called an* evolutionarily stable state (ESS) *if it satisfies the following equivalent conditions:*
   (a) *For all* p *in* Δ

$$(6.40) \qquad\qquad a_{pe} \leq a_{ee} .$$

Furthermore, if for some $p \neq e$ *equality holds, then*

$$(6.41) \qquad\qquad a_{pp} < a_{ep} .$$

   (b) e *is an equilibrium and for all* $p \neq e$ *in* Δ *close enough to* e *inequality (6.41) holds.*
   (c) e *is an equilibrium and for all* $p \neq e$ *close enough to* e *the inequality*

$$(6.42) \qquad\qquad \frac{dL^e}{dt} < 0$$

holds. Here $L^e(p) = -\sum e_i \ln(p_i/e_i)$, *with summation over* i *in the support of* e . *So* $L^e(p)$ *is defined for all* p *in* Δ *such that* $\text{supp}(e) \subset \text{supp}(p)$.

An ESS is a Nash equilibrium and it is internally strongly stable. Conversely, if e *is a regular Nash equilibrium which is internally strongly stable, then* e *is called a* regular ESS *which, as the name suggests, is an ESS.*

In particular, for an interior equilibrium, the concepts ESS and regular ESS both coincide with strong stability.

Proof. Condition (a) is essentially Maynard Smith's original definition of an ESS. Condition (6.40) is exactly (6.36) and so an ESS is Nash. For $J = \text{supp}(e)$ and p in Δ_J the equilibrium condition says exactly $a_{pe} = a_{ee}$, and so by (6.41) the strong stability condition (6.30) holds on Δ_J. Thus, an ESS is strongly stable when the system is restricted to its support strategies.

Conversely, if e is a regular Nash equilibrium, then (6.40) is true and by (6.37) equality holds only for p in Δ_J. For such p, (6.41) follows from (6.30) provided e is internally strongly stable.

Now we go back to the proof of Theorem 6.4 where we saw that:

$$\frac{dL^e}{dt} = a_{pp} - a_{ep} = a_{p-e\ p}$$

provided L^e is defined at p, i.e., $\text{supp}(e) \subset \text{supp}(p)$. It is then obvious that (b) is equivalent to (c).

So it remains to prove the result of Hofbauer and Sigmund that (a) is equivalent to $a_{p-e\ p} < 0$ for $p \neq e$ close enough to e.

Following the previous proof we write:

$$(6.43) \qquad a_{p-e\ p} = a_{p-e\ p-e} + a_{p-e\ e} .$$

Last time e was an interior equilibrium and so $a_{p-e\ e} = 0$ for all p. Now we have the weaker Nash condition (6.40) which says $a_{p-e\ e} \leq 0$. Furthermore, in this case the inequality $a_{p-e\ p-e} < 0$ holds only for those $p \neq e$ where $a_{p-e\ e} = 0$. For other points p this term can even be positive.

To deal with this complication fix any metric on $\mathbb{R}^I$, say the standard one, and define:

$$S_e = \{Y \in \mathbb{R}_o^I : Y_i \geq 0 \text{ for } i \notin \text{supp}(e) \text{ and } \|Y\| = 1\} .$$

Notice that because $p_i \geq 0$ for all i, the difference $p - e$ is the positive multiple of a vector in S_e. Conversely, if Y is in S_e, then $p = e + \varepsilon Y$ lies in Δ provided $\varepsilon \geq 0$ is small enough (check this!). So we can rewrite (6.43) as

$$(6.44) \qquad a_{p-e\ p} = \varepsilon^2 a_{YY} + \varepsilon a_{Ye} \equiv h(\varepsilon) ,$$

and condition (a) can be rewritten as

$$a_{Y_e} \leq 0 \text{ for } Y \text{ in } S_e,$$

(6.45)   and when equality holds, then

$$a_{YY} < 0.$$

Assume (6.45) is true and think of the right side of (6.44) as a real-valued function h of $\varepsilon \geq 0$ with Y fixed. If $a_{Y_e} < 0$, then $h(0) = 0$ and $dh/d\varepsilon < 0$ a 0. Hence, for $\varepsilon > 0$ and small enough $h(\varepsilon) < 0$. Even though a_{YY} may be positive in this case, it is multiplied by ε^2 and so is dominated by the linear term εa_{Y_e} for ε small enough. On the other hand, if $'a_{Y_e} = 0$ and $a_{YY} < 0$, then $h(\varepsilon) < 0$ for all nonzero ε. Finally, because the set S_e is compact, we can find a positive ε_o so that $h(\varepsilon) < 0$ for all $0 < \varepsilon < \varepsilon_o$ and all Y in S_e. The set $p = e + \varepsilon Y$ with $0 \leq \varepsilon < \varepsilon_o$ describes all points close enough to e. So from (6.44) (b) follows.

Conversely, (b) says exactly that for all Y in S_e, $h(\varepsilon) < 0$ for positive ε small enough. Hence, $dh/d\varepsilon \leq 0$ at $\varepsilon = 0$, i.e., $a_{Y_e} \leq 0$, and if equality holds, then $a_{YY} < 0$. Thus, from (b) we recover (6.45) which is equivalent to (a).   □

Notice that in contrast with an interior ESS, i.e., a strongly stable interior equilibrium, for a boundary ESS we no longer know that $a_{YY} < 0$ for all nonzero Y in $\mathbb{R}_o^I$. We only know it for Y in $\mathbb{R}_o^J$, where J is the support of the ESS. Also, we no longer have global stability. Conditions (b) and (c) only hold for p close enough to e. These two problems are closely related as shown by the following:

6.10 Corollary. *Assume that for all nonzero Y in $\mathbb{R}_o^I$, $a_{YY} < 0$. The Taylor-Jonker dynamic for* (a_{ij}) *admits a unique Nash equilibrium e. e is an ESS and for all $p \neq e$ with* $supp(e) \subset supp(p)$, $dL^e/dt < 0$.

Proof. By Proposition 6.7 at least one Nash equlibrium e exists. In the notation of the previous proof $a_{Y_e} \leq 0$ for all Y in S_e. Furthermore, for all such Y, $a_{YY} < 0$. In (6.44) it follows that $h(\varepsilon) < 0$ for all Y in S_e and all $\varepsilon > 0$. This time we avoid the problem of $a_{YY} > 0$ which imposed the requirement that ε be small. Consequently, (6.41) holds for all $p \neq e$. In particular, when $supp(e) \subset supp(p)$ so that L^e and its derivative are defined at p, $dL^e/dt < 0$. It follows that every solution path starting at a point in $\overset{\circ}{\Delta}$ approaches e in the limit. This common destiny implies that only one such e can exist.   □

When the payoff matrix is symmetric, the equations are the natural selection equations, the Shahshahani gradient of mean fitness (times $\frac{1}{2}$). Let us see how the ESS concepts specialize in that case.

6.11 Proposition. *Assume the payoff matrix* a_{ij} *is symmetric. An interior equilibrium is then linearly stable if and only if it is strongly stable. In general, an equilibrium* *is linearly stable if and only if it is a regular ESS. A point is an ESS if and only if it is a strict local maximum point for the mean fitness function* $\tilde{a}$.

Proof. Because the Hessian of (4.21) is symmetric, definitions (4.19) and (4.9) imply that the linearization at an interior equilibrium e is a symmetric linear map with respect to $(\,,\,)_e$. As observed in Propositon 4.1 the eigenvalues are all negative if and only if the Hessian is negative definite. So an interior equilibrium is linearly stable if and only if it is strongly stable.

A regular ESS is always linearly stable by Corollary 6.8 which also shows that a linearly stable equilibrium is regular Nash. Now apply the above symmetry results to go from linear stability to strong stability for the restricted system.

For the mean fitness result use the notation of the proof of Proposition 6.9. In particular, with e fixed, write $p = e + \varepsilon Y$ for Y in S_e there, and expand $\tilde{a}(p) = a_{pp}$:

$$(6.46) \qquad \tilde{a}(p) = \tilde{a}(e) + 2\varepsilon a_{Ye} + \varepsilon^2 a_{YY} .$$

We used symmetry to say that $a_{eY} = a_{Ye}$. Now just as with (6.44) we can argue to see that e is an ESS if and only if (6.45) holds, if and only if $\tilde{a}(p) < \tilde{a}(e)$ for all $p \neq e$ close enough to e . $\qquad\square$

We conclude this portion of our analysis by investigating the effect of changing the payoff matrix in various simple ways. In each case we replace (a_{ij}) by a new matrix $(\tilde{a}_{ij})$ of payoffs for the same set of strategies:

(i) $\qquad\qquad\qquad \tilde{a}_{ij} = a_{ij} + k$

(ii) $\qquad\qquad\qquad \tilde{a}_{ij} = a_{ij} + k_j$

(iii) $\qquad\qquad\qquad \tilde{a}_{ij} = a_{ij} + k_i$

(iv) $\qquad\qquad\qquad \tilde{a}_{ij} = ka_{ij} \qquad\qquad\qquad (k \neq 0)$

(v) $\qquad\qquad\qquad \tilde{a}_{ij} = k_j a_{ij} \qquad\qquad\qquad (k_j > 0)$.

Cases (i) and (ii), addition of constant or a term depending only upon the opponent's strategy, have no effect at all on the dynamical system. In (ii), for example,

$$\tilde{a}_{ip} - \tilde{a}_{pp} = (a_{ip} + k_p) - (a_{pp} + k_p) = a_{ip} - a_{pp} .$$

Case (iii) does change the dynamical system. Here k_i represents an additional cost or payoff for my use of strategy i independent of my opponent's strategy. This can clearly change the relative advantages among the strategies and so does affect the dynamical system. However, the Hessian form remains unchanged because, for Y^1, Y^2 in $\mathbb{R}_o^1$:

(6.47)
$$\sum \tilde{a}_{ij} Y_j^1 Y_i^2 = \sum (a_{ij}+k_i) Y_j^1 Y_i^2 = \sum a_{ij} Y_j^1 Y_i^2$$

(why?). The effect of the additive payoff term is to move the equilibrium. In fact, the system of equations (6.22) for an interior equilibrium $\tilde{e}$ with payoff matrix $(\tilde{a}_{ij})$ becomes:

$$\sum_j (a_{ij}-a_{uj})\tilde{e}_j = -k_i + k_u$$

(6.48)
$$\sum_j \tilde{e}_j = 1$$

where $u_j = 1/n$ for all j (n is the number of strategies) and so $k_u = n^{-1}\sum_j k_j$.

In particular, if $a_{YY} < 0$ for all nonzero Y in R_o, then Corollary 6.10 applies to all of the case (iii) payoff matrices, and so there is then a unique ESS whose position in Δ can be moved about arbitrarily by changing the choice of k_i's.

Multiplication by a positive constant, case (iv) with $k > 0$, leaves the solution paths unchanged but the speed of motion along the paths is multiplied by the constant k. On the other hand, if k is negative, then motion along the paths reverses direction. In particular, with $k = -1$, so that the new payoff matrix is $-a_{ij}$, the new motion is exactly the inverse of the original motion. Thus, moving t time units into the future in the new system is exactly the same as moving t units backwards into the past of the original one.

Case (v) is very interesting because the following proposition of Zeeman shows that it is dynamically equivalent to the original systsem with respect to a so-called projective transformation.

6.12   Proposition. *Define the map* $R : \Delta \to \Delta$ *by*

(6.49)
$$R(p)_i = (p_i/k_i)/\sum_j (p_j/k_j) .$$

R *maps a solution path of the original system* (6.18) *to a solution path of the new system with* a_{ij} *replaced by* $\tilde{a}_{ij} = k_j a_{ij}$. *The time parametrizations are related by a nonlinear time change.*

Proof. Recall from section 3 that our dynamical system on Δ is derived from one on the subset R_+^I of nonnegative vectors in R^I (see (3.7)):

$$\frac{dx_i}{dt} = x_i a_{ip} = \frac{1}{|x|} x_i \sum_j a_{ij} x_j .$$

This means that if we take a solution path and project it to Δ by the map

$$P : R_+^I \to \Delta \qquad P(x)_i = x_i/|x| ,$$

then we get a solution path of (6.18).

Now change variables, defining

$$\tilde{x}_i = x_i/k_i \quad \text{or} \quad x_i = \tilde{x}_i k_i$$

and substituting. Divide by k_i and get:

$$\frac{d\tilde{x}_i}{dt} = \tilde{x}_i a_{ip} = \frac{1}{|x|} \tilde{x}_i \sum_j \tilde{a}_{ij} \tilde{x}_j = \frac{|\tilde{x}|}{|x|} \tilde{x}_i \tilde{a}_{i\tilde{p}_i}$$

where $\tilde{p}_i = \tilde{x}_i/|\tilde{x}|$, i.e., $\tilde{p} = P(\tilde{x})$. Introduce the new time variable τ by

$$\frac{d\tau}{dt} = \frac{|\tilde{x}|}{|x|}$$

(see section 5 for a discussion of this kind of nonlinear time charge). The system has become

$$\frac{d\tilde{x}_i}{d\tau} = \tilde{x}_i a_{i\tilde{p}}$$

which projects by P to the system (6.18) with a_{ij} replaced by $\tilde{a}_{ij}$.

So if we start with a solution path $p(t)$ of the original system, converting to $\tilde{x}$ coordinates and projecting again to Δ , we get a solution for the $\tilde{a}$; system but with time τ . R is just the projection P applied to the $\tilde{x}$ coordinates of $p(\tau)$.   □

The propositon implies that e is an equilibrium for the original system if and only if $R(e)$ is an equilibrium for the new one. Notice that R maps each face of Δ into itself, i.e., $\text{supp}(R(p)) = \text{supp}(p)$. In particular, e is interior if and only if $R(e)$ is. By continuity of R and its inverse e is stable for a_{ij} if and only if $R(e)$ is stable for $\tilde{a}_{ij}$. Because R is smooth, linear stability is preserved as well. However, strong stability need not be. We call e a *generalized ESS* for a_{ij} , if $R(e)$ is an ESS with respect to $(k_j a_{ij})$ for some positive choice of k_j 's . A generalized ESS need not be an ESS and L^e need not be increasing on orbits of (6.18). Instead, one uses the function $L^{R(e)}(R(p))$ of p .

We apply this to Corollary 6.11, obtaining a more general condition for the existence of a globally stable equilibrium.

6.13 Corollary. *If there exist* $k_j > 0$ *such that for all nonzero* Y *in* $\mathbb{R}_0^I$:

$$\sum_{ij} a_{ij} k_j Y_i Y_j < 0 ,$$

then the system (6.18) *admits an equilibrium* e *such that every solution path in the interior of* Δ *approaches* e *in the limit.*   □

* * *

We now look at the analogues of some of these results for discrete time models.

A discrete time model is defined by a continuous mapping F of the state space to itself. $F(x)$ is the successor to state x, that is, the state one time unit later when the initial state is x. The subsequent states are obtained by iterating F. If t is a positive integer, then $F^t(x)$, meaning $F \circ F \circ ... \circ F$ (t times) applied to x, is the position, $x(t)$, when $x(0) = x$. If the map is invertible, then F^t is defined for negative integers as well by $(F^{-1})^{|t|}$. By convention, F^0 is the identity map I.

Given a vector field, solving the corresponding differential equation defines a map F^t for all real values of t. Here $F^t(x)$ is the value at time t of the solution path with initial position x (I am avoiding some technical problems by assuming these solution paths are defined for all real times t). It can be shown by using the uniqueness theorem for differential equations that these mappings satisfy the composition identity:

$$(6.50) \qquad\qquad F^t \circ F^s = F^{t+s} .$$

So, in particular, the maps F^t for integral t are just the iterates of the time-one map F^1.

Thus, from a vector field continuous time model we can always obtain, in theory, a discrete time model, and the solution maps for real t interpolate mappings between the integer time maps of the discrete model.

In particular, for the real-valued linear model

$$\frac{dx}{dt} = ax \qquad \text{with} \qquad F^t(x) = e^{at}x$$

the discrete time model is:

$$F(x) = Ax \qquad \text{with} \qquad F^t(x) = A^t x ,$$

related by

$$(6.51) \qquad\qquad a = \ln A .$$

Notice that $A = e^a$ is positive. If we begin with negative A, the discrete time model exhibits behavior for which there is no continuous time analogue: jumping back and forth across 0 due to the alternating sign. The map from a vector field never reverses orientation in this fashion.

An equilibrium for a discrete time system is a fixed point for the map, a point e at which $F(e) = e$. As in the continuous time case, stability can be investigated by linearizing at equilibrium. This requires that the state space be a manifold, for example a vector space V. Then we are replacing F by the linear map $d_eF : V \rightarrow V$. As before stability for the linearized system implies stability, locally, for the original equilibrium e.

If $F : V \rightarrow V$ is linear, then the equilibrium 0 is stable precisely when the eigenvalues lie inside the unit circle, that is, each eigenvalue A has absolute value $|A| < 1$. This is precisely the condition that the sequence of powers $A, A^2,...$ approach 0.

The relation between discrete and continuous time stability is illustrated by (6.51): the real part of the log of a complex number is the log of its absolute value. So $a = \ln A$ has real part negative if and only if A has absolute value less than 1.

To write down the discrete time map which is implicit in a vector field requires that we solve the differential equations explicitly. An alternative procedure is to get an approximation by replacing the original differential equation on a vector space by a difference equation.

Define for $F : V \rightarrow V$ the change in x :

$$(6.52) \qquad\qquad \Delta x = F(x) - x .$$

Given a differential equation we can define F formally by replacing the tangent vector dx/dt by the secant vector $\Delta x/\Delta t$ for some fixed time step Δt. For the simple real variable case replace:

$$\frac{dx}{dt} = ax \qquad \text{by} \qquad \frac{\Delta x}{\Delta t} = ax ,$$

and so with $\Delta t = 1$ we get

$$F(x) = (1+a)x \qquad \text{with} \qquad F^t(x) = (1+a)^t x .$$

This replaces the exact formula (6.50) with the approximation

$$(6.53) \qquad\qquad A = 1 + a .$$

Notice that if a is close to 0, or, equivalently, A is close to 1, then $\ln(1+a)$ is close to a.

Also, if we use $\Delta t = 1/n$, then to get the result after a single time unit, the map $F_n(x) = (1+(a/n))x$ has to be iterated n times. So the value at time t is given by

$$(1+(a/n))^{nt} x$$

and as n approaches infinity, i.e., Δt approaches 0, this becomes $e^{at} x$.

May gives a nice picture which relates the stability conditions for a linear differential equation to those for its difference equation analogue. Begin by replacing:

$$\frac{dx}{dt} = L(x) \qquad \text{with} \qquad \frac{\Delta x}{\Delta t} = L(x) ,$$

where $L : V \rightarrow V$ is a linear mapping. With $\Delta t = \tau$, the fixed step size, we define, via (6.52):

(6.53) $$F_\tau = I + \tau L ,$$

where I is the identity map.

Now 0 is stable for the differential equation if the eigenvalues of L have negative real parts, that is, they all lie in the left half of the complex plane.

On the other hand, 0 is stable for the map F_τ if its eigenvalues lie inside the unit circle. Notice that:

$$A \text{ is an eigenvalue for } F_\tau = I + \tau L$$

$$\Updownarrow$$

$$\frac{A-1}{\tau} \text{ is an eigenvalue for } L.$$

A lies in the unit circle if and only if $A - 1$ lies in the circle of radius 1 with center at -1 if and only if $(A-1)/\tau$ lies in the circle of radius τ^{-1} with center at $-\tau^{-1}$. As Figure 8 shows, the latter circles lie inside the left side of the complex plane getting larger as τ gets smaller. Hence, if 0 is stable for the map F_τ, then 0 is stable for the vector field L. The converse is only true once the step size τ is small enough.

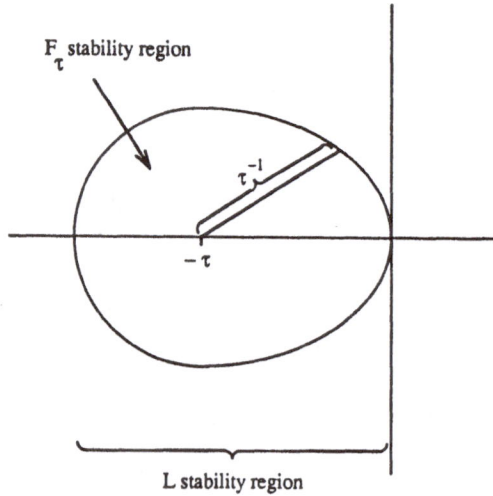

Figure 8

To apply this to the game dynamic model, go back to the beginning of section 3, where in the original definition (see (3.6) and (3.7)) we used a_{ip} as the growth rate of x_i, the size of the i strategy cohort. Replace

$$\frac{dx_i}{dt} = x_i a_{ip} \qquad \text{by} \qquad \frac{\Delta x_i}{\Delta t} = x_i a_{ip} .$$

Provided Δt is fixed at τ, the resulting discrete time dynamical system is given by the map $F : \mathbb{R}_+^I \to \mathbb{R}_+^I$:

(6.54) $$F(x)_i = x_i A_{ip} ;$$

where the matrix (A_{ij}) is:

(6.55) $$A_{ij} = 1 + \tau a_{ij} = \tau(\tau^{-1} + a_{ij}) .$$

Notice that the term added to τa_{ij} is not the identity matrix but the matrix with all entries equal to 1.

In order for the image of F to lie among the positive vectors $\mathbb{R}_+^I$ some positivity conditions are needed on the payoff matrix (A_{ij}). The most general ones which are convenient are:

(6.56)
$$A_{ij} \geq 0$$
$$\text{with} \qquad \text{for all i,j in I}$$
$$A_{ii} > 0$$

No matter what the entries a_{ij} are, the matrix of (6.55) will satisfy these conditions provided τ is small enough.

Summing on i in (6.54) we get the new population size:

(6.57) $$|F(x)| = |x| A_{pp} .$$

Now dividing (6.54) by (6.57) we get a formula for the successor's frequency distribution, and we observe that it depends only upon the old frequency distribution. So the map F induces a map $\overline{F}$ on the simplex Δ defined by:

(6.58) $$\overline{F}(p)_i = \frac{p_i A_{ip}}{A_{pp}} .$$

The conditions of (6.56) imply that $A_{ip} \geq 0$ for p in Δ with strict inequality when $p_i > 0$. Hence $A_{pp} > 0$ for all p in Δ and

(6.59) $$\text{supp}(\overline{F}(p)) = \text{supp}(p) .$$

The analogy between the discrete time dynamic given by $\bar{F}$ and the original (6.18) is easier to see if we rewrite (6.58) in difference equation form:

$$(6.60) \qquad (\Delta p)_i = \frac{p_i(A_{ip} - A_{pp})}{A_{pp}} .$$

When the payoff matrices are symmetric, the system (6.59) becomes Wright's discrete time equations for natural selection, in which A_{ij} is usually written W_{ij} . The symmetric version of (6.18) is, as we have seen, Fisher's differential equations for selection with the a_{ij}'s usually written m_{ij} . Thus, (6.51) shows why the m_{ij}'s are approximately the log of the W_{ij}'s or when the m_{ij}'s are small (equivalently the W_{ij}'s are near 1) (6.53) suggests W_{ij} is approximately $1 + m_{ij}$.

As a general rule of thumb, the discrete time systems often, but not always, behave in a fashion dynamically similar to their continuous analogues, but the discrete time results are harder to prove. Thus, for example, when the payoff matrix is symmetric, the mean fitness function $\bar{A}(p) = A_{pp}$ is a Lyapunov function for Wright's natural selection dynamic, i.e.,

$$(6.61) \qquad \bar{A}(\bar{F}(p)) \geq \bar{A}(p)$$

but the proof is harder than Fisher's Fundamental Theorem for the (6.18) symmetric case.

Returning to the general case, it is possible to prove that the system given by (6.58) or (6.60) is invertible, i.e., that the map $\bar{F}$ is a diffeomorphism of Δ . In the vector field case we saw that inverting the system, moving backwards in time, is equivalent to replacing the payoff matrix by its negative $-a_{ij}$. In contrast I know of no explicit formula for $\bar{F}^{-1}$. Instead, invertibility is proved by computing the derivative map at each point and using the inverse function theorem (see section 2). We will omit the proof, and compute the derivative only at equilibrium.

From (6.58) or (6.60) we see that the conditions that e be an equilibrium, $\bar{F}(e) = e$, are the same as for the continuous case: $e_i(A_{ie} - A_{ee}) = 0$ for all i . From now on we will only consider an interior equilibrium e so that $A_{ie} = A_{ee}$ for all i .

In computing the linearization $d_e\bar{F} : \mathbb{R}_o^I \to \mathbb{R}_o^I$ we proceed in a now familiar way. Replace the p_i's by x_i's and take the partials. It is easier to write $\bar{F}$ as the identity, I, plus the right side of (6.60). We then get for Y in $\mathbb{R}_o^I$

$$d_e\bar{F} = I + L , \quad \text{where}$$

$$(6.62)$$

$$L(Y)_i = \frac{1}{A_{ee}} \sum_j e_i(A_{ij} - A_{ej}) .$$

We omit the $-e_i A_{je} Y_j / A_{ee} = -e_i Y_j$ terms because $\sum Y_j = 0$.

Except for the factor $1/A_{ee}$ the map L is essentially the linearization of the vector field at e .

The first consequence of this result is:

6.14 Proposition. *If* e *is an equilibrium for the map* $\overline{F}$ *of* (6.58), *then the eigenvalues of* $d_e\overline{F}$ *are contained in the circle in the complex plane with center at* 1 *and radius equal* 1 .

Proof. From (6.62), z is an eigenvalue of $d_e\overline{F}$ if and only if z - 1 is an eigenvalue of L . We show that the eigenvalues of L are in the unit circle.

Define the linear map $\tilde{L}$ on $\mathbb{R}^I$ by

$$(6.63) \qquad\qquad \tilde{L}(x)_i = \sum_j \frac{A_{ij}}{A_{ee}} X_j .$$

Observe that since $A_{ie} = A_{ee}$ for all i , $\tilde{L}(e) = e$. That is, e is a positive eigenvector for $\tilde{L}$ with eigenvalue 1 . Furthermore, the conditions (6.56) allow us to apply the Frobenius theory of positive matrices to the linear map $\tilde{L}$. The positive matrix result we need is that 1 is the so-called dominant eigenvalue for $\tilde{L}$, and every other eigenvalue is at most 1 in absolute value.

(e_iA_{ij}/A_{ee}) is the matrix of $\tilde{L}$ with respect to the standard basis. Now we describe the matrix of $\tilde{L}$ with respect to the basis consisting of e , and then a basis for $\mathbb{R}_o^I$. Notice that for Y in $\mathbb{R}_o$

$$\tilde{L}(Y) = L(Y) - (\sum_j A_{ej}Y_j/A_{ee})e .$$

This means that with respect to this new basis the matrix of $\tilde{L}$ has the triangular form

$$\begin{pmatrix} 1 & C \\ 0 & D \end{pmatrix}$$

where D is the matrix for $L : \mathbb{R}_o^I \to \mathbb{R}_o^I$ with respect to the chosen basis of $\mathbb{R}_o^I$. The first column is 1 followed by 0 's because e is an eigenvector for $\tilde{L}$.

It follows that the eigenvalues of $\tilde{L}$ consist of 1 and those of L . But the eigenvalues of $\tilde{L}$ consist, we have seen, of 1 and the remaining eigenvalues with absolute value at most 1 .   □

While this result is interesting in a peculiar way, it is the formula (6.62) itself which is important. To interpret it let us see how simple changes of the payoff matrix affect the discrete time system.

Because we are now dealing with multiplicative systems it is multiplication by a positive constant which leaves the system unchanged, that is, if we replace A_{ij} by kA_{ij} with $k > 0$, then conditions (6.56) continue to hold and the common factor k cancels out in the definition (6.58).

This time the addition of a common constant, replacing A_{ij} by $A_{ij} + K$ does have an effect. Observe in formula (6.60) the difference in the numerator remains unchanged but in the denominator A_{pp} is replaced by $A_{pp} + K$. Thus, as K increases, the direction of the vector Δ_p at every point remains unchanged but its length shrinks. This fits also with the second equation of (6.55). Because the common factor of τ there has no effect, decreasing the step size τ amounts to increasing the common constant τ^{-1} which is added to a_{ij} in the definition of A_{ij}. This suggests that while the addition of K leaves the set of equilibria unchanged, increasing K should tend to stabilize the system. To make this precise, observe first that adding K changes the linear map L of (6.62) by multiplying by the constant $A_{ee}/(A_{ee}+K) = (1+(K/A_{ee}))^{-1}$.

6.15 Proposition. *Suppose* e *is an interior fixed point for the map* $\overline{F}$ *associated with* A_{ij}. *If the eigenvalues of* $d_e\overline{F}$ *are contained in the unit circle, then the eigenvalues of* L *have negative real parts. Conversely, if the eigenvalues of* L *have negative real parts, then for* $K > 0$ *large enough eigenvalues of* $d_e\overline{F}_K$ *are contained in the unit circle, where* $\overline{F}_K$ *is the map associated with* $A_{ij} + K$.

Proof. By the previous proof $d_e\overline{F} - I$ is the linear map of (6.62) with $A_{ij} + K$. Since this is a constant times L we can write this as:

$$(6.64) \qquad\qquad (1+(K/A_{ee}))[d_e\overline{F}_k - I] = L.$$

Now observe that a complex number z lies within the unit circle if and only if $\alpha_K(z-1)$ lies within the circle centered at $-\alpha_K$ with radius α_K. This circle lies in the negative half-plane and so multiplicative stability of $d_e\overline{F}$ implies that the eigenvalues of L have negative real part. The weak converse follows because as K increases the α_K circle grows to include any finite set of points in the half plane. The picture is like Figure 8 again. □

Remarks.   (a) Notice that if $d_e\overline{F}_K$ is stable, then it remains stable as K is increased. This is because each α_K circle remains inside the larger ones.

(b) If the eigenvalues of L are real and negative, then $d_e\overline{F}$ is stable. Observe that the α_K circle with $K = 0$, the unit circle centered at -1, includes the intersection of the negative real axis with the unit circle centered at the origin. Now apply Proposition 6.14. If the matrix A_{ij} is symmetric, then L is a symmetric linear map with respect to the Shahshahani metric $(\ ,\ )_e$. This is just the Hessian calculation (4.21). Hence, L has real eigenvalues in the symmetric case.

Now discrete or continuous, the definition of an ESS is the same. It is condition (a) of Proposition 6.9. Here we use A_{ij} instead of a_{ij}, but if they are related by (6.55), then the conditions that e be an ESS with respect to A_{ij} or a_{ij} are equivalent. That is to say, adding a positive constant K to A_{ij} does not affect the ESS conditions, nor does multiplication by a positive constant k.

In particular, if e is interior the ESS condition is just the strong stability condition

(6.65) $$(L(Y),Y)_e = A_{YY} < 0$$

for all nonzero Y in $\mathbb{R}_o^I$. This implies that the eigenvalues of L have negative real parts, but does not necessarily imply they are real. So we only get stability of $\bar{F}_K$ for K large enough. Cannings' detailed analysis of the paper-rock-scissors game provides a beautiful example of this phenomenon.

To see what is going on in general we assume that e is an interior ESS , so that (6.65) holds, and consider the function $L^e(p) = \Sigma \, e_i \ln(p_i/e_i)$ defined on $\mathring{\Delta}$.

6.16 Propositon. *If e is an interior ESS , then for K large enough, L^e is a local Lyapunov function. That is, there is a neighborhood of e on which $L^e(\bar{F}_K(p)) \leq L^e(p)$ with equality only at p = e .*

Proof. We proceed by analogy with the proof of Theorem 6.1. Define on $\mathring{\Delta}$ the function $\hat{L}^e$ by:

$$\hat{L}^e(p) = L^e(\bar{F}(p)) - L^e(p)$$

(6.66) $$= -\Sigma \, e_i \ln(A_{ip}/A_{pp})$$

$$= \ln A_{pp} - \Sigma \, e_i \ln A_{ip} .$$

Clearly, $\hat{L}^e(e) = 0$ and we try to prove that $\hat{L}^e$ has a strict local maximum at e . We take partial derivatives treating p_i 's as x_i 's as usual:

$$\frac{\partial \hat{L}^e}{\partial x_j} = \frac{A_{jp} - A_{pj}}{A_{pp}} - \Sigma_i \, e_i \frac{A_{ij}}{A_{ip}} .$$

When p equals e , $A_{je} = A_{ee} = A_{ie}$ for all i and j , and so this reduces to the constant 1 . So the differential of $\hat{L}^e$ applied to any Y in $\mathbb{R}_o$ is 0 and e is a critical point for $\hat{L}^e$ on $\mathring{\Delta}$.

$$\frac{\partial^2 \hat{L}^e}{\partial x_j \partial x_k} = \frac{A_{jk} + A_{kj}}{A_{pp}} - \frac{(A_{kp} + A_{pk})(A_{jp} + A_{pj})}{A_{pp}^2}$$

$$+ \Sigma \, e_i \frac{A_{ik} A_{ij}}{(A_{ip})^2} .$$

Now when we multiply by $Y_j Y_k$ and sum, for Y in $\mathbb{R}_o^I$, we note that with p = e : $(A_{ie})^2 = A_{ee}^2$ and $\Sigma A_{ke} Y_k = \Sigma A_{je} Y_j = A_{ee} \Sigma_j Y_j = 0$. Hence:

$$\sum_{jk} \frac{\partial^2 \hat{L}^e}{\partial x_j \partial x_k} Y_j Y_k = \frac{2A_{YY}}{A_{ee}} - \frac{(A_{eY})^2}{A_{ee}^2} + \sum_i e_i \frac{A_{iY}^2}{A_{ee}^2}$$

(6.67)

$$= \frac{2A_{YY}}{A_{ee}} + \frac{1}{A_{ee}^2} \sum_i e_i (A_{iY} - A_{eY})^2 .$$

We have combined the latter two terms noting that together they are the variance of A_{iY} with respect to the distribution e on I. In particular, this variance term is positive definite. (That it is nonnegative is clear. It is positive definite because for an ESS, the system (6.22) with A_{ij} replacing a_{ij} and e replacing u has e as unique solution.) But the Hessian term is negative definite by condition (6.65).

When the sum in (6.67) is negative definite, the critical point e is a relative maximum for $\hat{L}^e$ and the conclusion of the proposition will follow. If we replace A_{ij} by $A_{ij} + K$ the numerators of the two terms are unaffected. In the denominator A_{ee} is replaced by $A_{ee} + K$. Notice that $(A_{ee}+K)^2$ grows faster that $A_{ee} + K$. By restricting Y to the –compact– unit sphere in $\mathbb{R}_o$ we find that for K large enough the first term dominates and the sum is indeed negative definite. $\square$

Examples where the positive definite variance term dominates the sum in (6.67) can be constructed in general by using zero-sum games. In that case the payoff matrix a_{ij} is anti-symmetric:

(6.68)
$$a_{ij} = -a_{ji}$$

and so for every vector X in $\mathbb{R}^I$, $a_{XX} = 0$. In particular, the mean fitness a_{pp} is constantly 0, and the symmetrized version of the Hessian, a_{YY}, is also identically 0.

Suppose now that a_{ij} has been chosen antisymmetric with $|a_{ij}| < 1$ and having a unique interior equilibrium, i.e., a unique positive solution e to $a_{ie} = 0$ for all i. This requires that there be an odd number of strategies I and in that event, the required uniqueness will be the usual case.

Then with $A_{ij} = 1 + a_{ij}$ we have $A_{pp} = 1$ and $A_{YY} = 0$ for all p in Δ and Y in $\mathbb{R}_o^I$. Because $A_{ej} = 1 + a_{ej} = 1 - a_{je} = 1$, $A_{eY} = A_{Ye} = 0$, and so the variance term is $\sum_i e_i (A_{iY})^2$. If this were 0 for any nonzero Y in $\mathbb{R}_o^I$, then for ε small $e + \varepsilon Y$ would consist of equilibria, contradicting the assumed uniqueness of e. Hence, the variance term is positive definite and $\hat{L}^e$ has a local maximum at e. Thus, on a neighborhood of e, $L^e(\bar{F}(p)) \geq L^e(p)$ with equality only at 0. That is, e repels nearby points instead of attracting them.

Of course, because A_{YY} is identically 0, the equilibrium e is not an ESS. We cure this by adding $-\varepsilon I$ to A_{ij} for small positive ε, i.e., replace a_{ij} by $a_{ij} - \varepsilon \delta_{ij}$. Denoting this new matrix by A_{ij}^ε, we have that $A_{YY}^\varepsilon = -\varepsilon \sum_i y_i^2$ which is negative definite. The introduction of the positive parameter ε changes everything in a continuous way depending upon ε. e may move and so the value of A_{ee}^ε changes as does

$A^{\varepsilon}_{iY} = A_{iY} - \varepsilon Y_i$. But by continuity in ε, the positive definite variance term will dominate the Hessian term provided $\varepsilon > 0$ is small because the latter tends to zero with ε. Finally, for $\varepsilon > 0$ the interior equilibrium is an ESS because the Hessian is negative definite even though small.

* * *

We conclude by considering the effect of mixing strategies in the continuous time dynamic. We begin with a game having two strategies i_1 and i_2. Suppose now that a new strategy i_3 is introduced. The Taylor-Jonker dynamic is still a selection among relative frequencies of cohorts of pure strategists, each type, including i_3, breeding true. It happens, however, that i_3 is really a mixed strategy, playing i_1 $\frac{1}{3}$ of the time and i_2 the remaining $\frac{2}{3}$. As far as the dynamics are concerned the effect of this fact is completely described by the equations:

(6.69)
$$a_{i_3 j} = \tfrac{1}{3} a_{i_1 j} + \tfrac{2}{3} a_{i_2 j}$$

$$a_{j i_3} = \tfrac{1}{3} a_{j i_1} + \tfrac{2}{3} a_{j i_3}$$

for all strategies j including the new one i_3.

Now look at the matrix:

(6.70)
$$\left. \begin{matrix} 1 & 0 & \tfrac{1}{3} \\[4pt] 0 & 1 & \tfrac{2}{3} \end{matrix} \right\} A$$
$$\begin{matrix} \tfrac{1}{3} & \tfrac{2}{3} & -1 \end{matrix} \quad - \quad B.$$

We are going to apply the Product Theorem 3.1 with the subspace A of $\mathbb{R}^3$ generated by the first two rows and the subspace B consisting of multiples of the third row. Observe that the sum of the first two rows is $(1\ 1\ 1)$ and so the constant vectors lie in A. Also, the third row is perpendicular to the other two with respect to the usual metric. Letting Δ and Δ_S be the simplices in $\mathbb{R}^3$ and $\mathbb{R}^2$, respectively, we define for the product theorem two maps: the linear map $E^A : \Delta \to \Delta_S$ with

(6.71)
$$E^A(p_1, p_2, p_3) = (p_1 + \tfrac{1}{3} p_3 \,,\, p_2 + \tfrac{2}{3} p_3)$$

and the log-linear map $L^b : \Delta \to \mathbb{R}$ with

$$L^b(p_1, p_2, p_3) = \tfrac{1}{3}\ln p_1 + \tfrac{2}{3}\ln p_2 - \ln p_3$$

(6.72)

$$= \ln \frac{p_1^{\frac{1}{3}} p_2^{\frac{2}{3}}}{p_3}.$$

The map E^A is the *mean strategy* map. Its two components represent the relative frequency with which the two original strategies appear in play when the population is at p. Strategy 1 is used when an i_1 player occurs (probability $= p_1$) and $\tfrac{1}{3}$ of the time when an i_3 player occurs (probability $= p_3$), and similarly for strategy 2.

To understand the map L^b go back to the definition of domination (6.24). The first equation of (6.69) looks like $\tfrac{1}{3}i_1 + \tfrac{2}{3}i_2$ dominates i_3, except that the inequalities have been replaced by equalities. The map L^b is exactly the map defined in the proof of Theorem 6.3, but this time –using the same proof– we have, instead of (6.26), that

(6.73) $$\frac{dL^b}{dt} = 0 \qquad \text{at all } p \text{ in } \overset{\circ}{\Delta}.$$

To interpret this result we go back to the horizontal and vertical foliations defined by the maps E^A and L^b as shown in Figure 1, which is drawn exactly for the case at hand.

Think of the original system on Δ_S as the pure strategy dynamic and the new system on Δ as the mixed strategy dynamic. Over each point q of Δ_S lies a line segment, a leaf of the vertical foliation of Δ, consisting of all population states with mean strategy the given point q. A point of Δ is an equilibrium if and only if its mean is an equilibrium for the pure strategy dynamic. This follows from the second equation of (6.69) which says, in effect, that for each j the value of a_{jp} depends only on the mean $E^A(p)$.

Thus, if e is an equilibrium in Δ_S for the pure strategy dynamic, then the entire vertical leaf over e consists of equilibria in Δ. This implies that even if e is an interior ESS for the pure strategy dynamic below, the equilibria above it are not linearly stable. In fact, they are degenerate. Recall from section 4 that a nondegenerate equilibrium is always isolated, it is the only equilibrium among nearby points. This degeneracy is cured by restricting to the leaves of the horizontal foliation defined by L^b = constant. Each of these nonlinear curves intersects a vertical leaf, such as the equilibrium leaf, exactly once. Furthermore, (6.73) says exactly that each horizontal leaf is an invariant manifold for dynamic in Δ. Thus, the ratio $p_1^{1/3}p_2^{2/3}/p_3$ remains constant as the relative frequencies of the three types change in $\overset{\circ}{\Delta}$. Furthermore, if p^* is an equilibrium in $\overset{\circ}{\Delta}$, a point in the vertical leaf over the ESS e in Δ_S, then p^* is linearly stable for the restriction of the dynamic to the horizontal leaf containing p^*. In addition, we can understand how the degeneracy works. If we perturb p^* to a nearby nonequilibrium point p, then the solution path beginning at p tends back toward an equilibrium p^{**} with the same mean strategy e as p^*. If p happened to lie on the same horizontal leaf as p^*, i.e., $L^b(p^*)$ = $L^b(p)$, then $p^{**} = p^*$. Otherwise, p^{**} is the point on the same vertical leaf as p^* but on the new horizontal leaf determined by p.

This picture describes rather well the general situation. Once mixed strategies are allowed, the population state is a distribution not over the finite set I but over the entire simplex Δ thought of as possible mixed strategies. Thus each population state is a measure on Δ and the state space will be some, usually infinite dimensional, space of probability measures on Δ, which we will denote M without describing it. There is a nice linear map from M to Δ associating to each measure μ its mean strategy:

$$(6.74) \qquad\qquad E(\mu)_i = \int_\Delta p_i \mu(dp) \, ,$$

in other words, the mean frequency of strategy i is the frequency of i use by a p player (= p_i) times the probability of a p player in the population (= $\mu(p)$) summed over all p. The integration is just the rigorous version of this idea.

Over each point p of $\overset{\circ}{\Delta}$ there is a convex set of measures with mean p. Think of this as the vertical leaf over p. Each vertical leaf is again infinite dimensional if M is. But the analogue of the horizontal foliation also exists. In fact, there is a version of the Shahshahani metric on the space of measures with respect to which the Product Theorem holds. Each horizontal leaf intersects each vertical leaf exactly once and so maps onto $\overset{\circ}{\Delta}$ (There are some hidden technical assumptions here that I don't want to get into).

Zeeman introduced a definition of the mixed strategy dynamic interpreted as a vector field on such a space of measures. Just as in our simple example, μ is an equilibrium for the Zeeman mixed strategy dynamic if and only if its mean $E(\mu)$ is an equilibrium for the original Taylor-Jonker pure strategy dynamic. In particular, if e is an interior ESS, then the eqlibria for the mixed strategy system consist of the vertical leaf over e. Furthermore, the horizontal leaves are invariant manifolds for the Zeeman dynamic. This is a strong result because the definition of the horizontal foliation, like that of the map E, is prior to the introduction of any particular payoff matrix on the set I of strategies. The horizontal leaves are invariant whatever the choice of payoff matrix. In our simple example, what was needed was that i_3 was "really" the mix $\frac{1}{3}i_1 + \frac{2}{3}i_2$. Then equations (6.69) hold regardless of the original payoff matrix on $\{i_1, i_2\}$.

If e is an ESS, then each equilibrium measure is linearly stable in its finite dimensional horizontal leaf. In fact, it is globally stable. Any solution path in M tends toward the unique equilibrium μ^* lying over e and on the horizontal leaf in which the path remains. But be careful here. If μ_1 and μ_2 are two initial positions with the same means, $E(\mu_1) = E(\mu_2)$, then both solutions approach equilibria with the same mean, namely e, with paths in different horizontal leaves because $\mu_1 \neq \mu_2$. However, the projection under E of the two paths will usually be different. The path of approach of the means to e will depend not only on the mean of the initial position, but upon which horizontal leaf it lies in. In particular, the following suggestive idea does not work: starting with μ, solve the pure strategy dynamic with initial position the mean $E(\mu)$. At time t, lift your mean answer to the horizontal leaf determined by μ. What you get is *not* the same as solving the Zeeman dynamic starting at μ and running along to time t.

The reason for the latter difficulty turns out to be a problem concerning Riemannian metrics. Since E maps each horizontal leaf onto $\overset{\circ}{\Delta}$ we can think of the inverse of E as giving a coordinate system upon the leaf using $\overset{\circ}{\Delta}$. I mentioned that in the space M there is a version of the Shahshahani metric. Restricting it to the horizontal leaf and using E to push it down to $\overset{\circ}{\Delta}$ we get a new Riemannian metric on $\overset{\circ}{\Delta}$. This metric will depend upon the choice of leaf and, in particular, it will usually not be the original Shahshahani metric on $\overset{\circ}{\Delta}$. As a (tedious) exercise, try comparing (i) the restriction of the Shahshahani metric of Δ to a horizontal leaf in the triangle of Figure 1 with (ii) the metric on Δ_S .

It was Gordon Hines who originally looked at the mixed strategy dynamic and motivated Zeeman's measure theory approach. Hines, however, looks at the mean dynamic, in Zeeman's terms the image under E . For each horizontal leaf the change in Riemannian metric induces a change from the pure strategy Taylor-Jonker dynamic by multiplying by a positive definite matrix, the covariance matrix. The difficulty here is that the covariance matrix varies from point to point and so is hard to compute. But at equilibrium it is quite useful and leads to Hines' special characterization of ESS's.

If an interior equilibrium e is linearly stable but not an ESS , i.e., not strongly stable, then it is no longer true that each of the equilibria of the vertical leaf over e is linearly stable in its horizontal leaf. Some may be unstable though all are nondegenerate in the horizontal direction. It follows that if you move, parametrically, from a stable to an unstable equilibrium over e , a Hopf bifurcation occurs. Thus, even though e is stable for the Taylor-Jonker dynamic, over it may lie unexpected cycles among mixed strategist populations. The criticality of these Hopf bifurcations has however not been studied. So it is not known if limit cycles can occur in this case.

7.   Bibliography

To review the linear algebra, any standard text will do, although virtually all have the emphasis on matrices that I described in the beginnning of section 1. At the other extreme, Lang's book [16] gives the mathematician's view of the subject with total −and unreadable− purity. Instead, I would recommend Halmos [12] as a good second book on linear algebra.

The early chapters of Gale's [11] are a good secret source for convexity, cones and positive solutions of equations.

The covariance metric application is what mathematicians call folklore, meaning it is hard to find a good reference. It is discussed in more detail in chapter 1 of my book [1].

Calculus on vector spaces is developed in chapter 8 of [9] and is applied to manifolds in [17]. But these are the standard references for professionals with everything done in Banach space. Luckily, most recent multivariable calculus books adopt the vector space approach, e.g. [10]. Spivak is a great expositor and his [24] goes from the beginning of calculus -vector version- through a nice introduction to manifolds.

The Shahshahani metric was introduced in [22] and is discussed in detail in [1]. In particular, the proof of the Product Theorem together with applications to epistasis occur there. In [3] the concept is generalized to spaces of measures, and in the process its relationship with Fisher's original information measure appears. As mentioned in section 3, Antonelli and Strobeck in [8] introduced essentially the same metric.

For differential equations in general and linar stability analysis, in particular, I strongly recommend Hirsch and Smale [15]. The Hopf bifurcation theorem is proved in [19] but for the lighter expository needs of our applications the introductory discussions from [1] and [4] may suffice.

The stuff about the Hessian and its application to gradients comes from [1]. In particular, the cycling theorem is proved there.

In [4] Hopf bifurcation is studied in the general TLTA model and a parametrization method is introduced to generate all examples of such bifurcation in the family of these models. Also, the symmetry group of the TLTA models, the square group, is introduced and discussed there. [5] concentrates upon and describes in detail the π_+ symmetric examples which were sketched out in section 5.

Hastings in [13] describes a direct procedure for generating the cycling examples in the analogous discrete time model.

The biologist's view of evolutionary game theory is given in [22] by its originator. Taylor and Jonker introduced their dynamic model and discovered the relationship between the ESS conditions and dynamic stability. Hines introduced the mixed strategy dynamic in [14]. The measure theory interpretation of this dynamic is due to Zeeman [26] and the invariant manifolds in the spaces of measures are described in [3].

The relation between domination and equilibrium comes from [2]. See also [6].

My discussion of discrete time models is motivated by chapter 2 of May's book [21]. In [19] we prove the invertibility of the game dynamic mapping and in [7] zero-sum games are studied in both the continuous and discrete time contexts.

References

[1]   Akin, E., *The Geometry of Population Genetics*, Lecture Notes in Biomathematics, Vol. 31, Springer, Berlin-Heidelberg-New York, 1979.

[2]   Akin, E., Domination or Equilibrium, *Math. Biosci.* **50** (1979), 239-250.

[3]   Akin, E., Exponential families and game dynamics. *Canad. J. Math.* **34** (1982), 374-405.

[4]   Akin, E., *Hopf Bifurcation in the Two Locus Genetic Model*, Mem. Amer. Math. Soc., Vol. 284. AMS, Providence, RI, 1983.

[5]   Akin, E., Cycling in simple genetic systems II: The symmetric case, in: *Dynamical Systems* (Kurzhanskii and Sigmund, eds.), Lecture Notes in Economics and Mathematical Systems, Vol. 287, Springer, Berlin-Heidelberg-New York, 1987, 139-154.

[6]   Akin, E. and Hofbauer, J., Recurrence of the unfit, *Math. Biosci.* **61** (1982), 51-62.

[7]   Akin, E. and Losert, V., Evolutionary dynamics of zero-sum games, *J. Math. Biol.* **20** (1984), 231-258.

[8]   Antonelli, P. and Strobeck, C., The geometry of random drift I. Stochastic distance and diffusion, *Adv. in Appl. Probab.* **9** (1977), 238-249.

[9]   Dieudonné, J., *Foundations of Modern Analysis*, Academic Press, New York, 1960.

[10]   Edwards, C.H., *Advanced Calculus of Several Variables*, Academic Press, New York, 1973.

[11]   Gale, D., *Theory of Linear Economic Models*, McGraw-Hill, New York, 1960.

[12]   Halmos, P., *Finite-Dimensional Vector Spaces*, Van Nostrand, Princeton, NJ, 1958.

[13]   Hastings, A., Stable cycling in discrete-time genetic models, *Proc. Nat. Acad. Sci. U.S.A.* **78** (1981), 7224-7225.

[14]   Hines, W.G.S., Strategy stability in complex populations, *J. Appl. Probab.* **17** (1980), 600-610.

[15]   Hirsch, M. and Smale, S., *Differential Equations, Dynamical Systems and Linear Algebra*, Academic Press, New York, 1974.

[16]   Hofbauer, J., Schuster, P. and Sigmund, K., A note on evolutionary stable strategies and game dynamics, *J. Theoret. Biol.* **81** (1979), 609-612.

[17]   Lang, S., *Introduction to Linear Algebra*, 2nd ed, Springer, Berlin-Heidelberg-New York, 1985.

[18]   Lang, S., *Differential Manifolds*, Addison Wesley, Reading, MA, 1972.

[19]   Losert, V. and Akin, E., Dynamics of games and genes: Discrete versus continuous time, *J. Math. Biol.* **17** (1983), 241-251.

[20]   Marsden, J. and McCracken, M., *The Hopf Bifurcation and Its Applications*, Springer, Berlin-Heidelberg-New York, 1976.

[21]   May, R., *Stability and Complexity in Model Ecosystems*, Princeton University Press, 1973.

[22]   Maynard Smith, J., *Evolution and the Theory of Games*, Cambridge University Press, 1982.

[23]   Shahshahani, S., *A New Mathematical Framework for the Study of Linkage and Selection*, Mem. Amer. Math. Soc. Vol. 211. AMS, Providence, RI, 1979.

[24]   Spivak, M., *Calculus on Manifolds*, Benjamin, New York, 1965.

[25]   Taylor, P. and Jonker, L., Evolutionarily stable strategies and game dynamics, *Math. Biosci.* **40** (1978), 145-156.

[26]   Zeeman, E.C., Population dynamics from game theory, in: *Global Theory of Dynamical Systems* (Z. Nitecki and C. Robinson, eds.) Lecture Notes in Mathematics, Vol. 819, Springer, Berlin-Heidelberg-New York, 1979, 471-497.

[27]   Zeeman, E.C., Dynamics of the evolution of animal conflicts, *J. Theoret. Biol.* **89** (1981), 249-270.

TOPICS IN THE THEORY OF ESS's

C. Cannings
Department of Probability and Statistics
The University
Sheffield, S3 7RH
United Kingdom

ABSTRACT. The notion of an ESS is discussed, and illustrated with the aid of certain mathematical models. After examining the definition of an ESS, the War of Attrition with 2 players, with random rewards and with n players is discussed. Certain aspects of the dynamic system driven by conflicts are illustrated using the Rock-Scissor-Paper game. Finally the notion of a pattern of ESS's is introduced.

1. Introduction

The notion of an ESS (evolutionarily stable strategy) was introduced by Maynard Smith and Price (1973) in an attempt to understand certain aspects of intraspecific conflicts. They considered a simple model of fighting which involved individuals of various types – mouse, hawk, bully, retaliator and probe-retaliator – whose behaviour in fights with each other resulted in payoffs which depended on the type of the two protagonists. Thus, for this conflict, with a finite number of possible strategies, they defined a payoff matrix in the usual spirit of game theory (e.g. Luce and Raiffa, 1957). Using this matrix they argued that "limited war" strategies could persist under individual selection. They argued that limitation on the use of dangerous weapons (e.g. horns or venom) could be favoured by individual selection, and that recourse to "group selection" arguments were not necessary.

The approach adopted by Maynard Smith and Price (1973), and the notion of an ESS introduced by them (explained and expanded upon more formally below) has been enormously valuable in discussing not only fighting behaviour, but many other situations where conflicts arise between members of the same species. It is probably true that whenever individuals interact they are in conflict, even when their action appears co-operative.

Some of the literature on this topic has been reviewed by Maynard Smith (1982) and by Hines (1987). The objective of this presentation is very different. In it some of the underlying mathematical ideas will be presented together with a number of models of conflicts (some new) which will be used to illustrate, hopefully also to illuminate, these ideas. If, as seems likely, more questions are raised in the reader's mind than are answered, then the object of the article will have been achieved.

95

S. Lessard (ed.), Mathematical and Statistical Developments of Evolutionary Theory, 95–119.

Before proceeding to the discussion of ESS's it is appropriate to make a few comments on the purpose in formulating mathematical models. The purpose, at least as envisaged by this author, is not to represent reality in a set of mathematical equations; that would be an unreasonable objective in almost all biological situations. The object is to capture at least some aspects of reality, while neglecting others, in the hope that some understanding of the interplay of differing assumptions will emerge, and perhaps some principles which are robust over models be found. That said, the mathematical aspects of the models of conflicts have an intrinsic interest, and as we shall see, albeit briefly, problems arise in dynamical systems theory, combinatorics, graph theory, and many other areas of mathematics.

2. The definition and notion of an ESS

Maynard Smith and Price (1973) suggested that one seek strategies which if played throughout the whole population (there are various ways of interpreting this last phrase) are non-invadable, i.e., are capable of resisting an influx of an alternative strategy. Such influxes are considered to be at a low frequency, arising perhaps by mutation of the genes coding for the behaviour involved in the conflict, or by migration from another population.

Suppose then that we are interested in a population whose members are involved in pairwise conflicts (a later section will introduce conflicts with $n \geq 2$ contestants), the payoffs from which will contribute to the individual's overall fitness. In any contest an animal (or a plant) can choose from some behavioural repertoire, i.e., choose a value x (which need not be univariate, but which will be throughout this paper) from some set U. The opponent also chooses a behaviour from U, y say, and as a result the conflict is resolved, the x-player receiving payoff E(x,y), the y-player E(y,x). These payoffs must ultimately be translated into reproductive success. The conflict is completely defined therefore when one has specified U and E(x,y) for all x,y $\in$ U. Now, as in classical game theory, we extend the conflict by introducing mixed strategies, which are just weight functions p(x), x $\in$ U; here just probability vectors, or probability densities which we shall abbreviate as p. The extension of the payoffs is done by assuming additivity (or linearity) so that

$$E(x,q) \quad = \int\limits_{y \in U} E(x,y)q(y)dy$$

(1)

$$E(p,q) \quad = \int\limits_{x \in U} E(x,q)p(x)dx$$

where we use the integral notation throughout, though we may be dealing with vectors.

A mixed strategy p must be interpreted with care according to the situation. It may be a specification of the way that an individual plays (i.e., choosing to play pure strategy x with probability p(x)), or it may correspond to the way that a group behaves on average though possibly all the members of the group are playing specific x values, the underlying pure strategies. To distinguish these two ideas we should refer to an individual playing **p** as a p-player, and to the group as a p-group. The equations in (1) are valid for either of these possibilities, $E(x,q)$ specifying the payoff to an x-player (an x-group necessarily contains only x-players) in conflict with a q-player, or in conflict against a q-group.

Suppose then that a population is playing **p** , and there is an invasion by a q-group (Maynard Smith, 1974). The population strategy is perturbed from **p** to $(1-\varepsilon)p + \varepsilon q$, where ε , $0 \le \varepsilon \le 1$, specifies the size of the influx. We can now compare the fitness of the **p**-player (or p-group) in this population with that of the q-player. The strategy **q** will not invade a population playing **p** if **p** is fitter (i.e., obtains a greater payoff) than **q** . Thus using (1) we have that **q** will not invade if

(2) $$E(p,(1-\varepsilon)p+\varepsilon q) > E(q,(1-\varepsilon)p+\varepsilon q) \ .$$

Now this will hold, for sufficiently small ε if and only if,

(i) $E(p,p) \ge E(q,p)$, and

(3)

(ii)   if $E(p,p) = E(q,p)$, then $E(p,q) > E(q,q)$.

There is a simple interpretation of the two parts of (3). The population is almost entirely made up of **p**-players (or is approximately a p-group), so that the most important comparison between **p** and **q** is how well they do against this p-group. If **p** is to resist invasion it must do at least as well as **q** against the p-group, condition (i), and when it does exactly as well we have to look at the much less frequent encounters with the q-group, condition (ii). The conditions (i) and (ii) are essentially similar to those for initial non-increase of a new allele at a single autosomal locus under selection.

Now if **p** is to be an ESS it must be resistant to invasion by any **q** $(\ne p)$, and so we have Maynard Smith's (1974) definition.

Definition 1. **p** is an ESS if for all **q** $\ne p$ (3) holds.

Now conditions (2) and (3) are equivalent, and we can use (2) to provide an alternative definition of an ESS:

Definition 2. **p** is an ESS if for any **q** $\ne p$ there exists some $\varepsilon_0(q) > 0$ such that for all ε , $0 < \varepsilon < \varepsilon_0(q)$,

$$E(p,(1-\varepsilon)p+\varepsilon q) > E(q,(1-\varepsilon)p+\varepsilon q) \ .$$

Vickers and Cannings (1987) have pointed out that an alternate definition should be considered, perhaps preferred to the two equivalent forms above.

Definition 3. **p** is an ESS if there exists $\varepsilon_o > 0$ such that for all ε, $0 < \varepsilon < \varepsilon_o$ and any $q \neq p$,

$$E(p,(1-\varepsilon)p+\varepsilon q) > E(q,(1-\varepsilon)p+\varepsilon q) .$$

This latter definition essentially imposes a uniformity condition. It is not sufficient to look at each **q** in turn, and to pick an ε_o for that specific **q**, one needs to pick ε_o before looking at any of the **q**'s, that ε_o sufficing for all $q \neq p$. If **p** is an ESS under Definition 2 but not under Definition 3, then there will for any given ε be a set of **q**'s which can invade the **p**-population, if the **q**'s are introduced at frequency ε or more. Adopting Definition 3 guarantees that any influx at a level less than ε_o will not permanently disturb the **p**-population. The question still arises whether that ε_o is sufficiently large compared to possible perturbations, and perhaps the notion of **p** being an ε_o-ESS would be a useful one to explore. Riley (1979) discussed the notion of an ESS for a finite population (see also Hines, this symposium), and then, of course, if there are only n individuals, an influx must have ε at least $1/n$ (assuming that a new strategy is played by a single individual, though it may be possible to have $\varepsilon < 1/n$ if a deviant individual plays a mixture of **p** and a new **q**). There are also other problems with the ESS definition in this case as shown by Riley (1979).

Another possibility in which the ESS definition might need to be refined is when there is migration between contiguous populations as per Brown (1988).

As an illustration of a conflict which has an ESS under Definition 2 but not under Definition 3 we consider an example similar to one discussed in Vickers and Cannings (1987).

Example. Suppose $U = [1,\infty)$ and

$$E(x,y) = \begin{cases} 0 & \text{if } x = y = 1 \\ -1/x & \text{if } 1 = y < x \\ -1 & \text{if } 1 = x < y \\ 0 & \text{otherwise.} \end{cases}$$

Then $p = \{\text{play 1 always}\}$ has

$$E(p,p) = 0 , \; E(q,p) = \int (-\frac{1}{x})q(x)dx \in [-1,0) ,$$

$$E(p,q) = -1 , \; E(q,q) = 0$$

(assuming that **q** never involves playing 1 which, as we shall see later, is not an important restriction). Thus $E(p,p) - E(q,p) = \lambda(q)$, say, where $\lambda(q) > 0$, with $E(p,q) - E(q,q) = -1$, and **p** is an ESS by Definition 2, i.e., take

$$\varepsilon < \varepsilon(q) = \frac{\lambda(q)}{1+\lambda(q)} \, .$$

However, $\lambda(q)/(1+\lambda(q))$ can be made arbitrarily small by making q (= play x only) and letting $x \to \infty$, so p is not an ESS by Definition 3.

Vickers and Cannings (1987) specify a number of situations under which the distinction between the two definitions disappears. These are:

(i) if $U = \{1,2,...,n\}$ so that only a finite number of pure strategies are available, (ii) if the support of the ESS p, $R(p) = \{x; x \in U, p(x) > 0\}$, is such that $R(p) = U$, and (iii) in the War of Attrition with $U = [0,a]$, $U = [0,\infty)$, $U = \{1,2,...\}$ or $U = \{1,2,...,n\}$. Fortunately these cases cover all those to be considered in this paper, so we can adopt Definition 1 which is easier to work with, and is the one most commonly adopted in the literature.

3. The War of Attrition

In order to illustrate the various ideas to be presented here we consider a particular model, the so called War of Attrition introduced by Maynard Smith (1974). Subsequently a number of generalizations will be made.

In the War of Attrition

(4)
$$E(x,y) = \begin{cases} V - y & x > y \\ \dfrac{V}{2} - x & x = y \\ -y & x < y \, . \end{cases}$$

The rationale behind these payoffs is the following. There is a certain reward V; each individual selects a value from U and is prepared to contest (wait, display, fight, etc.) for only that length of time. When one of the contestants has reached his chosen value he leaves, and the other collects the reward V. The cost incurred by each individual is equal to the length of the contest $= \min(x,y)$. If they should choose x and y equal, then V is shared equally.

Initially we take $U = [0,\infty)$, and seek p which is an ESS. We begin by deriving some general properties of ESS's (Results 3.1, 3.2, and 3.3).

Result 3.1.

If p is an ESS with support $R(p)$, then $E(x,p) = c$ (a constant) for almost all $x \in R(p)$.

This property was stated and used by Maynard Smith (1974), and demonstrated more formally by Bishop and Cannings (1976).

Essentially p must be an equilibrium, for if there were some $y \in R(p)$ with

$$E(y,p) > c = E(x,p) = E(p,p),$$

then q, which is play-y always, has $E(q,p) > E(p,p)$ which is not permitted. On the other hand, if for $y \in R(p)$ $E(y,p) < c$ that will be allowable provided the weight $p(y)$ is zero.

Thus we seek p such that $E(x,p) = c$ for all $x \in R(p)$. The problem remains that we have in general no clear idea what $R(p)$ is. However, if we define

$$S(p) = \{x; E(x,p) = E(p,p)\}$$

so that $R(p) \subset S(p)$, then Bishop and Cannings (1976) proved that:

Result 3.2.

If p and q are ESS's (we shall see later that many ESS can exist for a given conflict), then

$$R(q) \not\subset S(p) \text{ and } R(p) \not\subset S(q).$$

We shall see that this is an immensely powerful result (though easy to demonstrate), and of great help when seeking to find the complete set of ESS's (see Section 8), for we can deduce (Bishop and Cannings, 1976) the following result.

Result 3.3.

If p is an ESS and $S(p) = U$, then p is the unique ESS.

Result 3.3 suggests a way forward in seeking an ESS for the War of Attrition in the case where $U = [0,\infty)$. We seek first on equilibrium p with $S(p) = U$ (using Result 3.1), and investigate whether this p is an ESS. If so, then it must be the unique ESS. Consider then $p = p(x)$, $x \in [0,\infty)$ with $p(x) \geq 0$, and

$$P(x) = \int_0^x p(z)dz,$$

so $p(x)$ is a density function and $P(x)$ a distribution function ($P(0) = 0$, $P(\infty) = 1$). Then, as per Maynard Smith (1974), if $E(x,p) = c$, the derivative with respect to x will be zero. Now

$$(5) \qquad E(x,p) = \int_0^x (V-y)p(y)dy - x\int_x^\infty p(y)dy.$$

If this is to be constant, then differentiating (5) we obtain

$$(6) \qquad 0 = (V-x)p(x) - (1-P(x)) + x\,p(x)$$

so

$$\frac{p(x)}{1-P(x)} = \frac{1}{V} .$$

We obtain therefore

(7) $$p(x) = \frac{1}{V} e^{-x/V} , \quad x \in [0,\infty) ,$$

the negative exponential density with rate $1/V$.

Since $E(x,p) = c$ for p defined in (7), and all $x \in U$, $E(p,p) = E(q,p)$. In Maynard Smith (1974) a check was then made of $E(p,y) - E(y,y)$ which proved negative for all $y \in [0,\infty)$, but this is not sufficient to ensure that p is an ESS. Bishop and Cannings (1976) addressed this issue and introduced a function $T(u,v)$ defined for any strategies u and v, by

$$T(u,v) = E(u,u) - E(v,u) - E(u,v) + E(v,v) ,$$

and proved this was negative provided $v \not\equiv u$. Taking $u = p$ so that $E(v,p) = E(p,p)$ implies that $E(p,v) > E(v,v)$ all v, so that p is an ESS, which by Result 3.3 is unique. Note that we have used the weaker form of the definition of an ESS, but since $R(p) = U$, it will also be an ESS in the stronger sense. We summarize the above as

Result 3.4.
    For the War of Attrition with $U = [0,\infty)$ the unique ESS is given by

$$p = p(x) = \frac{1}{V} e^{-x/V} , \quad x \in [0,\infty) .$$

Some comments on Result 3.4 are appropriate:
(i) the case where $x = y$ is unimportant as this has zero probability of occurring for the ESS, although it needs to be considered in proving that p is an ESS;
(ii) the negative exponential distribution is, in a sense, what we should have anticipated as the answer, for it has the lack of memory property (Feller, 1950), and so is essentially spyproof;
(iii) since each player plays negative exponential with rate $1/V$, the contest lasts a negative exponential time with rate $2/V$, i.e., on average $V/2$, and the expected time spent by the two individuals totals V, exactly matching the reward gained in each contest. The overall payoff for contesting is thus zero, and time expended exactly matches in value reward gained.

A DIFFERENT APPROACH

In the above we found a $p(x)$ equilibrium over U, and were able to prove that this was an ESS, and hence the unique ESS. More generally we may not be so fortunate. We introduce here a somewhat different approach to finding p and $R(p)$, not needed for the War of Attrition on $[0,\infty)$, but invaluable in other cases. These ideas are presented in Bishop and Cannings (1978) in a more general framework, and are only outlined here.

Result 3.5.

For the War of Attrition on an interval $[a,b]$, an ESS p can have no atom of probability, except possibly at b.

If there were an atom of probability for $x \in [a,b)$ in a strategy p, then a new strategy which was identical to p except for a slight shift of the atom to the right of x would beat p. On the other hand, if the atom is placed at b the same argument cannot be applied, and indeed there are cases (the rule rather than the exception) with an atom at b.

Result 3.6 ("Gaps" in $R(p)$).

For the War of Attrition on an interval $[a,b]$, there is no interval (c,d), $a \le c < d \le b$ with $(c,d) \cap R(p) = \emptyset$, unless there is an atom of probability at b (where $R(p) = \{b\} \cup [a,b-V/2]$ if $b > V/2$, and $R(p) = \{b\}$ if $b \le V/2$).

The proof is by contradiction. Suppose a gap exists in the support $R(p)$, i.e., no strategy in (c,d) is included in $R(p)$. Now compare $E(c,p)$ and $E(d,p)$ which are defined to be equal. Clearly a d-player wins no more contests than a c-player, but loses more "time" on average, unless there is an atom at d, for then the d-player will share the reward against another d-player. We know from Result 3.5 that there can only be an atom at b. Assuming there is an atom of weight r at b, we have $E(b,p) - E(c,p) = r\{V/2-b+c\}$, which is zero only if $c = b - V/2$.

Result 3.7.

For the War of Attrition on $[a,b]$ there is an atom of weight $V/2$ at b, and the remainder of the density is in the form $p(x)$ of (7) for $x \in [a,b-V/2]$ if $b > V/2$. If $b < V/2$ there is an atom of weight 1 at b.

OTHER U'S

(i) If $U = [a,s)$ there can be no ESS. As discussed above there would need to be an atom at s, but here $s \notin U$.

(ii) $U = \{m_1,m_2,...,m_k\}$. There is always an ESS, m_k is always included $(m_1 < m_2 < ...m_k)$, m_{k-1} is included only if $(m_k-m_{k-1}) > V/2$. More generally, if one knows that m_l is included in $R(p)$, for the ESS p, then the condition for m_{l-1} to be included also is of the form $(m_l-m_{l-1}) >$ some function of the m_i's in the support $R(p)$ for $i \ge l$, and V. Thus the m_i's included in the support must be sufficiently different from each other.

THE GENERALIZED WAR OF ATTRITION

Bishop and Cannings (1978) generalized the War of Attrition to allow a fairly general set up for the payoffs, while still retaining the "ordinal" feature, i.e., whoever is prepared to play longer wins. They defined

$$E(x,y) \; = \; \begin{cases} f(y) - g(y) & \text{if } x > y \\ \\ -g(x) & \text{if } x < y . \end{cases}$$

Essentially this implies a scaling of the time, and allows a variety a models (e.g. the graduated risk model) to be studied. The results above generalize fairly straightforwardly provided $f\nearrow$, $g\nearrow$ strictly and dg/dx bounded. One obtains, for $U = [0,\infty)$, the unique ESS

$$p(x) \; = \; \frac{g'(x)}{f(x)} \exp\left(-\int_0^x \frac{g'(m)}{f(m)} \, dm \right) , \quad x \in [0,\infty)$$

provided $g'(x)/f(x) \to \infty$ as $x \to \infty$. Note that this ESS has the lack of memory property discussed earlier, and that $E(x,p) = 0$. We shall return to the War of Attrition model later in order to introduce certain other generalizations. However it is worth mentioning here one generalization we shall not consider further. Cannings and Gardiner (1980) allowed relaxation of the conditions on f and g required above. In that case gaps and atoms may occur.

4. Dynamics

So far the ESS notion has been presented without reference to the dynamical perspective. It is important to add to the idea of conflicts some representation of the way in which strategies are determined for each individual, and the way in which the frequencies of strategies are modified through time.

The following is a brief discussion of certain salient features; Akin and Hines (this volume) explore certain aspects of the same problem. We shall assume here an essentially haploid model of inheritance, thus obtaining the frequency of a strategy in one generation by scaling that in the previous generation, by its payoff in that population. Accordingly if the population is playing v , and each individual in some subpopulation is playing u (possibly mixed), then we use E(u,v) to determine the scaling for that sub-population. If we are working with a continuous time model (i.e., overlapping generations), then take (where $x(u)$ = frequency for u-players)

(8) $$\dot{x}(u) \; = \; x(u)(E(u,v) - E(v,v))$$

(e.g. Taylor and Jonkers, 1978), where $\dot{x} = dx/dt$, $t = $ time, while for a discrete time model (i.e., non-overlapping generations), then take

$$(9) \qquad x'(u) = \frac{x(u)(k+E(u,v))}{(k+E(v,v))}$$

(e.g. Bishop and Cannings, 1978). An important feature of (9) is the appearance of k, a constant, which must be included, either directly into all the payoffs $E(x,y)$ when they are defined, or later to ensure that negative scaling factors never occur. The value of k should be regarded as the background fitness of the organism, to which the payoffs from the conflicts make a marginal contribution. Note that k has no effect on the definition of an ESS, nor on the continuous dynamic. On the other hand, R is vitally important in the discrete dynamic, as we see below.

CONTINUOUS DYNAMICS

Taylor and Jonkers (1978), Zeeman (1980), Hofbauer et al. (1981) have discussed the behaviour of (8) when only pure strategies are available to individuals. The behaviour is reasonably straightforward, though not trivial to delineate. Zeeman (1980) proved that
(i)   every ESS is an attractor (i.e., has some region of frequencies which lead to a sequence under (8) converging to the ESS);
(ii) if an internal ESS exists, then there is no other attractor, and global convergence to the ESS is assured (unless of course the initial population is missing a pure strategy); and
(iii) there may be attractors which are not ESS's.
Akin (1982) extends the discussion to consider cases where individuals are allowed to play mixtures. This necessitates the consideration of an ∞-dimensional space, as opposed to the previously finite dimensional one. Akin proves that, when there is an internal ESS, which is contained within the convex hull of the available strategies, then convergence to the ESS is assured.
The situation with the continuous dynamic is thus reasonably clear, at least whenever there is an internal ESS.

DISCRETE DYNAMICS

Under the dynamic (9) the behaviour is more complex, and one cannot even assert that ESS's are attractors. The following examples show that all combinations of being an ESS, and being stable, can occur.

Example 1 (Taylor and Jonkers, 1978). $U = \{1,2,3\}$,

$$E(i,j) = a_{ij} \quad \text{and} \quad A = (a_{ij}) = \begin{pmatrix} 2 & 1 & 5 \\ 5 & 2 & 0 \\ 1 & 4 & 3 \end{pmatrix}.$$

Then using the result of Haigh (1978) one can easily prove that there is no ESS. On the other hand, Taylor and Jonkers (1978) prove that $p = (13,11,10)/34$ is a stable equilibrium if $k \geq 10,999/5020$, and unstable otherwise.

Example 2 (Rock-Scissors-Paper, Bishop (1978)).

$$E(i,j) = a_{ij} \text{ and } A = (a_{ij}) = \begin{pmatrix} -\varepsilon & 1 & -1 \\ -1 & -\varepsilon & 1 \\ 1 & -1 & -\varepsilon \end{pmatrix}$$

then we have (note that we require $k \geq \max(1,\varepsilon)$) :

Tableau

ε	k	ESS	STABLE
$\varepsilon \leq 0$	$k \geq 1$	no	yes
$0 < \varepsilon \leq 1$	$k > \dfrac{1+\varepsilon^2}{2\varepsilon} > 1$	yes	yes
$0 < \varepsilon \leq 1$	$\dfrac{1+\varepsilon^2}{2\varepsilon} > k > 1$	yes	yes
$\varepsilon > 1$	$k \geq 1$	yes	yes

An important feature of the discrete dynamic is illustrated here (and discussed in greater detail in Bishop (1978)). The value of k is of importance in determining the stability of an equilibrium, though as pointed out earlier it does not enter into the consideration of ESS's. For certain values of ε (i.e., $\varepsilon \notin [0,1]$) the value of k does not influence stability, while for others, viz. $\varepsilon \in [0,1]$, there will be an ESS which is stable for larger k, unstable for smaller k. The reason for this is easy to appreciate. As k increases the discrete dynamic approaches the continuous one, the step size decreasing while the direction is unchanged. For the continuous dynamic ESS's are (at least locally) stable, and if internal are globally stable.

The behaviour of the system in rock-scissors-paper under the discrete dynamic is not too complex. When $\varepsilon > 0$ there is an ESS, at $p = (1/3,1/3,1/3)$. For $k > (1+\varepsilon^2)/2\varepsilon$ the system (under the discrete dynamic) will spiral inwards to that ESS, while for $k < (1+\varepsilon^2)/2\varepsilon$ it will spiral outwards. The critical case is when $k = (1+\varepsilon^2)/2\varepsilon$. For any initial value, the frequencies will lie on a closed curve. If one uses the term period (even though this will not necessarily be integer), one can say the behaviour is periodic, and that for initial values further from p, the period will be larger. This type of behaviour holds (when $k = (1+\varepsilon^2)/2\varepsilon$) except when $\varepsilon = k = 1$. Then the behaviour is periodic, with period six irrespective of the starting values (provided they are not on the boundary; i.e., $p > 0$).

We now turn our attention to the situation where individuals can play mixtures of strategies. General results, of the type demonstrated by Akin ((1982) and discussed above, are not available for the discrete dynamic.

It is easy to prove for the case where $U = \{1,2\}$, and only pure strategy players are available, that convergence is assured. One can exploit this simple case to prove that:

Result 4.1.

If $U = \{1,2\}$ and any subset of the set of mixtures is playable, then the system converges;

Result 4.2.

If $U = \{1,2,...,n\}$ and only two mixtures are available, then convergence is assured.

However, a much stronger result is possible which proves that for any reasonable dynamic an internal ESS (supposing one exists) is stable if playable. Suppose p is the internal ESS, and that p is playable. If the population frequency is $q \neq p$, then the fitness of the p-player will be a function of $E(p,q)$, and since p is an internal ESS, $E(p,p) = E(q,p)$ so $E(p,q) > E(q,q)$. Now a reasonable dynamic must have that if $E(p,q) > E(q,q)$, then the fitness of p in a q-population must be greater than the mean fitness. Accordingly the frequency of p-players must always increase from generation to generation. Provided the population is initiated with $q \neq p$ the system will converge to one with only p-players. If initially the population is at p, they any perturbations will lead to increases in the frequency of p-players. We thus have a very strong result.

Result 4.3.

If there exists an internal ESS p which is playable, and there are individuals in the population who play p, then irrespective of the dynamic applied, the system converges to one with only p-players.

In general of course the ESS will not be playable, but some mixtures may be. Some ideas regarding the outcome when mixtures are included in the rock-scissors-paper conflict are given below.

First suppose that only the three pures plus the ESS strategy $(1/3,1/3,1/3)$ are playable (i.e., we take $\varepsilon > 0$). Now for $k > (1+\varepsilon^2)/2\varepsilon$ the discrete dynamic with just the pures spirals inwards to the equilibrium. When $p = (1/3,1/3,1/3)$ is playable, as pointed out above, the p-players increase in frequency. Representing the frequency of the four available strategies in a tetrahedron, we observe the trajectory spiralling around, and towards the vertex "play p". The spiralling which is observed in the subspace from which p-players are excluded (i.e., the normalized triangle of pures) does not precisely match that if p-players are absent. The effect of the presence of p-players on the pure players dynamics is to successively reduce k in equation (9) as the frequency of the p-

players increases. When $k < (1+\epsilon^2)/2\epsilon$, then initially the spiral will be observed to expand, and then to begin to contract as it climbs up the tetrahedron.

We now look in some detail at the case where there are three mixed strategies $(0,\lambda,(1-\lambda))$, $((1-\lambda),0,\lambda)$ and $(\lambda,(1-\lambda),0)$ which we label r_1, r_2 and r_3. Now essentially we have replaced a system with three pure strategies by one with three different pure strategies but with a different A matrix. If the new payoff matrix is A*, then

$$A* = \begin{pmatrix} r_1 \\ r_2 \\ r_3 \end{pmatrix} A'(r_1,r_2,r_3) = RAR'$$

say, where $R = (\lambda J+(1-\lambda)J')$ and

$$J = \begin{pmatrix} 0 & 1 & 0 \\ 0 & 0 & 1 \\ 1 & 0 & 0 \end{pmatrix}$$

(J is a permutation matrix). Thus

$$A* = (\lambda J+(1-\lambda)J')(kU+A)(\lambda J'+(1-\lambda)J)$$

$$= (k-\epsilon\lambda(1-\lambda))U + (1-3\lambda(1-\lambda))A$$

$$= (1-3\lambda(1-\lambda)) \left[\frac{[k-\epsilon\lambda(1-\lambda]}{[1-3\lambda(1-\lambda)]} U + A \right].$$

The constant multiplying factor $(1-3\lambda)(1-\lambda)$ has no effect on the ESS, or on the dynamic. However the "background" fitness value has changed from k to $(k-\epsilon\lambda(1-\lambda))/(1-3\lambda(1-\lambda))$. The new value exceeds k provided $\epsilon < 3k$. This is clearly satisfied for $\epsilon \le 1$ since $k \ge \max(1,\epsilon)$, but need not be for $\epsilon > 1$. However in this latter case the ESS is always stable irrespective of the value of the background fitness. The only situation in which there is an ESS, whose stability might be affected is when $0 \le \epsilon \le 1$ and $(1+\epsilon^2)/2\epsilon > k > 1$, when the new background fitness may exceed $(1+\epsilon^2)/2\epsilon$, and the ESS becomes dynamically stable.

The important point to realize here is that a population of mixtures (i.e., players using mixtures) may well be stable, though one of pures is not. Admittedly the point is made here only in a very limited context, that of Rock-Scissors-Paper and for symmetric mixtures only. However, I believe it is a principle of a more general nature that when an internal ESS exists, the discrete dynamic with mixtures will often be stable while that with pures is not.

5. n-Person Wars of Attrition

In the previous discussion of the War of Attrition each contest involved precisely two protagonists. In the case where $U = [0,\infty)$ a particularly simple result arises, the ESS density is negative exponential with rate $1/V$. We now introduce a number of simple models in which n individuals compete for the reward (the detailed mathematical arguments are developed in Haigh and Cannings (1988)).

Model 5.1.
    Suppose there is a single reward value V , that at the outset of the contest each contestant chooses his value, and that value cannot be changed subsequently (for example when a contestant drops out). Individuals drop out form the contest as their value (in time) is reached, until only one remains (the one who chose the largest value initially) who then collects the reward. This model is relatively easy to solve by the following method: consider the "opponent" of an individual, which in reality is the amalgamation of the plays of his $(n-1)$ opponents. Thus if the times selected are $X_1, X_2, ..., X_{n-1}$ then the play of the "opponent" is $\max(X_1, X_2, ..., X_{n-1})$. If the density played by the "opponent" is p^*, then by Result 3.1 we have $E(x, p^*) = $ constant. Using the result of the two player contest we have that p^* is negative exponential with rate $1/V$. As the contest is symmetric, all individuals must play identical p's. If the p has distribution function $P(x)$, then the "opponent", the maximum of $(n-1)$ players, has distribution function $(P(x))^{n-1}$ (since $P[\max(X_1, X_2, ..., X_{n-1}) < x] = \prod_{i=1}^{n-1} P(X_i < x))$. Thus:

Result 5.1.
    For Model 5.1 the equilibrium p over $U = [0,\infty)$ has

$$P(x) = \left(1 - e^{-\frac{x}{V}}\right)^{\frac{1}{n-1}}.$$

Haigh and Cannings (1988) prove that this is indeed the ESS. It can be proved directly that the total cost per contest (i.e., total time expended by the n individuals) averages V , so that as for the case $n = 2$ reward per contest = average total cost per contest. This is clear also from the fact that $E(x, p) = 0$, by choosing $x = 0$.

Model 5.2.
    We now change one of the features of Model 5.1. In Model 5.1 each individual chooses his value, and is not permitted to reassess during the contest. In Model 5.2 reassessment is permitted. An individual is supposed to pick an initial value, but when the first individual drops out, the remaining $(n-1)$ can each choose a new value, and so each time someone is eliminated. Note that in a two person contest reassessment could be made at any stage without altering the "outcome" as the negative exponential has no memory. Here however the effect is dramatic. There is no ESS in this model. Consider the situation when r , the number of individuals remaining, has reached 2. Now clearly each

of these two will choose a value from the negative exponential distribution with rate $1/V$, since they are facing the War of Attrition of Section 3. Each therefore averages a loss of $V/2$ in time in this par of the contest, and a total loss between them of V. Thus the expected loss for the whole contest with $n > 2$ individuals will be ≥ 0. However, we can see that for an ESS the expected payoff cannot be negative as zero can be played. Since the last two contestants "use up" all the reward V, the others would need to pick 0, but no atoms are permitted. Thus follows (with a few more details as per Haigh and Cannings (1988):

Result 5.2.
Model 5.2 has no ESS.

Model 5.3.
The next generalization is to introduce multiple rewards. As individuals drop out, as in Model 5.2 with reassessment, they receive a reward. If they are the ith to drop out, they receive V_{n+1-i}, where $V_1 > V_2 > ... > V_n$. We shall argue that there is an ESS, though only on an intuitive basis (again details are given in Haigh and Cannings (1988)). Suppose that r individuals remain, so that an individual could guarantee V_r by dropping out immediately (i.e., playing $x = 0$). If an individual remains, he does so in order to win the right to contest in a contest in which V_{r-1} is guaranteed. The effective reward for this part of the contest is therefore $(V_{r-1}-V_r)$, the "opponent" (being the amalgamation of $(r-1)$ players) should play negative exponential with rate $1/(V_{r-1}-V_r)$, i.e., each individual plays negative exponential rate $1/(V_{r-1}-V_r)(r-1)$. This then specifies the ESS for this form of conflict, formalized in Result 5.3 below. As before, the total expected time played is equal to the total reward available $= V_1 + V_2 + ... + V_r$, and the expected lengths of the various parts being $(V_{r-1}-V_r)/r$ for $r = 1,2,...,n$.

Result 5.3.
For Model 5.3 an individual should play negative exponential with rate $1/(r-1)(V_{r-1}-V_r)$ when there are r individuals contesting. This constitutes an ESS.

Model 5.4.
Essentially as for Model 5.3 but with no reassessment permitted at any stage. The problem of investigating the occurrence of an ESS is considerably more difficult here than in any of the previous models. The reason is that it is difficult to specify the "opponent". At a time t the number of individuals remaining is random, whereas in previous models one could specify periods over which the number was well defined. In fact it turns out that an ESS may, or may not, exist; and the exact conditions, though relatively easy to state, are not particularly informative from a biological viewpoint (see Haigh and Cannings, 1988). We can assert again that the expected payoff will be V_0 for an ESS; again since zero can be played.

6.   The War of Attrition with random rewards

In the models discussed in the above sections the reward values were known, of fixed values, and identical for each contest. An important generalization is to allow the reward to vary between the two constestants; reflecting perhaps the state (of hunger, etc.) of the individual.   We consider here only contests which are symmetric, in the sense that contestants are stochastically identical. However, in any particular fight the individuals may differ in reward value, so that the contests are sometimes called asymmetric. We assume throughout that individuals know their own reward value exactly (even though it may not correspond precisely to the intrinsic value of the item to them), but that they do not know the precise value of their opponent's, but only have statistical information regarding the joint distribution of reward values for contestants. A number of cases are examined, all based on the War of Attrition, and all having $U = [0,\infty)$.

Model 6.1.

Suppose there are only two reward values possible, V_1 and V_2 , and that each contest involves a V_1-player and a V_2-player (the reward V_1 being assigned at random independently of the history of the contestants). Consider the V_1-player and denote the support of his strategy by $R_1(p)$, and similarly $R_2(p)$ for V_2 when he plays **p** . Then

$$E_{V_1}(x,y) = \begin{cases} V_1 - y & \text{if } x > y \\ -x & \text{if } x < y, \end{cases}$$

and analogously to Result 3.1 we have that

$$E_{V_1}(x,p) = \text{constant for all } x \in R_1 .$$

This implies that

$$p = \frac{k}{V_1} e^{-\frac{x}{V_1}}$$

is the appropriate play for the opponent of a V_1-player, i.e., for a V_2-player, for $x \in R_1$ (clearly we are making some assumption about R_1 at this stage, and this is justified below). It is possible to demonstrate here, as in cases above, that there can be no atoms, and careful argument regarding possible gaps in R_1 and R_2 leads to the conclusion that $R_1 = R_2 = [0,\infty)$. Accordingly the only possible candidate for an ESS is to take

$$p_1(x) = \frac{1}{V_2} e^{-\frac{x}{V_2}} \quad \text{and} \quad p_2(x) = \frac{1}{V_1} e^{-\frac{x}{V_2}} \quad \text{for } x \in [0,\infty) .$$

Denoting the strategy of an individual now as a pair (p_1, p_2), i.e., what is played when the reward is V_1 and V_2, respectively, we have

$$E((p_1,p_2),(p_1,p_2)) = E((q_1,q_2),(p_1,p_2)) = 0$$

for all (q_1,q_2) and (p_1,p_2) defined above. However, (p_1,p_2) is not an ESS, for an invader (q_1,p_2) (or (p_1,q_2)) has

$$E((p_1,p_2),(q_1,p_2)) = E_{V_1}(p_1,p_2 S) + E_{V_2}(p_2,q_1)$$

$$= E_{V_1}(q_1,p_2) + E_{V_2}(p_2,q_1)$$

$$= E((q_1,p_2))$$

so there is neutrality of (p_1,p_2) vis-a-vis any alternative which replaces one of the p_i's by some alternate strategy. The argument above is due to Selten (1980, who used it in the context of a general conflict where each individual is assigned one of two roles. The above argument then carries through in the same form, so that if (p_1,p_2) is such that $E((q_1,q_2),(p_1,p_2)) = $ constant for $R(q_1) \subset R(p_1)$ and $R(q_2) \subset R(q_2)$, then any (q_1,p_2) or (p_1,q_2) can invade. Thus the only possible ESS's are those for which there is no alternative candidate q_i, so p_1 and p_2 must be pure strategies. In the War of Attrition model on $[0,\infty)$ there can be no pure ESS, so there can be no ESS.

Result 6.1.
  If the reward value of the opponent, for any specific value of the reward V for a contestant, is uniquely determined and different from V, then there is no ESS.

Model 6.2 (Result 6.2).
  A trivial case is where the reward varies between contests but is the same for the two contestants, i.e., has correlation $+1$. Clearly then the ESS is play $p(x) = (1/V)e^{-x/V}$ when one's reward is V.

Model 6.3a: Finite number of Reward Values
  Bishop, Cannings and Maynard Smith (1978) considered the case where there were a finite number n of possible reward values, V_i, $i = 1,2,...,n$, say. In any contest each player is assigned a reward value V_i with probability p_i, independently of each other. Now as before,

$$E_{V_i}(x,p) = \text{constant} \quad \text{for all } x \in R_i$$

implies

$$p = \frac{1}{V_i} e^{-\frac{x}{V_i}}$$

for an opponent of an individual with reward value V_i, and for $x \in R_i$.

As discussed earlier in other models one can prove there are no atoms, and no gaps in R_i, any i. The final piece of the jigsaw is of a new type; the demonstration that R_i and R_j overlap by at most a single point.

No Overlap

Suppose $a \in R_i \cap R_j$. Then

$$E_{V_i}(a,p) = c_i = P_a(V_i - a^*) - (1-P_a)a ,$$

where $a^* =$ expected loss when winning with a, and $P_a =$ probability opponent plays less than a. Similarly

$$E_{V_j}(a,p) = c_j = P_a(V_j - a^*) - (1-P_a)a ,$$

the same P_a and a^* occurring since the opponent's play is independent of the individual's reward (it must be since the opponent has no information about the individual's reward).

Thus

$$(C_i - C_j) = P_a(V_i - V_j)$$

and

$$P_a = \frac{(C_i - C_j)}{(V_i - V_j)} ,$$

so P_a is uniquely determined, and since there are no gaps $P_x \nearrow x$, so there is a unique value a satisfying the above conditions.

These results imply that $R_i = [t_{i-1}, t_i]$ with $0 = t_0 < t_1 < t_2 < \ldots < t_{n-1}$ and "$t_n = \infty$". The final step is to pin down the precise form of the densities and of t_i. These are given by equations (8) and (9) in the paper cited.

There is no difficulty in extending the results to a countable number of reward values, and, by taking limits, to a continuum.

Result 6.3a.

For Model 6.3a an individual plays an exponential strategy over an interval defined according to its reward value.

Model 6.3b.

Bishop and Cannings (1986) extended the analysis to a continuum of reward values with a more general reward and loss structure. They proved that, under appropriate (and biologically meaningful) restrictions on the functions involved, if

$$E_V(x,y) = \begin{cases} w(v) - w(x,y) & x > y \\ l(v) - w(x,y) & x < y \end{cases}$$

then an ESS exists. A remarkable invariance result holds for these models. The expected payoff to an individual whose own reward is V, and who plays the ESS value (a unique value defined by a function $g(v)$) in an ESS population, is independent of the form of the functions $w(x,y)$ and $l(x,y)$.

The approach used to investigate the existence of an ESS is somewhat different here from earlier methods, and was used by Riley (1979a) in the context of the original War of Attrition.

It is clear that an individual whose reward value is v will, if there is an ESS, require to play a unique value $g(v)$, and that playing an alternative is just like getting your v value wrong while still using $g(v)$. An ESS will exist if and only if, $E_v(g(v^*),g)$ is maximized when $v^* = v$. The result follows from this maximization.

Result 6.3b.

The ESS for Model 6.3b is: play $g(v)$ when one's reward is v, where

$$[w(v)-l(v)]\phi(v) = G(v)$$

$\phi(v)$ being the density of reward value, and

$$g(v) = g'(v)\left[\int_0^v w'(g(v),g(u))\phi(u)du + \int_v^\infty l'(g(v),g(u))\phi(u)du\right].$$

There are certain interesting consequences for Models 6.3a and b. The first is that in a contest between individuals with reward values V_i and V_j, the individual with the higher reward value always wins. Secondly, and perhaps more importantly, the expected payoff is positive. There is no longer the exact translation of energy expended to energy received as in earlier models.

Model 6.4.

The models above address the cases where the opponent's reward is determined precisely by the contestant's reward, or is independent. These represent the extreme possibilities. An attempt to analyse the case with a more general correlation has been made by Hammerstein and Parker (1982). This model is similar to Model 6.1 but allows the possibility of incorrect assessment of their reward values by the contestants. An individual assesses himself as I, where I equals A or B (roles), and his opponent as J , also A or B , the payoff to the I-player is then defined as

$$E_{I,J}(x,y) = \begin{cases} V_{IJ} - C_{IJ}y & x > y \\ -C_{IJ}x & x < y. \end{cases}$$

Thus the reward part depends on the assessments made, as do the costs. We need to seek a strategy pair (p_A, p_B) defining what strategy should be played when assessing one's strategy as A , and as B . As in Bishop, Cannings and Maynard Smith (1978) it is proved that there are no atoms of probability in the ESS strategies. Hammerstein and Parker (1982) also prove under the so-called weak asymmetry conditions that there is no gap in R_A or R_B, that $R_A \cup R_B = U$, and that $R_A \cap R_B$ is a single point (though the proofs are rather difficult). The weak asymmetry conditions are

$$W_{AB}V_{AB} > W_{BB}V_{BB} \quad \text{and} \quad W_{BA}C_{BA} > W_{AA}C_{AA},$$

and an addition requirement is that

$$V_{AB}C_{BA} > V_{BA}C_{AB},$$

where

$$W_{IJ} = P[\text{Context is } I \times J], \quad I,J \in \{A,B\}.$$

With there restrictions they demonstrate that there is an ESS, and that the ESS is as given in Result 6.4.

Result 6.4.

For Model 6.4 (Hammerstein and Parker's model) the ESS is: play negative exponential with rate V_{II}/C when assess role as $I \in \{A,B\}$, with supports

$$R_A = \left[0, \frac{-V_{BB}}{C_{BB}} \ln\left(\frac{W_{BA}C_{BA}}{W_{BA}C_{BA} + W_{BB}C_{BB}}\right)\right] \quad \text{and} \quad R_B = [0,\infty) \backslash R_A.$$

Models 6.1 through 6.4 cover a range of possibilities. They must represent however only the tip of the iceberg. Even the most general model of Hammerstein and Parker has quite stringent restrictions imposed on the parameters, and more general models would be of considerable interest.

7.   Patterns in finite conflicts

We now turn to the consideration of the class of finite conflicts, i.e., those with $V = \{1,2,...,n\}$. As is well known, e.g. Haigh (1978), a conflict with a given payoff matrix A may have no ESS, a unique ESS, or a number of ESS's. Recently Vickers and Cannings (1988a) have begun to enumerate the various possibilities.

They introduce the idea of a pattern (of supports), as a set V of subsets of U, so $V = \{T_1,T_2,...,T_k\}$, where $T_i \subset U$, $i = 1,2,...,k$. A pattern V is then said to be attainable if there is a matrix A whose complete set of ESS's $p_1, p_2,...,p_k$ have precisely the supports $T_1,T_2,...,T_k$.

The conditions for an ESS p to exist for a matrix A (Haigh, 1978), while fairly easy to express mathematically, are rather difficult to work with, particularly if one is examining the possibility of coexistence of several ESS's. Moreover, we are working, as it were, from the "wrong end". It is a fairly mechanical task to evaluate the ESS's of a given matrix A, but to construct A for a given V, or demonstrate non-existence proves difficult. Nevertheless some progress has been made in proving various exclusion and existence results.

Result 7.1.
Bishop and Cannings (1976) prove that if p is an ESS with support $R(p)$, and q has $R(q)$, then if $R(q) \subset R(p)$, q is not an ESS.

This is a simple exclusion result, which immediately reduces the number of possible patterns considerably. A similar result was proved by Kingman (1961) in the case where the matrices were symmetric.

Subsequent results in this section (unless otherwise stated) are due to Vickers and Cannings.

Result 7.2.
No attainable pattern $V = \{T_1,T_2,...,T_k\}$ can have $T_1 = (1,2)$, $T_2 = (1,3)$, $T_3 = (2,3)$ (or any permutation thereof).

This result is in fact the only exclusion result which refers only to pairwise ESS's (i.e. ones with supports with two elements); the $T_4,...,T_k$ are not involved as such. Moreover, one can prove that no other exclusion result involving just pairs is possible.

Result 7.3.
If $V = \{T_1,...,T_k\}$ has all $|T_i| = 2$, then V is attainable provided there is no triple $T_i = (l,m)$, $T_j = (l,n)$ and $T_k = (m,n)$.

The easiest way to visualize the cases where $|T_i| = 2$ for all i, is to introduce a graph $G = (U,E)$, where $U = \{1,2,...,n\}$ is the set of nodes, and E the set of edges, so that nodes i and j are joined if and only if there is an ESS with support (i,j). Result (7.2) then asserts that for any attainable pattern with $|T_i| = 2$, the corresponding graph must be triangle-free, while Result 7.3 asserts that any triangle-free graph corresponds to an attainable pattern. In fact, given any graph $G = (U,E)$ which is triangle free we can construct a matrix A immediately which achieves the corresponding pattern (see section below "The special case $\mathbb{C}$"). It is known, Turán (1954) that the maximum number of edges in a triangle-free graph is achieved by the complete bipartite graph with equal (as nearly as possible) partition. Thus, if $n = 2m$, there can be m^2 pairwise ESS's, while for $n = 2m + 1$ there can be $m(m+1)$, but not more.

FUNDAMENTAL CONJECTURE

In attempting to delineate all patterns Vickers and Cannings (1988a) make a fundamental conjecture. They introduce the idea of a maximal attainable pattern, as being an attainable pattern whose set of supports is not a subset of that of any attainable pattern with the same value of n.

They then conjecture that if V is a maximal attainable pattern, then any subpattern of V (i.e., a pattern whose T_i's form a subset of those of V) is attainable. No proof has been found, and no counterexamples are known.

If true, this conjecture allows the specification of all attainable patterns through the specification of all maximal attainable patterns. It thus considerably simplifies the search procedure required. Using the theorems developed (examples given below) one can specify for a given r all those known to be impossible. One thus is left with a set of maximal patterns (i.e., not subpatterns of any others of the set). These need to be examined, and either shown to be impossible by special argument, or to be attainable by construction of a numerical example. In the former case one will then have new maximal patterns to examine. Using the various theorems, the cases $n = 3$ and $n = 4$ have been enumerated completely, but $n = 5$ is still incomplete. Before discussing any details of these cases we quote two results needed.

Result 7.4.
If $S \subseteq U\backslash\{1,2,3\}$, then if there are ESS's with supports $(1,2,S)$ and $(2,3,S)$ there is no ESS with support $(1,3,S)$.

Clearly Result 7.2 is a special case. Further we can deduce that if there are two ESS's with $(n-1)$ elements, then there are no other ESS's.

Result 7.5.
If $S \subseteq U\backslash\{1,2,\ldots,k\}$, then if there are ESS's with supports $(1,S)$, $(2,S),\ldots,(k,S)$ then there is no ESS with support $(1,2,\ldots,k)$.

The method of proof of these latter two results is to show that the existence of ESS's as specified implies the existence of an invader for an equilibrium over the final support.

These results enable one to find all the patterns for $n = 3$ ($V_1 = \{(1,2,3)\}$, $V_2 = \{(1,2),(2,3)\}$) where one only need list non-degenerate cases (a pattern $V = \{T_1,\ldots,T_k\}$ is called degenerate if for any $L \subset \{1,2,\ldots,k\}$,

$$\left(\bigcup_{i \in L} T_i\right) \cap \left(\bigcup_{i \in \bar{L}} T_i\right) = \emptyset,$$

and for $n = 4$,

$$V_1 = \{(1,2,3,4)\}, \qquad\qquad V_3 = \{(1,2,3),(2,4),(3,4)\},$$

$$V_2 = \{(1,2,3),(1,2,4)\}, \qquad\qquad V_4 = \{(1,2),(1,3),(1,4)\},$$

$$V_5 = \{(1,2),(2,3),(3,4),(4,1)\}$$

The case $n = 5$ is more difficult; the reader is referred to the papers cited.

THE SPECIAL CASE $\mathbb{C}$

The general class of matrices presents great difficulties. Accordingly Cannings and Vickers (1988) introduced a class of matrices $\mathbb{C}$ where the condition for an ESS reduces to a fairly simple one.

Define $\mathbb{C}$ as the class of $n \times n$ matrices $A = (a_{ij})$ which are symmetric and satisfy

$$a_{ij} = \begin{cases} 0 & \text{if } i = j \\ \pm 1 & \text{if } i \neq j. \end{cases}$$

Result 7.6.
    For any $A \in \mathbb{C}$ if $S \subset U$ is such that $a_{ij} = +1$ for any $i,j \in S$, $i \neq j$, and for each $k \notin S$ there exists $i \in S$ such that $a_{ik} = -1$, then there is an ESS with support S.

    Essentially there is an ESS with support S if between pure strategies in S the payoffs are $+1$ if different strategies, 0 if the same, and S is not contained in a larger set with the same property. We should also note that -1, 0 and $+1$ could be replaced by any c,d and e for which $c < d < e$.

    If we use the matrix A to define a graph $G = (U,E)$ where $(i,j) \in E$ if and only if $a_{ij} = +1$, then there will be an ESS with support S, whenever S is a clique (maximal complete subgraph) of G. There is an extensive literature regarding cliques, though often the questions addressed are somewhat different from those which arise in this context. Here we quote only one such result:

Result 7.7 (Moon and Moser, 1965).
    The maximum number of cliques (ESS's) for given r is

$$\begin{cases} 3^m & \text{, if } n = 3m \\ 4 \cdot 3^{m-1} & \text{, if } n = 3m + 1 \\ 2 \cdot 3^{m-1} & \text{, if } n = 3m - 1. \end{cases}$$

These maxima are achieved by creating a complete multipartite graph with partition of n as nearby as possible into 3's. We note, Vickers and Cannings (1988b), that this is not the maximum number of ESS's for general matrices.

A further topic of interest is to consider the generation of matrices in $\mathbb{C}$ at random, perhaps taking the entries to be +1 with probability p, and -1 with probability q = 1 - p (off-diagonal) and independently. A considerable body of knowledge exists on such random graphs, and its exploitation yields information regarding the evolution of ESS's. Various aspects of this problem are discussed in Cannings and Vickers (1988).

References

Akin, E. (1982), Exponential families and game dynamics, *Canad. J. Math.* 34, 374-405.

Bishop, D.T. (1978), Models in animal conflicts, Ph.D. Thesis, University of Sheffield.

Bishop, D.T. and Cannings, C. (1976), Models of animal conflicts, *Adv. in Appl. Probab.* 8, 616-621.

Bishop, D.T. and Cannings, C. (1978), A generalized war of attrition, *J. Theoret. Biol.* 70, 85-124.

Bishop, D.T. and Cannings, C. (1986), Ordinal conflicts with random rewards, *J. Theoret. Biol.* 122, 225-230.

Bishop, D.T., Cannings, C. and Maynard Smith, J. (1978), The war of attrition with random rewards, *J. Theoret. Biol.* 74, 377-388.

Brown, D.B. and Hansall, R.I.C. (1987), Convergence to the ESS in the two policy game, *American Naturalist* 130, 929-940.

Cannings C. and Gardiner, D. (1980), A dynamic programming approach to evolutionarily stable strategy theory, *Adv. in Appl. Probab.* 11, 3-5.

Cannings, C. and Vickers, G.T (1988), Patterns of ESS's. II, *J. Theoret. Biol.* 132, 409-420.

Feller, W. (1950), *An Introduction to Probability Theory and Its Applications*, Wiley, New York.

Haigh, J. (1978), Game theory and evolution, *Adv. in Appl. Probab.* 7, 8-11.

Haigh, J. and Cannings, C. (1988), The n-person war of attrition, in: *Proc. IASS Conference on Biomathematics* (K. Sigmund, ed.), Kluwer, Dordrecht.

Hammerstein, P. and Parker, G. (1982), The asymmetric war of attrition, *J. Theoret. Biol.* 96, 647-682.

Hines, W.G.S. (1987), Evolutionarily stable strategies: A review of basic theory, *Theoret. Population Biol.* 31, 195-272.

Kingman, J.F.C. (1961), A mathematical problem in population genetics, *Proc. Camb. Phil. Soc.* 57, 424-432.

Luce, R.D. and Raiffa, H. (1957), *Games and Decisions*, Wiley, New York.

Maynard Smith, J. (1974), The theory of games and the evolution of animal conflicts, *J. Theoret. Biol.* **47**, 209-212.

Maynard Smith, J. (1982), *Evolution and the Theory of Games*, Cambridge University Press.

Maynard Smith, J. and Price, G. (1973), The logic of animal conflicts, *Nature* **246**, 15-18.

Moon, J. and Moser, L. (1965), On cliques in graphs, *Israel J. Math.* **3**, 23-28.

Riley, J. (1979), Evolutionarily equilibrium strategies, *J. Theoret. Biol.* **76**, 109-123.

Selten, R. (1980), A note on evolutionarily stable strategies in asymmetric animal conflicts, *J. Theoret. Biol.* **84**, 93-101.

Taylor, P.D. and Jonkers, L.B. (1978), Evolutionarily stable strategies and game dynamics, *Math. Biosci.* **40**, 145-156.

Turán, P. (1954), On the theory of graphs, *Colloq. Math.* **3**, 19-30.

Vickers, G.T. and Cannings, C. (1987), On the definition of an Evolutionarily Stable Strategy, *J. Theoret. Biol.* **129**, 349-353.

Vickers, G.T. and Cannings, C. (1988a), Patterns of ESS's I, *J. Theoret. Biol.* **132**, 387-408.

Vickers, G.T. and Cannings, C. (1988b), On the number of stable equlibria in a one-locus, multi-allelic system, *J. Theoret. Biol.* **131**, 273-277.

Zeeman, E.C. (1980), Population dynamics for game theory, in: *Global Theory of Dynamical Systems* (Z. Nitecki, C. Robinston, eds.), Springer-Verlag, Berlin.

SELECTION IN NATURAL POPULATIONS

Freddy Bugge Christiansen
Department of Ecology and Genetics
University of Aarhus
DK-8000 Aarhus C
Denmark

ABSTRACT. The lectures focus on the development of selection component analyses based on population samples including mother-offspring combinations. The need for separation of selection into components is discussed and exemplified by experimental data. The genetic analysis of samples of both simple and complete mother-offspring combinations is developed for one-locus polymorphisms. This is subsequently used to develop an analysis of selection components using population samples including mother-offspring combinations, and this analysis is exemplified by data on natural populations of *Zoarces viviparus* (simple mother-offspring data) and *Gammarus oceanicus* (complete mother-offspring data). Finally, the mother-offspring analyses are generalized to cover polymorphisms at multiple loci, and as an introduction to this some population theory for multiple loci is discussed. Index set notation for multi-locus gamete types is introduced and applied to problems concerning population mixing. In addition, the effects of recurrent immigration into a population are investigated in a generalization of Wright's island model and in the stepping-stone cline model.

1. Inference on fitness

Natural selection is a process going on throughout the life cycle of a phenotypically varying species. Differences in survival among phenotypes from the time of birth to the time of reproduction is probably the most widely known aspect of natural selection. Survival, however, is in vain if the individual does not leave offspring, and fecundity, the number of offspring produced, is another aspect of selection.

This realization of natural selection as a process with many aspects is important when observations on selection are made (Christiansen, 1984). The focus of these lectures is to discuss the possibility of describing the action of selection based on observations of the genotypic composition of natural populations. Of interest is selection that influences the allelic variation at one or more loci, and initially we may discuss the simplest possible observation which historically has been of interest in the description of animal populations.

1.1. THE HARDY-WEINBERG PROPORTIONS

Consider an autosomal locus with two alleles, A_1 and A_2, in a diploid sexually reproducing species of animals. We will allow for two sexes, but in these initial discussions we will assume that the sexes are equal. Suppose that a sample of $X_{..}$

S. Lessard (ed.), Mathematical and Statistical Developments of Evolutionary Theory, 121–176.

individuals is taken from a population of the species, and that the three genotypes A_1A_1, A_1A_2 and A_2A_2 occur in the sample in the numbers X_{11}, X_{12} and X_{22}, respectively $(X_{11}+X_{12}+X_{22} = X_{..})$. The sample is assumed to be taken from a homogeneous population of individuals, so we may assume that (X_{11},X_{12},X_{22}) is multinomially distributed with number parameter $X_{..}$ and frequency parameter (x_{11},x_{12},x_{22}). The frequencies x_{11}, x_{12} and x_{22} $(x_{11}+x_{12}+x_{22} = 1)$ are referred to as the *genotypic frequencies* in the population, and they are estimated by the genotypic frequencies in the sample as

$$(1.1) \qquad \hat{x}_{11} = X_{11}/X_{..}, \hat{x}_{12} = X_{12}/X_{..}, \text{ and } \hat{x}_{22} = X_{22}/X_{..}.$$

The multinomial distributions form an exponential family of distributions where it is natural to consider the logarithm of the likelihood function instead of the likelihood function itself. This function is $L(x,X) = \sum X_{ij} \log(x_{ij})$.

The problem we are facing is to formulate expectations about the proportions among the genotypes in the population (and in the sample) under the assumption that no selection works on the considered variation. The formulation of such expectations is the basis of any analysis with the possibility of inferring selection in a population. The fundamental rule of population genetics is the law of the conservation of the gene frequency that applies in situations of no selection (see, e.g., Christiansen and Feldman, 1986, Ch. 2). The *gene frequencies* of A_1 and A_2 are given as $p_1 = x_{11} + x_{12}/2$ and $p_2 = x_{22} + x_{12}/2$, respectively, and these are the frequencies of allele A_1 and A_2 among the genes in the population. The conservation law then states that under Mendelian segregation and no selection these gene frequencies are conserved from generation to generation. This may be written as $p_1' = p_1$, where p_1' refers to the gene frequency among offspring of the population with gene frequency p_1 evaluated at the same development stage. The basis of this law is *Mendelian segregation*, where the two allelic genes carried by an individual has equal chance of being represented in a given (haploid) gamete. The conservation law in this form is a deterministic statement, but the principle of the law is really the basis for the process of random genetic drift in a finite population. However, we will be concerned with observations on short term phenomena where the statistical error covers any error due to random genetic drift.

This characterization of the absence of selection does not leave much chance for inference on selection from our sample. Our characterization of selection does not address the genotypic frequencies in an isolated sample, only the change in genetic composition through time.

The way out of this dilemma has been an assumption on the way in which the haploid gametes unite to form the diploid zygote which is the origin of the developed individual. In the analysis of animal populations the traditional form is to assume random union of gametes, or equivalently random mating of breeding individuals if each mating has equal fecundity. This assumption predicts that the genotypic frequencies among zygotes will be the *Hardy-Weinberg proportions*, i.e., the genotypic frequencies of A_1A_1, A_1A_2 and A_2A_2 are $x_{11} = p_1^2$, $x_{12} = 2p_1p_2$ and $x_{22} = p_2^2$. (Hardy-Weinberg equilibrium is the state where no selection occurs and reproduction is by random mating, so the genotypic frequencies are unchanged form generation to generation.)

The Hardy-Weinberg proportions describe the alleles as occurring independently in the zygotes. Thus, we may view our sample as a sample of $2X_{..}$ independent genes, so the estimates of the gene frequencies are simply the observed gene frequencies:

$$(1.2) \qquad \hat{p}_1 = \frac{2X_{11}+X_{12}}{2X_{..}} \quad \text{and} \quad \hat{p}_2 = \frac{X_{12}+2X_{22}}{2X_{..}} .$$

Now we can compare the genotypic proportions to the Hardy-Weinberg proportions. Deviations form these proportions may either be due to deviations from random mating or to selection. Really, as test for the Hardy-Weinberg proportions try the statistical hypothesis that the alleles occur independently in the individuals. The statistical analysis is fairly straightforward in that any genotypic distributions (x_{11},x_{12},x_{22}) may be uniquely determined as

$$(1.3) \qquad \begin{aligned} x_{11} &= p_1^2 + Fp_1p_2 , \\ x_{12} &= 2p_1p_2(1\text{-}F) , \\ x_{22} &= p_2^2 + Fp_1p_2 , \end{aligned}$$

where $p_1 = x_{11} + x_{12}/2$, $p_2 = x_{22} + x_{12}/2$ and F is the relative difference between the frequency of heterozygotes and the corresponding Hardy-Weinberg proportion (see Wright, 1969; for an extension to multiple alleles, see Nei, 1977). With the estimates (1.1) of the genotypic frequencies and with the gene counting estimates (1.2) of the gene frequencies we get

$$(1.4) \qquad \hat{F} = \frac{2\hat{p}_1\hat{p}_2 - \hat{x}_{12}}{2\hat{p}_1\hat{p}_2} = \frac{4X_{11}X_{22} - X_{12}^2}{(2X_{11}+X_{12})(X_{12}+2X_{22})}.$$

The observed genotypic frequencies will show an excess of homozygotes relative to the Hardy-Weinberg proportions when $\hat{F} > 0$ and vice versa for $\hat{F} < 0$.

As a test for the hypothesis: $F = 0$, we can use the classical χ^2-goodness-of-fit statistic: Σ (Observed-Expected)^2/Expected. This test statistic equals $X_{..}\hat{F}^2$ and it is approximately $\chi^2(1)$ distributed. Alternatively we may use the statistic $\hat{F}(X_{..})^{1/2}$ which is approximately normal $N(0,1)$ distributed (these tests are discussed by Brown, 1970). The latter test has the advantage of expressing in its value the sign of the deviation from Hardy-Weinberg proportions. The χ^2-statistic is just measuring the squared distance between the observed genotypic frequencies and the expected Hardy-Weinberg proportions (in the Shahshahani metric of the Hardy-Weinberg proportions corresponding to the observed gene frequencies, see Akin, this volume).

How is the presence or absence of a significant deviation from the Hardy-Weinberg proportions related to the presence or absence of selection on the observed genotypic variation? As already indicated we cannot expect the relation to be very strong, but it is important to argue that the relation is actually incredibly weak. Suppose for instance that breeding in the population indeed results in random union of gametes, and suppose that the zygotes differ in their probability of surviving to become adults, then we expect the genotypic proportions among adults to be given as in Table 1.1, where the average probability of the survival in the population is denoted by W, i.e., we have $W = p_1^2 W_{11} + 2p_1 p_2 W_{12} + p_2^2 W_{22}$. The adult population will then have the gene frequencies

$$(1.5) \qquad p_1' = p_1 \frac{p_1 W_{11} + p_2 W_{12}}{W} \;,\; p_2' = p_2 \frac{p_1 W_{12} + p_2 W_{22}}{W} \;,$$

so from (1.4) the deviation from Hardy-Weinberg proportions is given by

$$(1.6) \qquad F = \frac{4p_1^2 p_2^2}{4(p_1')^2 (p_2')^2} \frac{W_{11} W_{22} - W_{12}^2}{W^2} \;.$$

Therefore, if $W_{11} W_{22} = W_{12}^2$ no deviation from the Hardy-Weinberg proportions is expected (Frydenberg, 1956; Wallace, 1958; Lewontin and Cockerham, 1959). This pattern of selection (called geometric fitnesses because the fitness of the heterozygote, the probability of survival, is the geometric mean of the homozygote fitnesses) corresponds to directional selection. i.e., $W_{11} < W_{12} < W_{22}$ or $W_{11} > W_{12} > W_{22}$, and directional selection changes the gene frequency in a polymorphic population every generation. Thus, a lack of significant deviation from the Hardy-Weinberg proportions may cover strong geometrical selection, and in finite samples this extends to other patterns of selection (Lewontin and Cockerham, 1959). In addition, the observation of an excess of heterozygotes, $F < 0$, does not suggest overdominant selection, i.e., $W_{11} < W_{12}$ and $W_{22} < W_{12}$, it only points towards super-geometric selection, i.e., $W_{11} W_{22} > W_{12}^2$ (Frydenberg, 1956; Wallace, 1958; Lewontin and Cockerham, 1959).

Table 1.1
Selection by differential survival of zygotes

Genotype	$A_1 A_1$	$A_1 A_2$	$A_2 A_2$
Zygotic frequencies	p_1^2	$2p_1 p_2$	p_2^2
Probability of survival	W_{11}	W_{12}	W_{22}
Adult frequencies	$p_1^2 W_{11}/W$	$2p_1 p_2 W_{12}/W$	$p_2^2 W_{22}/W$

If a significant deviation from Hardy-Weinberg proportions is observed, can we then conclude that selection is working on the polymorphism? Again the answer is no! simply because we had to assume random mating to produce the expectation of Hardy-Weinberg proportions. Thus, a deviation from random mating may be the cause for the observation. In fact, this is a much more immediate cause for deviations from Hardy-Weinberg proportions, in that the genotypic proportions among zygotes are a reflection of the mating in the population. There are various causes for deviations from randomness in mating. Unrelated to the genotypic variation of interest are two main factors inbreeding, caused by an increased tendency for relatives to mate, and the Wahlund effect, caused by the mixing of different populations bred by random mating. We will return to discussions of the Wahlund effect as a reflection of immigration to the studied population. For inbreeding F is the natural parameter for the description of the effect in the population, so the Hardy-Weinberg test has much nicer properties when viewed in this context (Brown, 1970).

1.2. PARENT-OFFSPRING TRANSITION

The change from generation to generation is included into the description of our population if samples are taken in subsequent generations so the transition from the population of parents to the population of their offspring is studied. Thus, we consider a sample of adults (X_{11}, X_{12}, X_{22}) and a sample of their offspring at the adult stage, (Y_{11}, Y_{12}, Y_{22}) say. This sampling procedure may be continued over several generations, but the main idea of the analysis may be discussed on the basis of the two samples describing a *parent-offspring transition* on the population level.

Assuming reproduction by random mating among adults the zygotes of the offspring generation is expected to be in the Hardy-Weinberg proportions $(p_1^2, 2p_1p_2, p_2^2)$ corresponding to the gene frequencies among adults in the parent population. The expected genotypic frequencies among adults in the offspring population may then be read from Table 1.2 as a function of the probability of survival from zygote to adult. However, the genotypic frequencies are given in terms of ratios between the probability of survival of the genotype and the average probability of survival, so we can only infer information on these relative survival probabilities. A convenient way to parameterize this is to make inference on the survival probabilities of the homozygote relative to the heterozygote, so we replace (W_{11}, W_{12}, W_{22}) with (w_{11}, w_{12}, w_{22}) where $w_{12} = W_{12}/W_{12} = 1$ and selection is described by the *fitness* values of the homozygotes relative to the heterozygote, $w_{11} = W_{11}/W_{12} > 0$ and $w_{22} = W_{22}/W_{12} > 0$.

The gene frequencies among the parents may be estimated by equations (1.2), and from Table 1.1 we get the estimated fitnesses as

(1.7)
$$\hat{w}_{11} = \frac{Y_{11}}{Y_{12}} \frac{2\hat{p}_2}{\hat{p}_1} \quad \text{and} \quad \hat{w}_{22} = \frac{Y_{22}}{Y_{12}} \frac{2\hat{p}_1}{\hat{p}_2}.$$

Table 1.2
Parent-offspring transition with early and late selection

Genotype	A_1A_1	A_1A_2	A_2A_2
Parents: Adults	x11	x12	x22
Late selection	v_{11}	1	v_{22}
Parents: Breeders	$x_{11}v_{11}/v$	x_{12}/v	$x_{22}v_{22}/v$
Offspring: Zygotes	p_1^2	$2p_1p_2$	p_2^2
Early selection	w_{11}	1	w_{22}
Offspring: Adults	$p_1^2w_{11}/w$	$2p_1p_2/w$	$p_2^2w_{22}/w$

The estimates (1.2) and (1.7) are the maximum likelihood estimates of the gene frequencies among parents and the fitnesses describing the survival from zygotes to adults in the offspring generation. This may be seen by a simple, but tedious solution of the likelihood equations.

This procedure is clearly considerably better for the analysis of selection than that based on one isolated sample, in that the fitnesses w_{11} and w_{22} rather than the index F are estimated. However, when this is used in experimental populations the observation is commonly as shown in Figure 1.1. The estimated relative fitness values vary as a function of the genetic composition of the population. Such variation in the fitness values may be true to the population, and then it is referred to as *frequency-dependent selection*. However, the pattern of Figure 1.1, where the rarest type tends to have the highest fitness occurs so often that suspicion is in place. Indeed, as we will see below, this is exactly the kind of bias that would be expected in these fitness estimates (Prout, 1965).

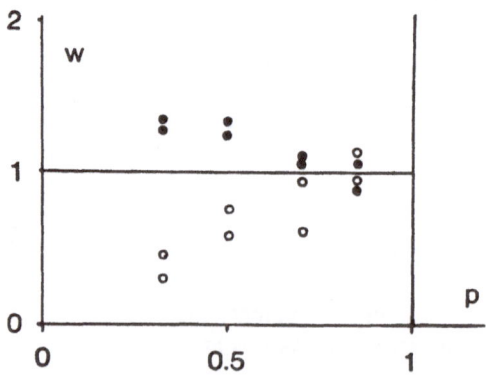

Figure 1.1. Fitness values estimated by (1.7) from parent-offspring transitions in an experimental population of *Drosophila melanogaster*. Two experiments were done for each of four genotypic compositions of the parental adult population (the gene frequency of allele A_1 is shown as the abscissa). o:w_{11}; o:w_{22}. After Bundgaard and Christiansen (1972).

Suppose that the adult individuals differ in their probability to breed, so the breeding population differs from the adults as shown in Table 1.2. Here, the average probability of becoming a breeder is $v = x_{11}v_{11} + x_{12} + x_{22}v_{22}$ relative to the probability that a heterozygote breeds. Similarly, $w = p_1^2 w_{11} + 2p_1p_2 + p_2^2 w_{22}$.

To simplify the analysis, assume that the zygote to adult survival is the same for all genotypes, i.e., we assume $w_{11} = w_{22} = 1$. With this we expect the genotypic frequencies among adults to be in Hardy-Weinberg proportions, so we may assume that $x_{11} = \pi_1^2$, $x_{12} = 2\pi_1\pi_2$ and $x_{22} = \pi_2^2$. So our gene frequency estimates (1.2) in reality estimate this adult gene frequency.

The adults in the offspring generation is expected to be in the proportions $(p_1^2, 2p_1p_2, p_2^2)$ because of the assumption $w_{11} = w_{22} = 1$ (Table 1.2). However, now we have

$$(1.8) \qquad p_1 = \pi_1 \frac{\pi_1 v_{11} + \pi_2}{v}, \quad p_2 = \pi_2 \frac{\pi_1 + \pi_2 v_{22}}{v},$$

so with very large samples we get approximately that $\hat{p}_1 = \pi_1$, and $\hat{p}_2 = \pi_2$ and therefore from (1.7) that

$$(1.9) \qquad \hat{w}_{11} = \frac{\pi_1 v_{11} + \pi_2}{\pi_1 + \pi_2 v_{22}} \quad \text{and} \quad \hat{w}_{22} = \frac{\pi_1 + \pi_2 v_{22}}{\pi_1 v_{11} + \pi_2}$$

(Prout, 1965). Thus, we get these spurious estimates of the fitness parameters w_{11} and w_{22} by erroneously assuming $v_{11} = v_{22} = 1$ when in fact $w_{11} = w_{22} = 1$. The estimates vary in the general way depicted in Figure 1.1, but in this simple situation we must have $\hat{w}_{11}\hat{w}_{22} = 1$ from (1.9). When A_1 is rare ($\pi_1 \approx 0$) then the estimated finesses (1.9) are $\hat{w}_{11} = 1/v_{22}$ and $\hat{w}_{22} = v_{22}$, and similarly for A_2 rare $\hat{w}_{11} = v_{11}$ and $\hat{w}_{22} = 1/v_{11}$. Thus, if the late selection is overdominant ($v_{11} < 1$ and $v_{22} < 1$), then the rare homozygote will show the highest fitness and the common homozygote the lowest.

These conclusions are due to Prout (1965), but some extensions are made by Christiansen, Bundgaard and Barker (1978), e.g., if the late selection Table 1.2 occurs in one sex only, then $\hat{w}_{11} + \hat{w}_{22} = 2$.

1.3. FITNESS ESTIMATION

The frequency dependence of the fitness estimates in the data of Bundgaard and Christiansen (1972), as shown in Figure 1.1 is indeed due to the effect disclosed by Prout (1965). The experimental population had discrete non-overlapping generations, and it was designed in a way that allowed an independent evaluation of late selection components. *Fecundity selection* was evaluated by counting the number of eggs laid by each female. *Sexual selection* is due to differences in the chance that adults have to become parents. This was evaluated by forcing the adult females in the population to mate with at most one male, and the genotype of this male was determined from the segregation among her offspring.

The data allowed information on the life stages shown in Figure 1.2. The knowledge of the population of breeders, their mating and their fecundity allows an unambiguous estimation of the genotypic frequencies in the population of zygotes by just assuming Mendelian segregation of the alleles. This allows the relative survival from zygotes to adults in the offspring generation to be estimated correctly; we will refer to this selection component as *zygotic selection*. The total selection during the life of the population from adult parents to offspring at the adult stage is therefore partitioned into the components of selection: zygotic selection (differential survival of zygotes to adults), sexual selection (differences in the propensity of adults to become parents) and fecundity selection (variation in number of offspring among mated pairs).

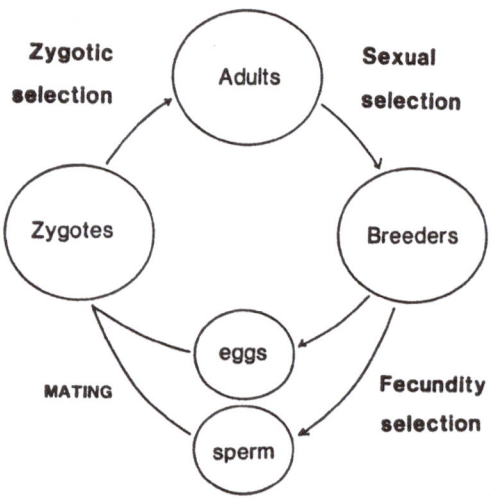

Figure 1.2. The life stages and the effects observable in the *Drosophila* experiment by Bundgaard and Christiansen (1972).

In this data the variation in fecundity was low, so we need only discuss zygotic selection and sexual selection. As we know the zygote population. we can mimic the estimation bias considered by Prout (1965) by calculating the estimated relative viabilities (1.7) with the parental adults and the offspring zygotes as the parent-offspring transition. This is obviously and evidently wrong, and therefore it might be difficult to envision what is going on. However, all that happens is that we force upon the population the assumption of no zygotic selection. These "estimates" are shown in Figure 1.3, and they show the expected frequency-dependent behavior. This supposed bias-prone estimation can be compared to more straightforward fitness estimates describing the sexual selection because the data contains information on the genotypic frequencies in the breeding population, (b_{11}, b_{12}, b_{22}) say. The sexual selection fitness values may therefore be estimated as

$$(1.10) \qquad \hat{v}_{11} = \frac{X_{11}}{X_{12}} \frac{\hat{b}_{12}}{\hat{b}_{11}} \quad \text{and} \quad \hat{v}_{22} = \frac{X_{22}}{X_{12}} \frac{\hat{b}_{12}}{\hat{b}_{22}} .$$

This estimation is done for the males and females separately (Figure 1.4), and for these estimates the spectacular frequency-dependence in Figure 1.3 vanishes. Similarly, no trace of this effect remains when the zygotic selection fitness values are estimated by a comparison between offspring zygotes and offspring adults (Figure 1.5). Thus, the pattern of the fitness estimates in Figure 1.1 has been resolved into virtually constant fitness estimates describing zygotic selection and sexual selection, and for these estimates the variation with initial frequency in the population becomes biologically interesting (Bundgaard and Christiansen, 1972).

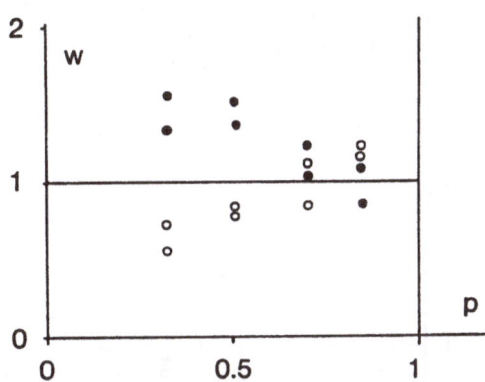

Figure 1.3. The bias caused by the influence of late selection components on the fitness values in Figure 1.1. The effect is illustrated by substituting the offspring zygote population in the place of the offspring adult population in the estimates (1.7). Otherwise as in Figure 1.1. After Bundgaard and Christiansen (1972).

The resolution of the *Drosophila* data comes from the compartmentalization of the selection effects, so the change in genotypic frequencies due to selection can be traced without disturbance from other causes. The experimental procedure, however, did not produce data on the segregation of gametes, and the population of zygotes was estimated assuming Mendelian segregation. Therefore, any deviation from Mendelian proportions among the gametes that unite to form the zygotes will be pushed forward and be evaluated as part of zygotic selection.

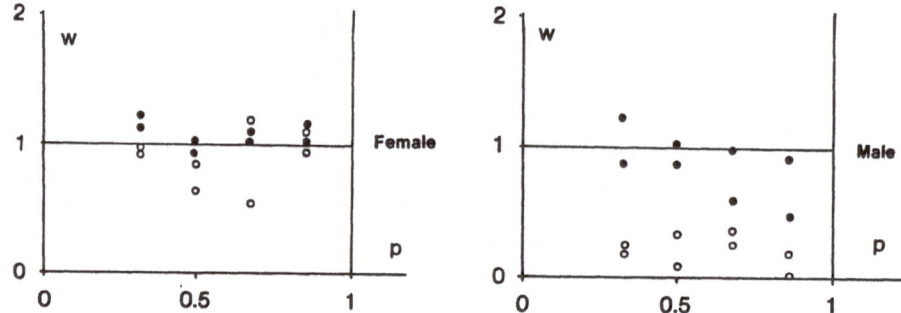

Figure 1.4. Sexual selection fitness values describing the transition from adults to breeders in the experimental population of *Drosophila* menalogaster. The fitness values are calculated for each sex separately using (1.10). o:v_{11}; o:v_{22} ; otherwise as in Figure 1.1. After Bundgaard and Christiansen (1972).

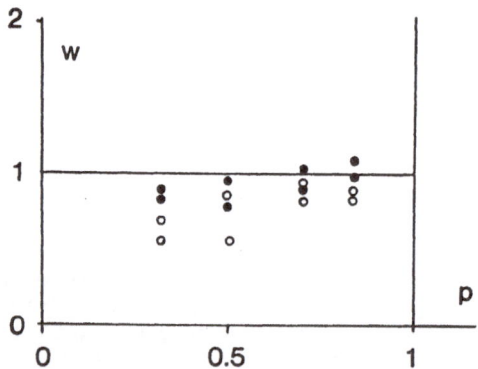

Figure 1.5. Zygotic selection fitness values describing the transition from zygotes to adults in the experimental population of *Drosophila* melanogaster. The fitness values are calculated for the sexes pooled as the genotypic frequencies in offspring adults did not differ between the sexes. o:w_{11}; o:w_{22} ; otherwise as in Figure 1.1. After Bundgaard and Christiansen (1972).

2.   Mother-offspring combinations

Genetic investigations of polymorphisms in natural populations may often be hampered by the difficulty of obtaining family data where segregation can be studied. The experimental production of such data may be difficult, tedious, or even practically impossible. However, for some groups of organisms it is possible to link the parents or a parent to the offspring, so the transmission and segregation of alleles can be studied in nature (Cooper, 1968; Christiansen and Frydenberg, 1973). A particularly interesting group of organisms are those where a parent broods the offspring, so data on parent-offspring combinations may be gathered from samples of adult individuals during the breeding season. The parent will often be the mother, and I will discuss data from two organisms where population samples of mother-offspring combinations can be gathered, the marine amphipod, *Gammarus oceanicus*, and the eelpout, *Zoarces viviparus*, which is a marine teleostean fish. Both of these organisms are common in shallow waters of northern Europe.

2.1. SIMPLE MOTHER-OFFSPRING COMBINATIONS

The sexually mature *Zoarces viviparus* mate in late summer, and the females go pregnant until they deliver their young in January. Thus, in late autumn a population sample will include pregnant mothers which on average carry between 50 and 100 feti. Data on *simple mother-offspring combinations*, i.e., genotypic combinations of mothers and one randomly chosen offspring, are shown in Table 2.1.   The data describes the EstIII polymorphism of the eelpout, but we may here just consider it as a data on the variation at an autosomal locus with two alleles A_1 and A_2 . The sample of mothers evidently describes the population of breeding females, but the mother-offspring combinations also contain information on the population of breeding males, on mating and on the segregation of female alleles.   Equally important, the mother-offspring combinations contain information about the population of zygotes, in that with no variation in fecundity the sum over female genotypes represents a random sample of the population of zygotes.

Table 2.1
Mother-offspring data on *Zoarces viviparus*
(Christiansen, Frydenberg and Simonsen 1977)

Mother genotype	Genotype of offspring A_1A_1	A_1A_2	A_2A_2	Sum
A_1A_1	305	516		821
A_1A_2	459	1360	877	2696
A_2A_2		877	1541	2418
Sum	764	2753	2418	5935

The A_1A_1 mothers transmit an A_1 allele to the offspring, the A_2A_2 mother transmit an A_2 allele to the offspring, whereas the heterozygote A_1A_2 mothers may transmit either an A_1 allele or an A_2 allele to the offspring. Thus, for homozygote mothers the two offspring classes correspond to observations of eggs fertilized by sperm carrying an A_1 allele and an A_2 allele, respectively. For heterozygote mothers the two classes of homozygote offspring correspond to observations of fertilization by an A_1 allele and an A_2 allele, respectively, whereas the origin of the alleles in the heterozygote offspring cannot be identified. The data therefore contain the observation of $305+459+877 = 1641$ male A_1 alleles and $516+877+1541 = 2934$ male A_2 alleles observed among $5935-1360 = 4575$ unambiguously determined male alleles.

In general for m alleles, $A_1,A_2,...,A_m$, a typical mother-offspring combination is (A_iA_j,A_iA_k) i,j,k = 1,2,...,m, of a mother A_iA_j and an offspring A_iA_k. If in this combination the mother is a homozygote, $i = j$, then we have an observation of a male A_k allele for all k = 1,2,...,m, and there will be m mother-offspring combinations for each homozygote genotype of the mother. If she is heterozygote, $i \neq j$, then we have an observation of a male A_k allele except when k = j, as the (A_iA_j,A_iA_j) combination can be formed by two distinct events, i.e., either by transmitting A_i from the mother and A_j from the father or by transmitting A_j from the mother and A_i from the father. Thus, the 2m distinct transmission events give rise to 2m - 1 offspring genotypes. Let F_{ij} denote the number of mothers of genotype A_iA_j, so F is multinomially distributed $(F_{..},\phi)$, and let $C_{ik}(ij)$ be the number of (A_iA_j,A_iA_k) combinations, so C(ij) is multinomially distributed $(F_{ij},\gamma(ij))$ with C(ij) independent for i,j = i1,2,...,m. Thus, the frequency of the offspring genotypes for a given mother genotype will be designated $\gamma_{ik}(ij)$, i.e., this is the frequency of the mother-offspring combination (A_iA_j,A_iA_k) within A_iA_j mothers.

An offspring zygote randomly selected from the mother represents a random egg produced by the mother and a random sperm collected from the father. Therefore, if the offspring may be considered to be newly united zygotes, then we would expect the frequency $\gamma_{ik}(ij)$ to be given as the product of the probability that the mother transmit A_i and the probability that the father transmit A_k. Let $\mu_k(ij)$, i,j,k = 1,2,...,m, designate the frequency of A_k carrying sperm that fertilize A_iA_j mothers, then by the above arguments we have

$$(2.1) \qquad \gamma_{ik}(ii) = \mu_k(ii) .$$

Let $\beta_i(ij)$ and $\beta_j(ij)$, $i \neq j$, i,j = 1,2,...,m, with $\beta_i(ij) + \beta_j(ij) = 1$ describe the segregation of the two alleles from the heterozygote A_iA_j mother, then

$$\gamma_{ik}(ij) = \beta_i(ij)\mu_k(ij) \text{ for } k \neq j ,$$

$$(2.2) \qquad \gamma_{jk}(ij) = \beta_j(ij)\mu_k(ij) \text{ for } k \neq i ,$$

$$\gamma_{ij}(ij) = \beta_i(ij)\mu_j(ij) + \beta_j(ij)\mu_i(ij) .$$

Thus, the $2m - 2$ free parameters that we need to describe the observed frequencies of offspring types form a heterozygote mother are by the product structure (2.2) reduced to m free parameters, i.e., one describing the segregation in the female and $m - 1$ describing the male gamete frequencies. The contention that the observed offspring represent newly formed zygotes may therefore be tested by a test of $m - 2$ degrees of freedom.

For two alleles ($m = 2$) the segregation from the only heterozygote A_1A_2 becomes particularly simple:

(2.3)

Offspring:	A_1A_1	A_1A_2	A_2A_2
Frequency:	$\beta_1\mu_1$	$\beta_1\mu_2 + \beta_2\mu_1$	$\beta_2\mu_2$

where the shorthand notation $\beta_i = \beta_i(12)$ and $\mu_k = \mu_k(12)$ is used. In this case it seems impossible to conclude a difference between the supposed product structure and the observed genotypic proportions among the offspring. However, 2/3 of all possible genotypic frequency vectors (z_{11}, z_{12}, z_{22}) cannot be written in the form (2.3). To see this, compare (2.3) to the Hardy-Weinberg proportions corresponding to the gene frequencies in (2.3), namely $p_1 = (\beta_1 + \mu_1)/2$ and $p_2 = (\beta_2 + \mu_2)/2$, to see that $\beta_1\mu_2 + \beta_2\mu_1 \geq 2p_1p_2$, so the product structure (2.3) corresponds to genotypic frequency vectors (z_{11}, z_{12}, z_{22}) which show an excess of heterozygotes when compared to the Hardy-Weinberg proportions with the same gene frequencies (Figure 2.1). (The conclusion is more general, in that random union of a population of female gametes with a different population of male gametes will always lead to $F \leq 0$.) Thus, deviation from the product structure may be detected even in the case of two alleles (Østergaard and Christiansen, 1981).

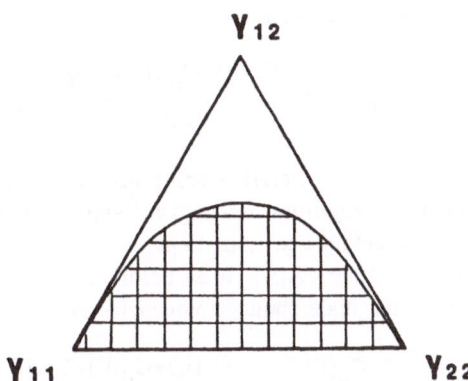

Figure 2.1. The range of the random union of gamete proportions (2.3) for the frequencies $\gamma_{11}(12)$, $\gamma_{12}(12)$ and $\gamma_{22}(12)$. The frequencies are depicted in a de Finetti diagram, and the hatched area delimits the frequencies which are not random unions of gamete proportions.

A more serious problem in the analysis of two allele mother-offspring combinations is that the two parameters (β_1,β_2) and (μ_1,μ_2) are completely symmetric in (2.3), so an interchange of β and μ will not change the genotypic proportions. This will considerably weaken the analysis of segregation, as Mendelian segregation, or $(\beta_1,\beta_2) = (1/2,1/2)$, is completely indistinguishable from the situation $(\mu_1,\mu_2) = (1/2,1/2)$. The Mendelian expectation is often formulated as the more unspecific relation $\gamma_{12}(12) = 1/2$. This means that the number of (A_1A_2,A_1A_2) combinations, $C_{12}(12)$, is expected to be binomially distributed with probability parameter $1/2$ and number parameter F_{12}, and this may then be used as a basis for a comparison of data to expectations from the Mendelian segregation (Christiansen and Frydenberg, 1973). A significant deviation from the hypothesis $\gamma_{12}(12) = 1/2$, however, may either be due to a deviation from Mendelian segregation or to a deviation from the product structure (2.3). If Mendelian segregation prevails, then the male gamete frequencies are estimated as the observed frequency among the unambiguouly determined male gametes.

For more alleles, $m > 2$, the information on the segregation of the female alleles is considerably improved. In a heterozygote mother A_iA_j, $i \neq j$, the relative frequency of the mother-offspring combinations (A_iA_j,A_iA_k) and (A_iA_j,A_jA_k) for $k \neq i$ and $k \neq j$ constitutes a direct observation of the relative segregation of the two female alleles. Explicit maximum likelihood estimates of $\beta(ij)$ and $\mu(ij)$ have not been found in general. This is due to the incomplete observation of the transmission classes, in that two are lumped together into the class (A_iA_j,A_iA_j). However, for frequencies of male gametes not carrying A_i or A_j explicit estimates are given by

$$(2.4) \qquad \hat{\mu}_k(ij) = \frac{C_{ik}(ij)+C_{jk}(ij)}{C_{..}(ij)}$$

for $k \neq i,j$, $k = 1,2,...,m$, so for the sum $\mu_i(ij) + \mu_j(ij)$ we get

$$(2.5) \qquad \hat{\mu}_i(ij) + \hat{\mu}_j(ij) = \frac{C_{ii}(ij)+C_{ij}(ij)+C_{jj}(ij)}{C_{..}(ij)}.$$

For $\beta_i(ij)$, $\beta_j(ij)$, $\mu_i(ij)$ and $\mu_j(ij)$ iterative solutions of the likelihood equations are obtained using the gene counting algorithm (Østergaard and Christiansen, 1981). An easy way of constructing the gene counting equations is given in appendix 2.4.

Mendelian segregation in female A_iA_j is as before characterized by $(\beta_1(ij),\beta_2(ij)) = (1/2,1/2)$. The estimates of the male gamete frequencies are then given by

$$(2.6) \qquad \hat{\mu}_i(ij) = \frac{C_{ii}(ij)}{C_{ii}(ij)+C_{jj}(ij)} \quad \frac{C_{ii}(ij)+C_{ij}(ij)+C_{jj}(ij)}{C_{..}(ij)},$$

$$(2.7) \qquad \hat{\mu}_j(ij) = \frac{C_{jj}(ij)}{C_{ii}(ij) + C_{jj}(ij)} \; \frac{C_{ii}(ij) + C_{ij}(ij) + C_{jj}(ij)}{C_{..}(ij)} \; ,$$

and (2.4) otherwise.

Finally a comparison of the male gamete frequencies among mother genotypes will disclose the mating pattern in the population. Random mating, in particular, means that $\mu_k(ij) = \mu_k$ for all $i,j,k = 1,2,...m$, and this provides a very simple description of the population of male gametes. With random mating, Mendelian segregation, and equal fecundity, the genotypic frequencies in the population of zygotes are $\nu_i\mu_i$ for the homozygote A_iA_i and $\nu_i\mu_j + \nu_j\mu_i$ for the heterozygote A_iA_j, where ν_i is the gene frequency among the mothers, i.e.,

$$(2.8) \qquad \nu_i = \phi_{ii} + \frac{1}{2} \sum_{j \neq i} \phi_{ij} \; .$$

2.2. DATA WITH MORE OFFSPRING PER MOTHER

The first offspring from a mother usually determines a male gamete. Similarly, a second randomly chosen offspring also determines a male gamete, but at the considered locus this second gamete may carry a gene which is identical to the gene carried by the first gamete. Alternatively, it may carry a copy of the other gene in the same male and thereby contain information on the genotype of the male. However, a third possibility may exist, namely that the gene in the gamete originates form another male that the female mated. Thus, if the mothers may carry broods with multiple sires additional offspring supply little information on the population of breeding males. In *Zoarces viviparus* the observation of three alleles segregating in broods from homozygote mothers establishes that females may mate more than once, and the frequency of multiple sired broods seems rather high.

The description of the segregation and transmission of the female alleles may be qualitatively enhanced from mother-offspring combinations with many offspring per female. For two alleles the expectation of half heterozygote offspring from heterozygote mothers also holds within broods, and this may be used for a more in depth analysis of Mendelian segregation (Hjorth, 1971; Simonsen and Frydenberg, 1972; Simonsen and Christiansen, 1981, 1984). For multiple alleles this expectation is valid within the offspring that only carries the two alleles of the heterozygote female. In addition, within the offspring that carries alleles not transmitted by the female, the male and female alleles should combine randomly, and the two female alleles should occur equally often.

2.3. COMPLETE MOTHER-OFFSPRING COMBINATIONS

If the possibility of multiple sired broods is excluded, then mother-offspring combinations with more than one offspring can provide an increase in the qualitative information on the breeding male population (Christiansen, 1980). In addition, information on segregation in males is gained (Nadeau, Dietz and Tamarin, 1981). However, to gain sufficient information for a full qualitative description of the male breeding population, male segregation, and female segregation, mother-offspring combinations including three or more offspring are needed, and we will refer to such data as *complete mother-offspring combinations* (Christiansen, 1980).

The breeding biology of Gammarids allows the collection of population samples of mother-offspring combinations and suggests that each brood is sired by one male only. Thus, they are suitable for the collection of population samples including complete mother-offspring combinations. Mating can occur only at the time when the female moult, and before mating the male carries the female around in a precopula. At moulting the female releases the eggs (about 50 to 150 for *Gammarus oceanicus*) into her brood pouch where they are fertilized. She then carries the eggs until they hatch and the juveniles leave the brood pouch.

Siegismund (1985) collected mother-offspring combinations of *Gammarus oceanicus* including five offspring per mother describing the Mpi polymorphism segregating two alleles, A_1 and A_2, say. Table 2.2 show the observed numbers of the six different mother-offspring combinations corresponding to the A_1A_1 mothers in a total sample of 361 females with broods (176 A_1A_1, 151 A_1A_2 and 34 A_2A_2). To the A_2A_2 mothers correspond in the same way six mother-offspring combinations, but the A_1A_2 mothers may show 21 different mother-offspring combinations with five offspring.

Table 2.2
Mother-offspring data on *Gammarus oceanicus*,
A_1A_1 females only (Siegismund, 1985)

| Genotype of offspring | | | | |
A_1A_1	A_1A_2	A_2A_2	Observed	Expected
5	0	0	92	92
4	1	0	12	12,0
3	2	0	24	21,4
2	3	0	14	19,1
1	4	0	11	8,5
0	5	0	23	23
Sum			176	176,0

The A_1A_1 females transmit only allele A_1 so the four segregating classes in Table 2.2 all unambiguously determine the father to be heterozygote A_1A_2. The two non-segregating classes, however, do not determine the father as they result either from a mating with a homozygote male or from a mating with a heterozygote male where by chance only one of the two expected offspring classes became represented among the five sampled offspring.

The mating of an A_1A_1 female with an A_1A_2 male segregates into A_1A_1 offspring with probability $\gamma_{11}(11,12)$, say, and A_1A_2 with probability $\gamma_{12}(11,12)$, so the number of A_1A_1 offspring among the five in the mother-offspring combination is expected to follow a binomial $(5,\gamma_{11}(11,12))$ distribution. Thus, the segregation parameters may be estimated from the doubly truncated binomial distribution leaving out the offspring classes $(5,0,0)$ and $(0,5,0)$ (Nadeau et al., 1981), and this may be done when the number of offspring per mother-offspring combination is three or more. Using these segregation parameters we can judge the contribution of offspring in the non-segregating classes that originated from a mating between a A_1A_1 female and a A_1A_2 male as a function of the frequency $\alpha_{12}(11)$, say, of A_1A_2 sires of A_1A_1 females. From this the frequencies $\alpha_{ij}(11)$ of A_iA_j sires of A_1A_1 females can be estimated using the relative proportions of the two non-segregating and the segregating classes of mother-offspring combinations (Siegismund and Christiansen, 1985). For the data on *Gammarus* these estimates are as follows

$$(2.9) \qquad \hat{\alpha}_{11}(11) = C(11;5,0,0)/F_{11} - \hat{\gamma}_{11}(11,12)^5\hat{\alpha}_{12}(11) ,$$

$$(2.10) \qquad \hat{\alpha}_{12}(11) = [1-\hat{\gamma}_{11}(11,12)^5 - \hat{\gamma}_{12}(11,12)^5]^{-1} \sum_{a=1}^{4} C(11;a,5-a,0)/F_{11} ,$$

$$(2.11) \qquad \hat{\alpha}_{22}(11) = C(11;0,5,0)/F_{11} - \hat{\gamma}_{12}(11,12)^5\hat{\alpha}_{12}(11) ,$$

where $C(ij;a,b,c)$ is the number of mother-offspring combinations with mother of genotype A_iA_j and the three offspring genotypes, A_1A_1, A_1A_2 and A_2A_2, in the numbers a, b and c, respectively.

The same procedure may immediately be used for the mother-offspring combinations corresponding to A_2A_2 females. For heterozygote females the offspring classes where both homozygotes have been observed, i.e., offspring class (a,b,c) where both a and c are non-zero, the male must have transmitted both allele A_1 and allele A_2, so the sire is unambiguously determined as heterozygote. For these classes we then estimate the segregation parameters $(\gamma_{11}(12,12),\gamma_{12}(12,12),\gamma_{22}(12,12))$ in a multiple truncated multinomial distribution with number parameter 5. Using these segregation parameters we can again judge the contribution of offspring in the classes $(a,b,0)$ and $(0,b,c)$ that originated from a mating between an A_1A_2 female and an A_1A_2 male as a function of the sire frequency $\alpha_{12}(12)$. Matings between an A_1A_2 female and an A_1A_1 male only contribute to the classes $(a,b,0)$, and their contribution is a function of the sire frequency $\alpha_{11}(12)$ and the segregation parameters $(\gamma_{11}(12,11),\gamma_{12}(12,11),0)$, and similarly for the contribution to the $(0,b,c)$ classes from matings between an A_1A_2 female and an A_2A_2 male. The estimation of all these segregation and sire parameters can be done with complete mother-offspring combinations (Siegismund and Christiansen, 1985), and the method is as everywhere else in this analysis the gene counting method. The general estimation procedure allows for a variable number of offspring per mother-offspring combination.

The description so far has relied only on the assumption that each brood has one sire and that the offspring represent newly formed zygotes. This model can be compared to the data, and the fit is nice (Table 2.2), so now we can commence on the analysis corresponding to the analysis of the simple mother-offspring combinations. First, we can check whether the segregation among offspring of the matings can be described as segregation in the heterozygote female, (β_1,β_2), and in the heterozygote male (δ_1,δ_2), separately. This corresponds to homogeneity of the segregation proportions (Table 2.3), and it simplifies the description of segregation from six free parameters to two. Secondly, we can check whether Mendelian segregation prevails. This is conveniently done by asking wether segregation is the same in males and females, i.e., $(\beta_1,\beta_2) = (\delta_1,\delta_2)$, and then suppose Mendelian segregation, i.e., $(\beta_1,\beta_2) = (1/2,1/2)$.

Table 2.3

Homogeneous segregation proportions in complete mother-offspring combinations

Female genotype	Male genotype A_1A_1	A_1A_2	A_2A_2
A_1A_1	$(1,0,0)$	$(\delta_1,\delta_2,0)$	$(0,1,0)$
A_1A_2	$(\beta_1,\beta_2,0)$	$(\beta_1\delta_1,\beta_1\delta_2+\beta_2\delta_1,\beta_2\delta_2)$	$(\beta_1,\beta_2,0)$
A_2A_2	$(0,1,0)$	$(\delta_1,\delta_2,0)$	$(0,0,1)$

With Mendelian segregation the estimated sire genotypic frequencies are shown in Table 2.4. A comparison of the sire frequencies among mother genotypes will disclose the mating pattern in the population. Random mating means that $\alpha_{kl}(ij) = \alpha_{kl}$ for all $i,j,k,l = 1,2$, and this provides a very simple description of the population of breeding males (bottom line of Table 2.4). With random mating, Mendelian segregation, and equal fecundity the genotypic frequencies in the population of zygotes is again given by $v_1\mu_1$ and $v_2\mu_2$ for the homozygotes A_1A_1 and A_2A_2, respectively, and $v_1\mu_2 + v_2\mu_1$ for the heterozygote A_1A_2, where $v_1 = \phi_{11} + \phi_{12}/2$ is the gene frequency of A_1 among the mothers, and $\mu_1 = \alpha_{11} + \alpha_{12}/2$ is that among the fathers.

The *Gammarus* data contained five offspring per female even though many more would have been possible. However, the information gained by taking an extra offspring rapidly diminishes with the number of offspring already analysed. If in addition to the offspring of simple mother-offspring combinations n extra offsprings are analysed then with Mendelian segregation the probability that both male alleles have been detected is of the order of magnitude $1 - (1/2)^n$. The uncertainty of the evaluation is due to the incomplete information on the mating type, and this uncertainty will lower the probability, so extra offspring is expected to provide more information for broods of heterozygote females (Cooper, 1968). In *Gammarus oceanicus* this may be illustrated by comparing the confidence regions of the estimates in Table 2.4 with confidence regions of population samples with similar genotypic composition. The sire frequencies estimated within A_1A_1, A_1A_2 and A_2A_2 mothers correspond to population samples of 163, 87 and 32 individual males, respectively (Siegmund and Christiansen, 1985). This may be compared to the number of mother-offspring combinations, 176, 151 and 34, respectively. Thus, the efficiency in the heterozygote females is only little more than half the efficiency in the homozygote females, so double as many offspring from heterozygote females would have been desirable. In laboratory applications it may be convenient to analyse the offspring in batches which are fractions of a chosen maximum number of offspring and then stop the analysis of a brood as soon as two male alleles have been detected (Watt et al., 1985).

Table 2.4
Genotypic frequencies among sires in *Gammarus oceanicus*
(After Siegismund, 1985)

Female genotype	Male genotype A_1A_1	A_1A_2	A_2A_2
A_1A_1	0,51	0,37	0,12
A_1A_2	0,49	0,37	0,14
A_2A_2	0,49	0,44	0,08
Combined	0,50	0,38	0,12

2.4. APPENDIX: THE GENE COUNTING ALGORITHM

The estimation of the parameters in the model (2.2) is developed here using the gene counting algorithm or the EM-algorithm (reviewed by Østergaard and Christiansen, 1981). The procedure follows the general procedure that can be used for estimation in the multinomial distribution concerning hypotheses of statistical independence where the observations are incomplete.

Suppressing the reference to the mother type, the likelihood function may be written as

(2.A1)
$$
\begin{aligned}
L(\gamma(ij),C(ij)) = {} & C_{ii}\log(\beta_i\mu_i) + C_{jj}\log(\beta_j\mu_j) \\
& + C_{ij}\log(\beta_i\mu_j + \beta_j\mu_i) \\
& + \sum_{k \neq i,j} [C_{ik}\log(\beta_i\mu_k) + C_{jk}\log(\beta_j\mu_k)] .
\end{aligned}
$$

The maximum likelihood estimates is found by maximizing this function under the constraints that $\beta_i + \beta_j = 1$ and $\sum \mu_k = 1$. A convenient way to do this is by means of Lagrange multipliers. That is, maximize the function

(2.A2)
$$L(\gamma(ij),C(ij)) + (1-\beta_i-\beta_j)\Lambda_\beta + (1-\sum_k \mu_k)\Lambda_\mu$$

with β and μ varying freely and then determine the constants Λ_β and Λ_μ so that $\beta_i + \beta_j = 1$ and $\sum \mu_k = 1$.

Differentiation with respect to μ_i provides the equation

(2.A3)
$$\frac{C_{ii}}{\mu_i} + \frac{C_{ij}\beta_j}{\beta_i\mu_j+\beta_j\mu_i} = \Lambda_\mu ,$$

with respect to μ_j provides

(2.A4)
$$\frac{C_{jj}}{\mu_j} + \frac{C_{ij}\beta_i}{\beta_i\mu_j+\beta_j\mu_i} = \Lambda_\mu ,$$

and with respect to μ_k for $k \neq i,j$ gives the equations

(2.A5)
$$\frac{C_{ik}+C_{jk}}{\mu_k} = \Lambda_\mu .$$

Multiplying each equation with the μ_k of the differentiation and subsequently summing the equations yields $C_{..} = \Lambda_\mu \sum \mu_k$, so the constant becomes $C_{..}$ when $\sum \mu_k = 1$. Differentiation with respect to β_i and β_j yields

(2.A6)
$$\frac{C_{i\cdot}-C_{ij}}{\beta_i} + \frac{C_{ij}\mu_j}{\beta_i\mu_j+\beta_j\mu_i} = \Lambda_\beta ,$$

(2.A7)
$$\frac{C_{j\cdot}-C_{ij}}{\beta_j} + \frac{C_{ij}\mu_i}{\beta_i\mu_j+\beta_j\mu_i} = \Lambda_\beta ,$$

and as before $\Lambda_\beta = C_{..}$ when $\beta_i + \beta_j = 1$.

The gene counting procedure is initiated by choosing initial values of β and μ in the interior of their domain of variation, i.e., none of the components is allowed the value zero. New values are then provided by the gene counting equation. This equation is obtained simply by multiplying each of the above equations with the differentiation variable and equating this variable on the right side of the equation to the new value of the variable times the constant Λ of the equation. Thus, we get for μ the equations

(2.A8)
$$C_{..}\mu'_i = C_{ii} + \frac{C_{ij}\mu_i\beta_j}{\beta_i\mu_j + \beta_j\mu_i},$$

(2.A9)
$$C_{..}\mu'_j = C_{jj} + \frac{C_{ij}\mu_j\beta_i}{\beta_i\mu_j + \beta_j\mu_i},$$

(2.A10)
$$C_{..}\mu'_k = C_{ik} + C_{jk}$$

for $k \neq i,j$, $k = 1,2,...,m$. Equation (2.A10) immediately provides an explicit expression (2.4) for the maximum likelihood estimate of μ_k for $k \neq i,j$. In addition we only need to iterate either (2.A8) or (2.A9) because the sum of the two equations provides an explicit estimate of $\mu_i + \mu_j$ (2.5). For β we get

(2.A11)
$$C_{..}\beta'_i = (C_{i.} - C_{ij}) + \frac{C_{ij}\beta_i\mu_j}{\beta_i\mu_j + \beta_j\mu_i}$$

and $\beta'_j = 1 - \beta'_i$. The gene counting iteration now proceeds by choosing β' and μ' as the new initial values and repeating the procedure.

As soon as the gene counting equations (2.A8) and (2.A11) are produced the reason for their name becomes apparent. The right side of equation (2.A11), for instance, is first the number of unambiguously observed female allele A_i in the offspring, and secondly the expected number of female A_i alleles in the incompletely observed class of A_iA_j offspring given the segregation β and the male gamete frequencies μ. Thus, this is our best guess on the number of A_i alleles of female origin in the data, and this is used to provide a supposedly better guess on the transmission probability β'_i on the left side of the equation. The attractive nature of the iteration procedure is that this very intuitive form is combined with a sure convergence property, in that the likelihood function increases for every iteration cycle.

The construction of the gene counting equations using the intuitive partitioning of the incompletely observed classes functions works whenever the frequencies of these classes are given as homogeneous polynomials. This secures that the multiplier on the left side of the equation is a nice integer.

3.   Selection component analysis

In my first lecture the possibilities of disclosing selection from observations on the genotypic frequencies in a population was discussed. We saw that zygotic selection, or variations in probability of survival from zygote to adult, could be evaluated only if information on the genotypic frequencies in the population of zygotes was available.

Population samples of mother-offspring combinations provide this kind of information, so for this reason alone they lend considerable support to attempts to describe selection in natural populations. However, zygotic selection is only one aspect of selection, others are sexual selection, fecundity selection and gametic selection. These should all be evaluated when the goal is to describe the selection working on a polymorphism. Zygotic selection removes individuals from the population, so its effect is obvious as a change in genotypic composition of the population. Sexual selection works in a very parallel way in that it removes adult individuals form the breeding population, but the individuals remain in the population and the process can only be studied by a characterization of the population of breeders. The components fecundity selection and gametic selection share that property, in that they are intimately connected to the process of procreation. However, they differ from the two other components because fecundity selection works on mated pairs of individuals and gametic selection works on a haploid phase in the life cycle. This will influence the evolutionary interpretation of selection effects, and for fecundity selection these are more varied than with zygotic selection (Feldman, Christiansen and Liberman, 1983). Sexual selection is also more intricate than zygotic selection in that it often is related to mating and results form interactions among individuals (O'Donald, 1980). As we shall see population samples including mother-offspring combinations may be used for very detailed analyses of selection that may in principle approach a complete description of selection in a natural population.

   In the following I will restrict attention to data on genotypic frequencies and neglect data on fecundity counts. Therefore, the fecundity of all types of mated pairs will be assumed to be equal, so fecundity selection will be assumed absent.

3.1. COMPLETE MOTHER-OFFSPRING DATA

The population sample of mother-offspring combinations in *Gammarus oceanicus* provided information on the segregation of alleles in male and female heterozygotes, on the breeding populations of males and females, and on the mating pattern in the population. This information can be utilized to estimate the genotypic frequencies in the population of offspring zygotes. Thus, the lower half of the life-cycle in Figure 3.1 is described by the mother-offspring data and the upper half is included simply by analysing the adults collected in the population at the same time as the mother-offspring combinations (Table 3.1).

   The analysis of mother-offspring combinations can now be viewed as an analysis trying to disclose evidence of selection. Deviations from homogeneity among segregation proportions (Table 2.3) may be due to selection between the time of zygote formation and the time of observation. As this would formally be zygotic selection we may name this component *fetal zygotic selection*. However, as a test for the absence of fetal zygotic selection the comparisons between the observed segregation proportions and the homogeneous segregation proportions of Table 2.3 share the same weakness as the Hardy-Weinberg test seen as test for the absence of zygotic selection. Therefore, this initial test, just as the comparison between the observed and expected distributions of mother-offspring combinations (Table 2.2), should be viewed as a goodness-of-fit test for the basic conception of the structure of the data.

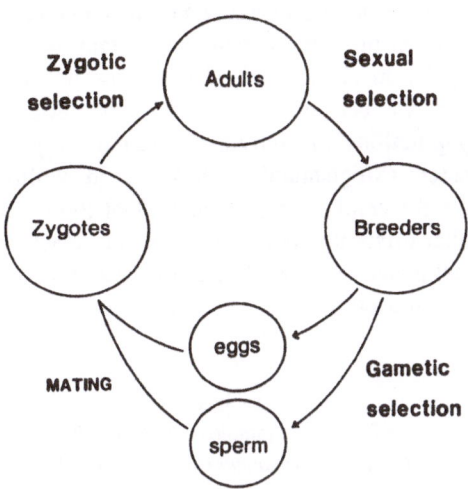

Figure 3.1. The life stages and the effects observable in the genotypic data on *Gammarus*.

Table 3.1
Genotypic data on *Gammarus oceanicus* (Siegismund, 1985)

Population section	Genotype A_1A_1	A_1A_2	A_2A_2	Sum
Mothers	176	151	34	361
Other females	20	12	3	35
Adult females	196	163	37	396
Adult males	204	173	36	413
Adults	400	336	73	809

Next, the comparison of the segregation in female and male heterozygotes to the expectation of Mendelian segregation may be viewed as an analysis of gametic selection. Again Mendelian segregation is part of our basic description of the polymorphism, so deviations from Mendelian segregation can either mean that our genetic description of the polymorphism is erroneous or that gametic selection is working. Thus, in general it will be hard to obtain evidence for gametic selection from mother-offspring data.

With Mendelian segregation the mother-offspring combinations describe the breeding male population and the mating between genotypes, and the mating is statistically indistinguishable from random mating (Table 2.4). In *Gammarus*, however, the mating, as concluded from the mother-offspring combinations, may be compared to the precopula that is found in the population (Table 3.2). These also occur in proportions consistent with random mating, so we can make a simple comparison between the breeding populations and the populations of individuals found in precopula. No significant differences were detected (Siegismund, 1985), so in addition to other biological conclusions we can make the comforting conclusion that the population of breeding males estimated from the mother-offspring combinations seems reasonable. In particular, there exists no sign of an estimated excess of heterozygote breeders (Table 3.3), the bias expected if multiple paternities occur.

Table 3.2
Population sample of precopula in
Gammarus oceanicus (Siegismund, 1985)

Female genotype	Male genotype A_1A_1	A_1A_2	A_2A_2	Sum
A_1A_1	114	81	17	212
A_1A_2	77	59	13	149
A_2A_2	24	9	6	39
Sum	215	149	36	400

Table 3.3
Genotypic frequencies in the male populations of
Gammarus oceanicus (After Siegismund, 1985)

Population section	Genotype A_1A_1	A_1A_2	A_2A_2	Sample Size
Breeders	0,50	0,38	0,12	(278)
Precopula	0,54	0,37	0,09	400
Free Adults	0,49	0,42	0,09	413
Total	0,51	0,39	0,10	(1091)

As soon as the genotypic frequencies in the breeding male population are found, then sexual selection may be disclosed by a comparison between the genotypic composition of the breeders and the adults (in females between breeders and non-breeders).   Finally, zygotic selection may be disclosed by a comparison between the population of zygotes and the adult population.

The analysis is in principle given by Christiansen and Frydenberg (1973), and the details concerning the complete mother-offspring data is given by Siegismund and Christiansen (1985).   In the data on the Mpi polymorphism in *Gammarus oceanicus* no unequivocal evidence of selection appeared, and the biological conclusions of the analysis are discussed by Siegismund (1985).

3.2. SIMPLE MOTHER-OFFSPRING DATA

For the *Zoarces viviparus* more extensive data is available (Tables 2.2 and 3.4), but the quality of the information on selection is not as good. The breeding male population is only represented by the transmitted male gametes, so the effects of sexual selection and gametic selection in the males are lumped together (Figure 3.2).   However, the main concern is that sexual selection in the males is only given by the genic effect, so if only male sexual selection is working on the polymorphism and if the polymorphism is at a stable equilibrium due to selection, then the gene frequencies among male gametes is expected to be the same as among adult males, no matter how strong the selection is.

Due to this incomplete resolution of the events in the life-cycle of the organism the selection component analysis is conceptually less straightforward, but statistically simpler. The analysis is described and discussed by Christiansen and Frydenberg (1973) with the estimation methods changed to the gene counting method by Christiansen et al. (1977).

Table 3.4

Genotypic data on *Zoarces viviparus*

(Christiansen, Frydenberg and Simonsen, 1977)

Population section	Genotype			Sum
	A_1A_1	A_1A_2	A_2A_2	
Mothers	821	2696	2418	5935
Other females	43	161	160	364
Adult females	1008	3344	3029	7381 [a]
Adult males	693	2332	2201	5226 [a]
Adults	1701	5676	5230	12607

a: Includes a sample of adults after the breeding season.

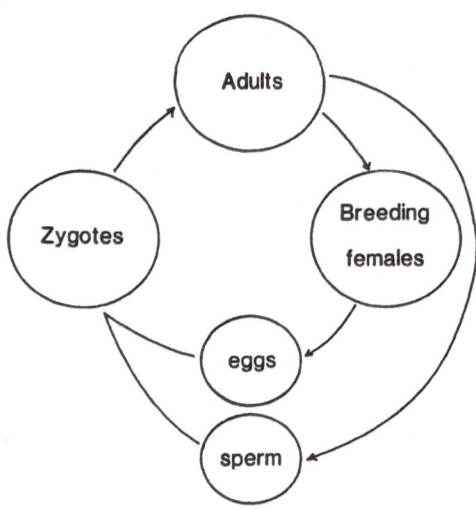

Figure 3.2. The life stages observable in the genotypic data on *Zoarces*.

3.3. POPULATIONS WITH OVERLAPPING GENERATIONS

The conceptualization of the life-cycles in Figures 3.1 and 3.2 implicitly assumes that the organism has discrete non-overlapping generations. This is true for *Gammarus oceanicus*, but certainly not for *Zoarces viviparus*, where the individuals mature at an age of about two years with breeding every year thereafter. In terms of Figure 3.2 the adult populations are "recycled", and the breeders return to the adult population after breeding. Evidently, a comprehensive description of the selection working on the EstIII polymorphism has to take these events into account.

The data of Tables 2.2 and 3.4 is sampled in the years 1969 through 1974, and from 1971 the individual age was determined. The highest age observed was ten years, but most fish in the samples had an age between two and five years. Recording the year of birth of the various age classes in the samples reveals the cohort structure of the total of the samples (Table 3.5).

Maturity is reached at age 2 and the survival of mature adults from year to year may be followed through the cohort (Table 3.5). This component of selection will be called *adult zygotic selection* and the component describing the survival of zygotes to mature adults will be called *juvenile zygotic selection*. Adult zygotic selection is formally among the selection components involved in breeding, as it is an important determinant of an individual's life time production of offspring. Table 3.5 shows, in addition, that the data allows an analysis of stability in the population by comparing the various cohorts.

Table 3.5
Cohorts in the data on *Zoarces viviparus*

Age class	Year of sample 1971	1972	1973	1974
0	71	72	73	74
1	70	–	–	73
2	69	70	71	72
3	68	69	70	71
4	67	68	69	70
5	66	67	68	69
6+				

The proper breeding components of selection are most naturally analysed within the sample for a given year, which involves vertical comparisons in Table 3.5. Adult zygotic selection is most naturally analysed within cohorts, which is diagonally in Table 3.5. To reconcile these two conflicting interests in the analysis we need to know the age structure of the population. However, this age structure is poorly represented by the sample as the method of catch introduces a size bias. The analysis is developed and further discussed by Christiansen and Frydenberg (1976).

The data on the EstIII polymorphism is analysed by Christiansen, Frydenberg and Simonsen (1977). They concluded that the population reproduced by random mating and Mendelian segregation, and that no selection except juvenile zygotic selection worked on the polymorphism. The zygote population is then expected to be in Hardy-Weinberg proportions every generation, and on the basis of this and the observed population of adults we can calculate the fitness values corresponding to juvenile zygotic selection (Table 3.6).

Table 3.6
Juvenile zygotic selection in *Zoarces viviparus*
(Christiansen, Frydenberg and Simonsen, 1977)

	Genotype A_1A_1	A_1A_2	A_2A_2
Zygotes	0,129	0,461	0,410
Fitness	1,065	1,000	1,037
Adults	0,135	0,450	0,415

Gene frequency of A_1 : $0,360 \pm 0,003$

3.4. APPLICABILITY OF THE SELECTION COMPONENT ANALYSIS

The key to the selection component analysis developed in this lecture is the population sample of mother-offspring combinations. This allows the description of the breeding components of selection and the genotypic frequencies in the population of offspring formed by the population of parents sampled. In *Zoarces viviparus* this is indeed the population of zygotes for the cohort produced in the year of sample, so the population of feti sampled is a natural entity in the population. This identification between the sampled population and the natural initiation of a cohort stems from the discreteness of the event of breeding in the population. Further, the population of mothers sampled in a given year is indeed the population of females that breed in that year, and the number of feti in a female reflects her fecundity that year.

In *Gammarus oceanicus*, on the other hand, the identification of the sampled population of offspring with the population of zygotes initiating the next generation is more questionable, because here the breeding is not discrete. The females may breed repeatedly in the final few months of their life, so the population of breeding females is not well defined in relation to the sample. Females without progeny in their brood pouch may just have released an earlier batch of offspring, they may breed later, or they may not breed at all. In addition, variation in the number of eggs per brood may be a poor representation of the fecundity as this number may be correlated to the total number of broods produced.

Discrete breeding is therefore important for a comprehensive selection component analysis based on changes in genetic composition of the population within a generation. Overlap between generations can be handled as long as the age structure of the population can be determined. Continuous breeding through a breeding season is enough to cause difficulties for the analysis, and the only way around these is to determine the number of broods of the analysed individuals. This may require that demographic data based on the individual survival and reproduction is gathered (Bodmer, 1968), and in natural populations this will often be impractical.

4. Population mixing

Deviations of genotypic frequencies form Hardy-Weinberg proportions may occur for many reasons in a locus that does not influence the individual mating behavior. Inbreeding in the population will produce an excess of homozygotes (as mentioned in my first lecture), and differences between the sexes in gene frequency will produce an excess of heterozygotes (as mentioned in my second lecture). In addition, the mixing of populations bred by random mating will produce genotypic proportions that show an excess of homozygotes, and this phenomenon is the so called *Wahlund effect* of population mixing (Wahlund, 1928).

4.1. THE WAHLUND EFFECT FOR ONE LOCUS

Consider an autosomal locus with alleles A_i, $i = 1,2,...,m$, in a population partitioned into N randomly mating subpopulations of relative sizes c_k, $k = 1,2,...,N$, with $c_1 + c_2 + ...+ c_N = 1$, and suppose that no selection influences this locus. Let p_{ik} denote the gene frequencies of A_i in the k'th population, so the genotypic frequencies of A_iA_i and A_iA_j in population k are p_{ik}^2 and $2p_{ik}p_{jk}$, respectively. Now suppose that these subpopulations in a given generation are mixed and breed by random mating. After mixing the gene frequencies become

$$(4.1) \qquad p_i = \sum_{k=1}^{N} c_k p_{ik} \, ,$$

and the frequency of genotype A_iA_i becomes

$$(4.2) \qquad \sum_{k=1}^{N} c_k p_{ik}^2 = p_i^2 + \sum_{k=1}^{N} c_k (p_{ik}\text{-}p_i)^2 \, .$$

The last term in this expression is the variance, V_i, in gene frequency of allele A_i among the subpopulations (for two alleles $V_1 = V_2$). Similarly we get the frequency of the heterozygote A_iA_j, $i \neq j$, as

$$(4.3) \qquad \sum_{k=1}^{N} c_k (2p_{ik}p_{jk}) = 2p_ip_j + 2\sum_{k=1}^{N} c_k (p_{ik}\text{-}p_i)(p_{jk}\text{-}p_j) \, .$$

The last term is here the covariance, C_{ij}, between the gene frequency of allele A_i and the gene frequency of allele A_j among the subpopulations (for two alleles $C_{12} = -V_1$). The genotypic frequencies in the mixed population therefore is given by $p_i^2 + V_i$ for the homozygote A_iA_i and by $2p_ip_j + C_{ij}$ for the heterozygote A_iA_j. Thus, the Wahlund effect is characterized by an excess of homozygotes relative to the Hardy-Weinberg proportions.

In the case of two alleles the excess of homozygotes is necessarily accompanied by a deficit of heterozygotes (Equation 1.3). Thus, the pattern of the Wahlund effect is as in the adult population of *Zoarces viviparus* in Table 3.6. This deficit was interpreted as due to selection in the selection component analysis, but that analysis is not able to differentiate between the effect of underdominant juvenile zygotic selection and immigration into the population during the juvenile stage. The size of the effect seen corresponds to the situation where the population is an equal mixture of two populations that differ by 0.13 in gene frequency. Although geographical variation in EstIII gene frequencies are known (Frydenberg et al., 1973), this size of the Wahlund effect seems highly improbable (Christiansen et al. 1977). A well described occurrence of population mixing and Wahlund effect is in the cod of the western Baltic Sea (Sick, 1965).

4.2. THE WAHLUND EFFECT AND MULTIPLE LOCI

The Wahlund effect immediately generalizes to multiple loci, in that a mixed population will show an excess of homogametic types compared to the Hardy-Weinberg proportions corresponding to the mean gametic frequencies in the mixed population. It should indeed be the same, because it is a property of the structure of the genotypic proportions. The separation between one locus gamete types and gamete types for multiple loci is in the transmission, not in the zygote formation.

The Wahlund effect really reflects the mating structure among the parents of the observed population, so another property of the Wahlund effect is that reproduction by random mating in the mixed population produces offspring in Hardy-Weinberg proportions. For one locus the offspring population will be in Hardy-Weinberg proportions with the gene frequencies p_i, $i = 1,2,...,m$, but for multiple loci breeding reshuffles the genes in the gametes by recombination, so the exact genotypic proportions cannot as immediately be predicted from the gamete frequencies.

4.3. MIXING AND TWO LOCI

Consider first two loci each with two alleles, A, a and B, b, and let r denote the frequency of recombination between the two loci. The four gametes AB, Ab, aB and ab are usually numbered 1, 2, 3 and 4, respectively. Suppose in a population breeding by random union of gametes that the zygotes are formed by the union of gametes with the four types in the frequencies x_1, x_2, x_3 and x_4. When these zygotes become mature adults and produce gametes, then the gene frequencies is these gametes are unchanged, i.e., $p'_A = p_A = x_1 + x_2$ and $p'_B = p_B = x_1 + x_3$. Thus, to determine the new gamete frequencies we only have to find x'_1, and to do so imagine all the gametes produced to be partitioned into those that were produced without recombination between the two loci, a fraction $1-r$, and those that were produced by recombination between the two loci, a fraction r. The frequency of the AB gamete in the first lot is evidently x_1 because those gametes are just the parental gametes transmitted unchanged. The frequency of the AB gamete among the recombinant gametes is equal to the frequency that one gamete of the parent carries allele A and the other gamete carries B, and because we assumed random union of gametes, this frequency is $p_A p_B$. We have therefore shown

$$(4.4) \qquad x'_1 = (1-r)x_1 + rp_A p_B .$$

Rearranging this yields

$$(4.5) \qquad x'_1 - p_A p_B = (1-r)(x_1 - p_A p_B) ,$$

showing that $x_1^{(t)} \to p_A p_B$ as $t \to \infty$ when $r > 0$. Thus, the gametic frequencies converge to a state where the alleles at the two loci are distributed independently in the gametes (Robbins, 1918). This state is called linkage equilibrium, and the characteristic form of the gamete frequencies are called *Robbins proportions* (Table 4.1). The difference between the gametic frequencies and the Robbins proportions is diminished by a factor $1 - r$ every generation, and this factor varies from $1/2$ for freely recombining loci to very close to one for tightly linked loci.   The difference used in (4.5) is usually called the *linkage disequilibrium* and is referred to as

$$(4.6) \qquad\qquad D = x_1 - p_A p_B \,,$$

so with this we can write (4.5) as $D' = (1-r)D$.

Table 4.1

The Robbins Proportions for two loci

Gamete type	AB	Ab
aB	ab	
Frequency	$p_A p_B$	$p_A q_b$
$q_a p_B$	$q_a q_b$	

Suppose as before that we have N subpopulations with genotypic frequencies in Hardy-Weinberg proportions , and suppose that we mix these populations. Then we get the Wahlund effect on the genotypic proportions, and we might expect a similar effect on another structure of independence, namely the Robbins proportions (4.6). The gametes produced in the mixed population are given by

$$(4.7) \qquad\qquad x_1' = \sum_{k=1}^{N} c_i[(1-r)x_{k1} + rp_{Ak}p_{Bk}] \,,$$

because the formation of gametes occurs inside the individual, so the gametes produced are just the total of the gametes produced by the individuals of the various compartments of the population.   Introducing the linkage disequilibrium $D_k = x_{k1} - p_{Ak}p_{Bk}$ in the k'th population before mixing we get

$$(4.8) \qquad D' = x_1 - p_A p_B = (1-r)D^* + \sum_{k=1}^{N} c_k(p_{Ak} - p_A)(p_{Bk} - p_B) \,,$$

where

$$(4.9) \qquad\qquad D^* = \sum_{k=1}^{N} c_k D_k$$

is the average linkage disequilibrium in the population before mixing, and where p_A and p_B are the gene frequencies at the two loci in the mixed population given as the average gene frequency (Equation 4.1). Thus, Equation (4.8) says that the linkage disequilibrium among the gametes produced in a mixed population equals the average disequilibrium before mixing, discounted by the effect of recombination, plus a term which is the covariance of gene frequencies at the two loci over the subpopulations. The mixing of populations in Robbins proportions therefore results in a population that produces gametes with a deviation from Robbins proportions equal to the covariance in gene frequencies at the two loci (Nei and Li, 1973; Prout, 1973; Feldman and Christiansen, 1975).

4.4. TRANSMISSION RULES FOR MULTIPLE LOCI

The principle (4.8) has a generalization to multiple loci, but before formulating this principle we need to consider the generalization of Equation (4.4) to an arbitrary number of loci.

Consider n loci each with two alleles designated allele number 0 and 1. We can view this description as focusing on a particular allele at each locus, allele number 1, and designating the alternative 0 (this alternative may consist of the collection of other alleles at the particular locus). The 2^n different types of gametes are described as n-tuples, $(g_1, g_2, ..., g_n) \in \{0,1\}^n$, where g_a is the number of the allele carried at the a'th locus in the gamete. The indexing of gametes used above for two loci corresponds to viewing the n-tuples as binary numbers, in that gametes are indexed $1, 2, ..., 2^n$ by the definition that the index is given by $g_1 2^{n-1} + g_2 2^{n-2} + ... + g_n + 1$.

Considerable simplification of this formulation is reached by representing the gametes of this two-allele model in terms of index sets of loci, i.e., subsets of $N = \{1, 2, ..., n\}$ the set of locus numbers, rather than the n-tuples (Christiansen, 1987, 1989). A gamete is uniquely determined by the subset A of loci in N, where the gamete carries allele number 1. That is, the gamete corresponding to the set A has allele number 1 at the loci $a \in A$ and allele number 0 at the loci $a \in N \backslash A$ ($\backslash$ designates set difference), and the gametes $(0,0,...,0)$ and $(1,1,...,1)$ correspond to the empty set, $\emptyset$, and to the set N, respectively. The gamete complementary to A is therefore $N \backslash A$. The set of all subsets of N is designated by $S(N)$, and each n-locus gamete corresponds to exactly one of these sets, so we can consider $S(N)$ as the set of gametes. To address a subset, $A \in S(N)$, of the n loci we use $S(A)$ for the set of subsets of A, i.e., $S(A)$ is the set of gametes corresponding to the loci in A.

For two loci, $n = 2$, the four gamete types are 00, 01, 10 and 11, where the first number designates the allele at locus 1 and the second number designates the allele at locus 2. In terms of index sets these gametes are $\emptyset$, $\{2\}$, $\{1\}$ and $\{1,2\}$, and the usual indexation used in Section 4.3 of these four gametes is 1, 2, 3 and 4. For three loci the gametes are 000, 001, 010, 011, 100, 101, 110 and 111 usually indexed as 1 through 8 and now indexed as $\emptyset$, $\{3\}$, $\{2\}$, $\{2,3\}$, $\{1\}$, $\{1,3\}$, $\{1,2\}$ and $\{1,2,3\}$.

The genotypes are given by ordered pairs, AB, of gametes, $A, B \in S(N)$. The genotype AB is a heterozygote at all the loci $a \in A*B$, where $A*B$, $A, B \in S(N)$, is the disjoint union (or symmetric difference) of A and B ($A*B = (A \backslash B) \cup (B \backslash A)$), where $\cup$ designates set union). The genotype AB is a homozygote for allele number 0 at all the loci $a \in (N\backslash A) \cap (N\backslash B)$ and a homozygote for allele number 1 at all the loci $a \in A \cap B$ (where $\cap$ designates set intersection).

The segregation of gametes is described by Geiringer's recombination distribution which may be viewed as a probability distribution on $S(N)$ (Geiringer, 1944; Schnelll, 1961). The probability $R(K)$, $K \in S(N)$, is the probability, that the genotype AB , say, segregates the gamete with alleles from gamete A at the loci $a \in N\backslash K$ and alleles from gamete B at the loci $a \in K$. The recombination event complementary to K is the complementary set $N\backslash K$, so the assumption of Mendelian segregation is now that $R(N\backslash K) = R(K)$. The marginal recombination distribution R_A describes the segregation process with respect to a subset $A \in S(N)$ of the loci:

$$(4.10) \qquad R_A(B) \;=\; \sum_{C \in S(N\backslash A)} R(B \cup C), \quad B \in S(A) \,.$$

Further discussions of the recombination distribution are given by Karlin and Liberman (1978, 1979 abc).

We will consider the change in gamete frequencies from generation to generation due to the process of recombination in a homogeneous random mating population. Let $\pi(A)$ be the frequency of gamete A in a given generation, and the frequency $\pi(A)'$ of gamete A in the offspring generation will be calculated in two different ways to illustrate the formulation. First, we will consider the generalization of the argument used to prove Equation (4.4). Thus, the frequency of gamete A in the offspring generation is calculated from a classification of gametes according to the recombination event that produced the gamete. Consider therefore a gamete produced by the event K , $K \in S(N)$. This gamete is of type A if the parent is formed by the union of gametes, one of which is of type $A \cap K$ among the loci in K with the other being of type $A\backslash K$ among the loci in $N\backslash K$. The frequency $\pi(A)'$ is therefore given by

$$(4.11) \qquad \pi'(A) \;=\; \sum_{K \in S(N)} R(K)\,\pi_{N\backslash K}(A\backslash K)\,\pi_K(A \cap K) \,,$$

where $\pi_A(B)$, $B \in S(A)$, are the marginal gamete frequencies for the loci in A defined like in Equation (4.10):

$$(4.12) \qquad \pi_A(B) \;=\; \sum_{C \in S(N\backslash A)} \pi(B \cup C), \quad B \in S(A) \,.$$

Next, let us calculate the frequency $\pi(A)'$ of gamete A in the offspring generation from the probabilities that a particular parent segregates gametes A . First let us characterize for a given $B \in S(N)$ the gametes $X \in S(N)$ that allow an individual of genotype BX to produce gamete A . The requirement is simply that BX must carry allele number 1 at all the loci in A and allele 0 at all the loci in $N\backslash A$. Thus, X must contain all the loci in $A\backslash B$ and $N\backslash X$ must contain all the loci in $B\backslash A$. Potential segregation of A therefore requires that $X = (A\backslash B) \cup C$ for some $C \in S(N\backslash (B*A))$ (see

Fig. 4.1). Secondly, the genotype BX is a heterozygote at the loci B * ((A\B)∪C) = (A∪B) * C, so only recombination among these loci is relevant. To produce A the recombination event should pick the allele number 1 from X at the loci in A\B and the allele number 0 from X at the loci in B\A. The frequency $\pi(A)'$ of gamete A in the offspring generation is therefore given by

$$(4.13) \qquad \pi'(A) = \sum_{B \in S(N)} \sum_{C \in S(N\setminus(B*A))} R_{(A\cup B)*C} (A*B)\pi(B)\pi(C\cup(A\backslash B)),$$

as the frequency of genotype AB (ordered) by the assumption of random union of gametes is given by $\pi(A)\pi(B)$.

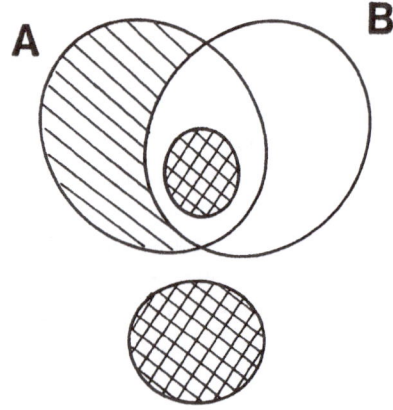

Figure 4.1. A parent of genotype BX may produce gamete A when X ⊇ A\B (symbolized by singly hatched area) and N\X ⊇ B\A. Thus X = (A\B) ∪ C for some C ∈ S(N\A * B)), and the set C is symbolized by the union of the two cross-hatched areas.

The difference between Equation (4.11) and Equation (4.13) is that Equation (4.11) focuses on the event of recombination that formed the gametes and Equation (4.13) focuses on the segregation of gametes from a given parent.

A linear transformation of the gamete frequencies simplifies theoretical investigations of the development of the gamete frequencies in the population (Karlin and Feldman, 1970; Feldman, Franklin and Thomson, 1974; Christiansen, 1987, 1988, 1989). The transformation is given by:

$$(4.14) \qquad \theta_A = \sum_{B \in S(N)} \gamma(A \cap B)\pi(B), \quad A \in S(N),$$

where $\gamma(C) = (-1)^{\#C}$, and $\#C$ is the number of elements in C. Thus, $\theta_\emptyset = 1$, $\theta_{\{a\}} = p_a - q_a$, where p_a and q_a are the gene frequencies of allele 0 and allele 1 at locus a. In general θ_A is determined from the marginal frequencies $\pi_A(B)$, $B \in S(A)$, so $\theta_{\{a\}} = \pi_{\{a\}}(\emptyset) - \pi_{\{A\}}(\{a\})$. (The linear transformation is usually defined in terms of tensor powers of the transformation

$$\begin{pmatrix} 1 & 1 \\ 1 & -1 \end{pmatrix}$$

applied on the vector of gametic frequencies with the indexation defined above in terms of binary numbers, see Feldman et al., 1974.) The reverse transformation given by:

$$(4.15) \qquad \pi(A) = 2^{-n} \sum_{B \in S(N)} \gamma(A \cap B)\, \theta_B \, ,$$

for all $A \in S(N)$. This is seen by the calculation

$$\sum_{B \in S(N)} \gamma(A \cap B)\theta_B = \sum_{B \in S(N)} \sum_{C \in S(N)} \gamma(A \cap B)\gamma(B \cap C)\pi(C)$$

$$(4.16)$$

$$= \sum_{C \in S(N)} \pi(C) \sum_{B \in S(N)} \gamma(B \cap (A * C)) = 2^n \pi(A) \, ,$$

because $\gamma(A \cap B)\gamma(B \cap C) = \gamma(B \cap (A * C))$ and the last sum is different from zero if and only if $A * C = \emptyset$ because

$$(4.17) \qquad \sum_{C \in S(A)} \gamma(B \cap C) = \begin{cases} 2^{\#A} & \text{for } A \cap B = \emptyset \, , \\ 0 & \text{otherwise.} \end{cases}$$

This transformation is conserved for the marginal gametic frequencies, e.g., the gene frequency of allele number 0 in locus a is $p_a = (1+\theta_{\{a\}})/2$ and the frequency of allele number 1 is $q_a = (1-\theta_{\{a\}})/2$.

The recurrence equations (4.11) or (4.13) are simplified to

$$(4.18) \qquad \theta'_A = \sum_{B \in S(A)} R_A(B)\theta_B \theta_{A \backslash B}$$

in the transformed variables (Christiansen, 1987, 1988, 1989). This may be seen by the following calculations. The transformation (4.14) of Equation (4.11) may be written as

$$(4.19) \qquad 4^n \theta'_A = \sum_{B \in S(N)} \sum_{C \in S(N)} \Omega_A(B,C)\theta_B \theta_C \, ,$$

where

$$\Omega_A(B,C) = \sum_{Z \in S(N)} \sum_{K \in S(N)} \sum_{X \in S(K)} \sum_{Y \in S(N\setminus K)} \gamma(A \cap Z)R(K) \cdot$$

(4.20)

$$\cdot \; \gamma([(Z\setminus K)*X] \cap B) \; \gamma([(Z \cap K)*Y] \cap C) .$$

Now make the substitution $X' = X * Y$, $X' \in S(N)$, and subsequently the substitution $Y' = X' * Z$, $Y' \in S(N)$, and rearrange to get

$$\Omega_A(B,C) = \sum_{K \in S(N)} \sum_{Y \in S(N)} \gamma(B \cap K \cap Y)\gamma(C \cap [Y \setminus K])R(K) \cdot$$

(4.21)

$$\cdot \sum_{Z \in S(N)} \gamma(Z \cap [A*B*C]) .$$

The sum over Z from (4.17) equals zero except in case $A = B * C$ when it equals 2^n. Then substituting $N\setminus K$ for K allow summation of $R(K)$ over $N\setminus A$ to provide $R_A(K)$, $K \in S(A)$. Finally, the sum over Y is from (4.17) equal to zero except when $B = K$. Thus, the only non-zero value of $\Omega_A(B,C)$ is $\Omega_A(B,A*B) = R_A(B)$ for $B \in S(A)$.

4.5. MIXING AND MULTIPLE LOCI

As for two loci the gametic frequencies converge to linkage equilibrium, where the alleles at the different loci are distributed independently in the gametes if recombination can occur between any pair of loci (Robbins 1914; Geiringer 1944). The equilibrium gametic proportions, the Robbins proportions, are given by

(4.22)
$$\pi(A) = \prod_{a \in N\setminus A} p_a \prod_{b \in A} q_b$$

which in the transformed variables are

(4.23)
$$\rho_A = \prod_{a \in A} \theta_{\{a\}} .$$

Thus, at linkage equilibrium $\theta_A = \rho_A$ for all $A \in S(N)$.

The deviation from linkage equilibrium, the linkage disequilibrium, may be measured by the deviation, $\delta_A = \theta_A - \rho_A$, the deviation from linkage equilibrium on the level A. The deviations are non-trivial for two or more loci, in that $\delta_A = 0$ when $\#A < 2$. For two loci this is the usual linkage disequilibrium, in that $D_{\{ab\}} = 4\delta_{\{ab\}}$. For more loci the usual Bennett disequilibrium differs from these linear measures (Christiansen, 1987, 1989). The recurrence equations in the δ's are

$$(4.24) \qquad \delta_{A'} = \sum_{B \in S(A)} R_A(B) \delta_B (\delta_{A \backslash B} + 2\rho_{A \backslash B}) \,.$$

The terms corresponding to $B = \emptyset$ and $B = \{a\}$, $a \in A$, vanish in this sum.

The eigenvalues for the convergence to linkage equilibrium are simply given as $2R_A(A)$, $A \in S(N)$, which is the probability of no recombination among the loci in A. Thus, $\delta_A^{(t)} \to 0$ as $t \to \infty$ when $2R_A(A) < 1$ (Geiringer, 1944). In the limit the higher order deviations show the fastest convergence, in that $R_A(A) \leq R_B(B)$ for for $B \in S(A)$, but only the two- and three-locus linkage disequilibria decrease monotonically to zero.

The generalization of Equation (4.8) to multiple loci turns out to be natural in terms of the linear measures of deviation from linkage equilibrium (Christiansen, 1989). Consider again our population which is divided into N subpopulations that in a given generation are mixed, and consider as before the gametes produced by this population. The parameters θ_{Ak}, $A \in S(N)$, describe the gametic proportions in subpopulation k, and as the recombination process occurs within the individuals bred in the subpopulations, then from (4.18) the proportions among the gametes forming the offspring generation is

$$(4.25) \qquad \theta_A' = \sum_{i=1}^{k} c_i \sum_{B \in S(A)} R_A(B) \theta_{Bi} \theta_{A*Bi} \,.$$

This simple argument works because the transformation (4.14) is linear. In terms of the deviation from Robbins' proportions among the gametes forming the offspring generation this equation is given by

$$(4.26) \qquad \delta_A' = \sum_{i=1}^{k} c_i \delta_{Ai}^* + \left(\sum_{i=1}^{k} c_i \rho_{Ai} - \rho_A' \right) \,,$$

where δ_{Ai}^*, obtained from (4.24), is the deviation from Robbins' proportions among the gametes produced by the individuals that originated from i'th subpopulation (Christiansen, 1989).

As for two loci populations mixing produces a deviation from linkage equilibrium that only depends on the variation in gene frequencies, and the produced deviation on level A depends only on the loci in A. The amount of deviation that is produced equals the difference between the average of the Robbins proportions in the subpopulations before mixing and the Robbins proportions in the population after mixing calculated from the average gamete frequencies in the population.

The interpretation of this multi-locus result is not quite as straightforward as the two-locus result, because if a locus, a say, does not vary among the subpopulations, then the second term in (4.26) is not necessarily zero for all A with $a \in A$. However, this is in accordance with the properties of the linear measures of linkage disequilibrium. For instance, if for locus $a \in N$ we have that $\pi(A) = \pi_{N \backslash \{a\}}(A)p_a$ and $\pi(A \cup \{a\}) = \pi_{N \backslash \{a\}}(A)q_a$ for all $A \in S(N \backslash \{a\})$, then $\theta_{A \cup \{a\}} = \theta_{\{a\}}\theta_A$ and $\delta_{A \cup \{a\}} = \theta_{\{a\}}\delta_A$ for all $A \in S(N \backslash \{a\})$. Thus, for a constant locus $a \in A$ the second factor in equation (4.26) has a factor $\theta_{\{a\}}$ which corresponds to no production of linkage disequilibrium on the level A.

This property of the linear linkage disequilibria makes them less well suited for the interpretation of results than the usual measures of linkage disequilibrium introduced by Bennett (1954). These may in a slightly modified form (Slatkin, 1972) be defined by

$$(4.27) \qquad D_A = \sum_{B \in S(A)} \gamma(A\backslash B) \pi_B(\emptyset) \prod_{a \in A\backslash B} p_a$$

for $A \in S(N)$. Note that the definition includes $D_\emptyset = 1$ and $D_{\{a\}} = 0$ for all $a \in N$. The correspondence between these and the linear linkage disequilibria are discussed elsewhere (Christiansen, 1987, 1989). The recurrence equations (4.18) or (4.24) in Bennett's measures are given by

$$(4.28) \qquad D'_A = \sum_{B \in S(A)} R_A(B) D_B D_{A\backslash B}$$

which is the simple well-known equation for two or three loci.

Different aspects of linkage disequilibrium are measured by the two ways of expressing the deviation of the gametic frequencies from the Robbins proportions. If the alleles of the loci in $A \in S(N)$ are distributed in the gametes independent of the alleles of the loci in $N\backslash A$, then for all $B \in S(N): \pi(B) = \pi_A(A \cap B)\pi_{N\backslash A}(B\backslash A)$, which is equivalent to $\theta_B = \theta_{A \cap B}\theta_{B\backslash A}$, $D_B = D_{A \cap B}D_{B\backslash A}$, and $\delta_B = \delta_{A \cap B}\delta_{B\backslash A} + \delta_{A \cap B}\rho_{B\backslash A} + \rho_{A \cap B}\delta_{B\backslash A}$. If, in addition, the alleles of the loci in A are distributed independently in the gametes, i.e., the π_A's are in Robbins' proportions, then $\theta_B = \rho_{A \cap B}\theta_{B\backslash A}$ and $\delta_B = \rho_{A \cap B}\delta_{B\backslash A}$ for all $B \in S(N)$ in the present parametrization, but in Bennett's parametrization $D_B = 0$ if $B \cap A \neq \emptyset$.

Equation (4.26) in Bennett's multiplicative linkage disequilibria is more complicated (Christiansen, 1989). It depends on higher order central product moments among the gene frequencies and the Bennett linkage disequilibria in the subpopulations before mixing. However, when mixing of subpopulations in linkage equilibrium occurs, then the Bennett linkage disequelibrium is the central product moment in the gene frequencies

$$(4.29) \qquad D'_A = \sum_{i=1}^{k} c_i \prod_{a \in A} (p_{ai} - p'_a)$$

which corresponds to the covariance in gene frequencies in the two-locus case.

5. Population mixing due to immigration

Immigration into a population is effectively mixing of populations in the sense that we considered in the last lecture. With N subpopulations, where migration is described by the backward migration matrix $M = (m_{kl})$, population k after migration is a mixture of the N subpopulations given by $c_l = m_{kl}$, $l = 1,2,...,N$. If reproduction occurs by random mating and if there is no selection, then this immigration produces a Wahlund effect on the genotypic proportions. In addition, the mixing will produce deviations from linkage equilibrium, but in contrast to the Wahlund effect this deviation will fade away at a finite rate, so the recurrence of mixing every generation will cause this deviation to build up. For two loci this phenomenon was studied by Feldman and Christiansen (1975).

5.1. THE ISLAND MODEL

To study this effect of recurrent immigration into a large population on the deviation from linkage equilibrium we need persistent gene frequency variation among the compartments of the mixed population. This may be achieved by incorporating two constant source populations into an island model instead of one in the classical island model (Wright, 1931, 1943), so we consider an island population that receives immigrants from two mainlands. Let the island population be population number 1, and let the two source populations have numbers 0 and 2, then the model is given by

(5.1)
$$M = \begin{pmatrix} 1 & 0 & 0 \\ m_0 & 1-m_0-m_2 & m_2 \\ 0 & 0 & 1 \end{pmatrix}.$$

The gene frequencies of the island population will eventually settle on a value determined by the gene frequencies in the source populations and the relative amount of immigration from these two populations, $p_1 = (m_0 p_0 + m_2 p_2)/(m_0 + m_2)$. The result is that each generation the island population is a mixture with three parts originating from the two source populations and from the resident population. In this stable situation we can study the deviations from Robbins' proportions as a balance between production by immigration and erosion due to recombination.

The simplest model is that equal fractions $m_0 = m_2 = m$ originate as immigrants from the two source populations, so at equilibirum we have that

(5.2)
$$\theta_{\{a\}1} = (\theta_{\{a\}0} + \theta_{\{a\}2})/2$$

for all $a \in N$, and $\theta_{\{a\}1}$ converges to this equilibrium with rate $1 - 2m$. For simplicity we may further assume that the two source populations are at linkage equilibrium, i.e., $\theta_{A0} = \rho_{A0}$ and $\theta_{A2} = \rho_{A2}$ or $\delta_{A0} = \delta_{A2} = 0$ for all $A \in S(N)$. With this regularity we may drop the population index from δ_{A1} and simply write δ_A. Using Equation (4.26) we get

(5.3)
$$\delta'_A = (1-2m) \sum_{B \in S(A)} R_A(B)\delta_B(\delta_{A*B} + 2\rho_{A*B1})$$

$$+ \ [(1-2m)\rho_{A1} + m(\rho_{A0} + \rho_{A2})] - \rho'_{A1}.$$

Then at equilibrium in the island population we have

$$[1-2R_A(A)(1-2m)]\delta_A \;=\; m(\rho_{A0}+\rho_{A2}-2\rho_{A1})$$

(5.4)

$$+ \;\; (1-2m) \sum_{B\in S(A)\setminus\{A,\emptyset\}} R_A(B)\delta_B(\delta_{A*B}+2\rho_{A*B1}) \,,$$

and the convergence to this equilibrium will be dominated by the convergence to the gene frequency equilibrium, given by the eigenvalue $(1-2m)$, because from (5.3) the convergence of the δ's will be at rates $2R_A(A)(1-2m)$, $A \in S(N)$ (Holgate, 1976).

With the gene frequencies at equilibrium given by Equation (5.2), Equation (5.4) constitutes an iterative solution for the equilibrium in the island population as a function of the gene frequencies in the source populations. The first term in (5.4) is given by

(5.5)
$$\rho_{A0}+\rho_{A2}-2\rho_{A1} \;=\; \sum_{B\in S(A)\setminus\{\emptyset\}} \rho_{A*B1}[1+\gamma(B)] \prod_{b\in B} \Delta_b$$

where $\Delta_a = (\theta_{\{a\}0}-\theta_{\{a\}2})/2$ measures the difference in gene frequencies between the source populations at locus a, $a \in N$. As $1+\gamma(B)$ is zero for #B, odd every part of the first term in (5.4) contains an even number of Δ-factors. Thus, the equilibrium value of δ_A for #A odd may be written in terms of the lower levels of linkage disequilibria, δ_B for $B \in S(A)\setminus\{A\}$. It turns out, that if δ_A for #A odd is written in terms of the lower level even disequilibria only, then the relation becomes independent of the recombination-segregation distribution R and independent of the immigration fraction m:

(5.6)
$$\sum_{B\in S(A)\setminus\{\emptyset,A\}} \delta_B \rho_{A*B1} \, \psi(A*B)[1+\gamma(B)]/2 \,,$$

where the function ψ depends on the number of loci in the argument, i.e., $\psi(A) = \psi(\#A)$ for all $A \in S(N)$. The values of this function are determined recursively for sets $A \in S(N)$ with an even number of loci by $\psi(\emptyset) = -1$ and

(5.7)
$$\sum_{B\in S(A)} \psi(B)[1+\gamma(B)] \;=\; -[1-\gamma(A)]\psi(A)$$

(the right side is zero for #A even), and at the same time this equation defines the values of ψ for sets $A \in S(N)$ with an odd number of loci. For sets with one or two elements the value is 1, and for sets with $3, 4, 5, 6, 7$, and 8 elements the value is $-2, -5, 16, 61, -272$, and -1385, respectively. This relation has been established for up to nine loci (Christiansen, 1988), and it constitutes a nice pattern in the linkage disequilibria produced by recurrent population mixing which may have potential applications in the analysis of population data.

The initial inspiration to consider patterns of linkage disequilibria due to mixing and immigration comes from some observations in natural populations of *Zoarces viviparus*.

The observation of a Wahlund effect in a sample from nature only shows that the sample contains a mixture of populations, as, e.g., for the cod in the western Baltic Sea (Sick, 1964). The observation says nothing about the mixing and interbreeding of the populations in nature. In *Zoarces viviparus* a similar observation was made in a Danish inlet, Mariager Fjord, where the populations at the head of the fjord are genetically and morphologically differentiated from the populations at the mouth of the fjord (Schmidt, 1917; Christiansen, Frydenberg and Simonsen, 1984; Christiansen, Nielsen and Simonsen, 1981, 1988). The population mixing is evident in Schmidt's morphological data and in our data on biochemical polymorphisms, but in both cases the observation is the mechanical mixing that produces the Wahlund effect. The effects of population interbreeding on the deviations from Robbins' proportions could form the basis of observations of interbreeding in natural populations. Unfortunately, the eelpout data could not disclose significant linkage disequilibria in the region of population mixing.

5.2. THE STEPPING-STONE CLINE MODEL

The deviations from linkage equilibrium build up through time due to the gradual erosion of linkage disequilibrium by recombination, but by the same mechanism the linkage disequilibrium may build up through space in a zone of population mixing. Feldman and Christiansen (1975) studied this in a model, the stepping-stone cline model, which is a modification of the stepping-stone model (Kimura, 1953). By assuming two source populations linked by a stepping-stone chain of populations stable gene frequency clines may be maintained.

Consider a chain of L populations, $1,2,...,L$. This is a stepping-stone chain if migration is allowed between neighbours only, and we will consider the simplest version of the model where each population after migration is a mixture of a fraction m of immigrants born in the neighboring populations and a fraction $1-2m$ of residents. To get the stepping-stone cline model add two constant source populations with number 0 and $L+1$ at the ends of the chain. Then we get the following simple backward migration matrix:

$$(5.8) \qquad M = \begin{pmatrix} 1 & 0 & 0 & 0 & \cdots & 0 & 0 \\ m & 1\text{-}2m & m & 0 & \cdots & 0 & 0 \\ 0 & m & 1\text{-}2m & m & \cdots & 0 & 0 \\ 0 & 0 & m & 1\text{-}2m & \cdots & 0 & 0 \\ \cdot & \cdot & \cdot & \cdot & & \cdot & \cdot \\ \cdot & \cdot & \cdot & \cdot & & \cdot & \cdot \\ \cdot & \cdot & \cdot & \cdot & & \cdot & \cdot \\ 0 & 0 & 0 & 0 & \cdots & 1\text{-}2m & m \\ 0 & 0 & 0 & 0 & \cdots & 0 & 1 \end{pmatrix} .$$

The gene frequencies in the populations change as

$$(5.9) \qquad \theta'_{\{a\}k} = (1\text{-}2m)\theta_{\{a\}k} + m(\theta_{\{a\}k\text{-}1} + \theta_{\{a\}k+1}) ,$$

so they converge to a linear cline given by

(5.10) $$\theta_{\{A\}k} = \theta_{\{a\}0} - k\Delta_a = \theta_{\{a\}L+1} - (L-k+1)\Delta_a,$$

where $\Delta_a = (\theta_{\{a\}0} - \theta_{\{a\}L+1})/(L+1)$. The linkage disequilibria change according to (4.26) as

(5.11)
$$\delta'_{Ak} = [(1-2m)\delta^*_{Ak} + m(\delta^*_{Ak-1} + \delta^*_{Ak+1})]$$

$$+ \{[(1-2m)\rho_{Ak} + m(\rho_{Ak-1} + \rho_{Ak+1})] - \rho'_{Ak}\}.$$

where δ^*_A in a given population is given as a function of δ_B, $B \in S(A)$, in that same population by Equation (4.24).

The linkage disequilibria in the cline converge to an equilibrium at a rate dominated by the convergence to the gene frequency cline (Holgate, 1976). This equilibrium is found form (5.11) by putting $\delta'_{Bk} = \delta_{Bk}$ for all $B \in S(A)$ and all $k = 1,2,...,L$. Separating out the highest level produces the equilibrium equation as

(5.12) $$[1-2R_A(A)(1-2m)]\delta_{Ak} - 2R_A(A)m(\delta_{Ak-1} + \delta_{Ak+1}) = \Gamma_{ak},$$

where Γ_{Ak} is a function of the gene frequency cline, i.e., $\theta_{\{a\}l}$ for $a \in A$ and $l = k-1$, k, $k+1$, and the linkage disequilibria on the levels $B \in S(A)\backslash\{A\}$, i.e., δ_{Bl} for $l = k-1$, k, $k+1$. Equation (5.11) with (4.24) yields

(5.13)
$$\Gamma_{Ak} = (1-2m) \sum_{B \in S(A)\backslash\{A\}} R_A(B)\delta_{Bk}(\delta_{A\backslash Bk} + 2\rho_{A\backslash Bk})$$

$$+ m[\sum_{B \in S(A)\backslash\{A\}} R_A(B)\delta_{Bk-1}(\delta_{A\backslash Bk-1} + 2\rho_{A\backslash Bk-1})$$

$$+ \sum_{B \in S(A)\backslash\{A\}} R_A(B)\delta_{Bk+1}(\delta_{A\backslash Bk+1} + 2\rho_{A\backslash Bk+1})]$$

$$+ m(\rho_{Ak-1} + \rho_{Ak+1} - 2\rho_{Ak})$$

(Christiansen, 1987). The last term in Γ_{Ak} is from (5.10) given by a sum of products with an even number of Δ-factors:

(5.14) $$\rho_{Ak-1} + \rho_{Ak+1} - 2\rho_{Ak} = \sum_{B \in S(A)\backslash\{\emptyset\}} \rho_{A\backslash Bk}[1+\gamma(B)] \prod_{b \in B} \Delta_b,$$

so as in the case $L = 1$ the odd levels of linkage disequilibria will play a special role. The recurrence equation (5.12) can be written as $\alpha_A\delta_{Ak} - \beta_A(\delta_{Ak-1} + \delta_{Ak+1}) = \Gamma_{Ak}$, where

$\alpha_A = 1 - 2R_A(A)(1-2m)$ and $\beta_A = 2R_A(A)m$. The characteristic roots of this equation are $\lambda_A = [\alpha_A + (\alpha_A^2 - 4\beta_A^2)^{1/2}]/(2\beta_A)$ and λ_A^{-1}.

This form of Equation (5.12) is the basis for an iterative solution for the linkage disequilibria. Assuming δ_{Bk} known for all $B \in S(A)$ and all $k = 0,1,2,...,L+1$, we can solve for δ_{Ak}, $k = 1,2,...,L$, as a function of these and δ_{A0} and δ_{AL+1}. Feldman and Christiansen (1975) considered the stepping-stone cline model in the case of two loci, i.e., for $A = \{ab\}$. In this case $\Gamma_{\{ab\}k}$ only contains the term (5.14) which is the covariance in gene frequencies among the three parts of the breeding population in subpopulation number k, and it becomes $2m\Delta_a\Delta_b$ for all $k = 1,2,...,L$. The solution is illustrated in Figure 5.1 in a cline where both source populations are monomorphic, population 0 is fixed for allele number 0 at all loci and population L is fixed for allele number 1. The value of the linkage disequilibrium attains its maximum in the middle of the cline, and this maximal value is by a crude approximation equal to $\Gamma_{\{ab\}}/R_{\{ab\}}(\{ab\})$ which is considerably larger than the amount of deviation from Robbins' proportions produced every generation due to mixing, namely $\Gamma_{\{ab\}}$.

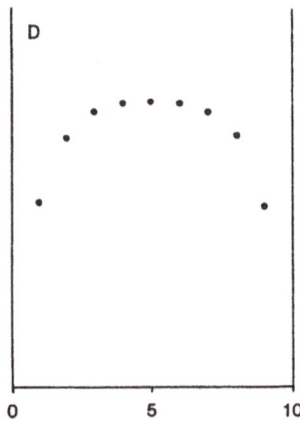

Figure 5.1. Typical variation of the two locus linkage disequilibrium, D, though a stepping-stone cline where the two source populations are in linkage equilibrium (L = 9).

For two loci the linear linkage disequilibrium measures and the usual Bennett linkage disequilibria coincide. For more than two loci the two measures differ, but as argued in Section 4 the interpretation of results is easier in terms of the Bennett disequilibria, so in the following we will use these measures in numerical examples.

For more than two loci the value of Γ_{Ak} depends on k, as it depends on two locus disequilibria in subpopulations $k-1$, k and $k+1$ (Figure 5.1). So the solution of the equilibrium equation becomes more tedious (Christiansen, 1987). The conclusion is that the buildup of linkage disequilibrium generalizes to any level A where #A is even. However, the range of variation for the linkage disequilibria is dependent on the gene frequencies.   For instance, in the case of two loci the range of variation is $\max(-p_a p_b, -q_a q_q) \leq D_{(ab)} \leq \min(p_a q_b, q_a p_b)$, so in the cline of Figure 5.1 the maximal range of $D_{(ab)}$ is in the middle of the cline where p_a and p_b are close to 1/2 (equal to if L is odd). Thus the maximal range and the maximal value coincide, so a judgement of the values in Figure 5.1 in terms of the strength of the linkage disequilibrium is difficult.

Another way of illustrating the variation in linkage disequilibria is to relate them to the maximal value that can be obtained in the considered cline, i.e., the value obtained when no recombinations between the loci are allowed. In this case the model degenerates into a multiple allele model and linear clines in the gamete frequencies result (Christiansen, 1987). For two, four and six loci in the same cline as in Figure 5.1 this relative measure is shown in Figure 5.2. The variation in the relative two locus linkage disequilibrium shows that the concern about the range of variation was just, as the linkage disequilibrium in the middle of the cline is a lower proportion of its maximal value than at the ends of the cline. With four loci, and more so for six loci, the trend is the opposite: the relative size of the linkage disequilibrium increases through the cline. Therefore, the buildup of linkage disequilibria through space is pronounced for more than two loci.

Further examples and discussions of this model are found in Christiansen (1987).

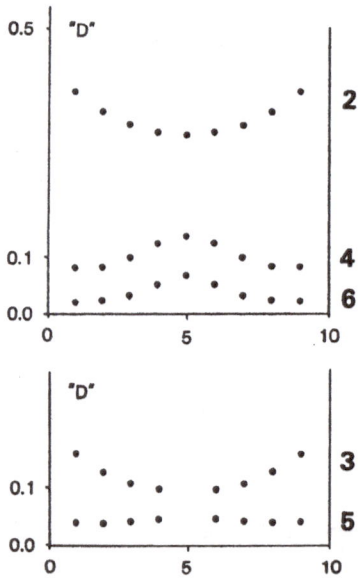

Figure 5.2. Strength of the deviation from Robbins' proportions in a stepping-stone cline with variation at six loci (with recombination frequency 0.02 between neighbouring loci). The figure shows the linkage disequilibrium at equlibrium relative the equilibrium value with no recombination. The level of the interaction is shown to the right, and in each instance the figure shows the highest values

among the six loci, e.g., the two locus measures correspond to neighboring loci. After Christiansen (1987).

5.3. THE SYMMETRIC VIABILITY MODEL

The transformation (4.14) was originally applied in the analysis of the symmetric viability model for three loci (Feldman et al., 1974), and it is indeed well suited for treatment of the n-locus symmetric viability model.

The symmetric viability model assumes that the fitness contribution of all homozygote genotypes are equal, so the fitness $w(AB)$ of genotype AB depends only on the loci in which it is heterozygote, i.e., we may write $w(AB) = v(A*B)$. With selection the recurrence equation (4.18) becomes more complicated, and it is given by

$$(5.15) \qquad 2^n W\theta'_A = \sum_{B \in S(N)} \theta_B \theta_{A*B} \sum_{C \in S(A)} R_A(C) e_{B*C} ,$$

where the average fitness is given by

$$(5.16) \qquad 2^n W = \sum_{A \in S(N)} \theta_A^2 e_A$$

and e_X, $X \in S(N)$, are epistatic parameters given by

$$(5.17) \qquad e_X = \sum_{Y \in S(N)} \gamma(X \cap Y) v(Y)$$

(Christiansen, 1988). Alternatively, the recurrence equation may be written as

$$(5.18) \qquad W\theta'_A = \sum_{B \in S(N)} \theta_B \theta_{A*B} \sum_{C \in S(A)} \gamma(B \cap C) v(C) L(A \cap C)$$

(Christiansen, 1988), where $L(X)$, $X \in S(N)$, are the linkage values given by

$$(5.19) \qquad L(X) = \sum_{Y \in S(N)} \gamma(Y) R_X(Y)$$

(Schnell, 1961; Karlin and Liberman, 1978, 1979abc).

The multi-locus symmetric viability model has been analysed by Karlin (1977) and by Karlin and Avni (1981). They focussed on the analysis of the central equilibrium, $\theta_A = 0$ for $A \in S(N)\backslash\{\emptyset\}$, and the eigenspaces for the local stability analysis are the coordinate axes in the present parametrization. Thus many of their results are produced extremely easily in the present formulation (Christiansen, 1988).

For total negative epistasis, $e_A < 0$ for $A \in S(N)\backslash\{\emptyset\}$, the central equilibrium is stable for all recombination distributions, and for free recombination, i.e., $R(A) = 2^{-n}$ for all $A \in S(N)$, it is stable for $e_{\{a\}} < 0$ for all $a \in N$ and unstable if $e_{\{a\}} > 0$ for an

$a \in N$. For no recombination, i.e., $R(N) = R(\emptyset) = 1/2$, and therefore for low recombination the central point is unstable if there exists an $A \in S(N)$ so $e_A > 0$, and then a pair of high complementarity equilibria, $\theta_A \neq 0$ and $\theta_B = 0$ for $B \in S(N)\backslash\{A,\emptyset\}$, exists and may be stable. In general, the various types of equilibria may be identified with the subgroups of the group $S(N)$ organized by the rule $*$ of multiplication.

6. Population data for multiple loci

In this section we will consider samples from a population which is polymorphic for n loci. Suppose in this population that among adult individuals the genotypic frequency of genotype AA is $Q(A,A)$ and the frequencies of the genotypes AB and BA are $Q(A,B)$ and $Q(B,A)$. Assuming $Q(A,B) = Q(B,A)$ produce the combined frequency of the genotypes AB and BA as $2Q(A,B)$. Among these individuals the frequency of gamete A is given by

$$(6.1) \qquad \pi(A) = \sum_{B \in S(N)} Q(A,B), \quad A \in S(N).$$

The marginal genotype frequencies with respect to the loci in $A \in S(N)$ are given by

$$(6.2) \qquad Q_A(B,C) = \sum_{X \in S(N\backslash A)} \sum_{Y \in S(N\backslash A)} Q(B \cup X, C \cup Y), \quad B,C \in S(A).$$

I will consider only the analysis of population samples of simple mother-offspring combinations from this population, because the differences between one locus and multiple loci will be aparent there.

6.1. GENOTYPIC MOTHER-OFFSPRING COMBINATIONS

In the simplest case the mother-offspring combinations will consist of counts of genotypic combinations (AB,CD). The mother AB is 11 homozygote in $A \cap B$, 01 heterozygote in $A * B$, and 00 homozygote in $N\backslash(A \cup B)$, so she segregates gametes of the form $(A \cap B) \cup X$ where $X \in S(A*B)$. The various offspring types from this mother are therefore obtained by fertilizing these eggs with sperm of all the possible types. As in one locus, we need to describe the female segregation frequencies $\beta((A \cap B) \cup X | AB)$, $X \in S(A*B)$, which are related to $R_{A*B}((X \cap A) \cup (B\backslash X))$, and the frequencies of the fertilizing male gametes $\mu(K|AB)$, $K \in S(N)$. The offspring frequencies for a given mother type is given in terms of the products, $\beta((A \cap B) \cup X | AB)\mu(K|AB)$, that provide the probability that the mother transmit $(A \cap B) \cup X$ and the father transmit K.

The problem of gamete identification in the offspring was in the one locus case restricted to the class where the offspring shows the same genotype as the mother. This is aggravated in the multiple locus case, because all the offspring genotypes CD where C and D may both be segregated by the mother, i.e., $C = (A \cap B) \cup X$ and $D = (A \cap B) \cup Y$ for $X, Y \in S(A*B)$, pose an identification problem when $X \neq Y$, so the frequency of this offspring genotype among mothers of genotype AB is

(6.3) $\beta((A \cap B) \cup X | AB)\mu((A \cap B) \cup Y | AB) + \beta((A \cap B) \cup Y | AB)\mu((A \cap B) \cup X | AB)$.

When $X = Y$ the determination of the origin of the gametes is of course unambigous, and if the offspring is heterozygote at a locus in $N(A*B)$, then the determination is also certain. The number of genotypic mother-offspring combinations for mother AB is shown in Table 6.1, and the total number of mother-offspring combinations becomes $2^n(6^n/2 - 5^n/4 + 3^n/4 + 2^n/2)$. As in the case of one locus with multiple alleles we can compare the observed frequencies of mother-offspring combinations for heterogametic mothers with the expectation from the product structure outlined here to produce a test for fetal zygotic selection.

Table 6.1
Number of offspring genotypic classes
from mothers of genotype AB

Offspring type	Number of classes
Homogametic	$2^{\#(A*B)}$
Heterogametic, ambigous	$2^{\#(A*B)}(2^{\#(A*B)}-1)/2$
Heterogametic, straight	$2^{\#(A*B)}(2^n - 2^{\#(A*B)})$
Total	$2^{\#(A*B)}[2^n - (2^{\#(A*B)}-1)]$

    The analysis now proceeds as before by an analysis of Mendelian segregation which for multiple loci is

(6.4) $\beta((A \cap B) \cup X | AB) = \beta((A \cap B) \cup ((A*B) \backslash X) | AB)$

which in terms of the recombination-segregation distribution is that $R_{A*B}((X \cap A) \cup (B \backslash X)) = R_{A*B}((X \cap B) \cup (A \backslash X))$. However, Hypothesis (6.4) only specifies Mendelian segregation for the recombination-segregation distribution for AB mothers, we also need that

(6.5) $\beta((A \cap B) \cup X | AB) = R_{A*B}((X \cap A) \cup (B \backslash X))$,

where R is a recombination-segregation distribution common for all mother genotypes. The analysis of mating may proceed by supposing random mating, i.e.,

(6.6) $\mu(K | AB) = \mu(K)$.

Thus, the analysis of multiple locus mother-offspring genotypic combinations is a fairly straightforward extension of the one locus analysis. The extension to a selection component analysis is equally easy, because the inclusion of data on genotypic frequencies in the population is no different. The analysis is developed by Østergaard and Christiansen (1981) where the gene counting estimates of the parameters are also given.

The only really different aspect of the multiple locus analysis is that if Mendelian segregation and random mating prevails and if no selection is working on the considered variation, then we can ask whether the gamete frequencies in the population are in the Robbins proportions and thereby compare the population to expectations under linkage equilibrium. This analysis is interesting in that selection effects that go unnoticed in the selection component analysis may produce significant linkage disequilibrium for tightly linked loci. Thus, a final test for Robbins' proportions may be viewed as a kind of consistency test for the conclusions of the selection component analysis. A significant deviation from the hypothesis of linkage equilibrium cannot, however, be taken as evidence for selection. Population interbreeding due to past immigration events may be responsible.

6.2. PHENOTYPIC DATA ON MULTIPLE LOCI

Genotypic data on multiple loci is rare and usually extremely difficult, if not impossible, to collect. A more typical data consists of determinations of the genotype at each of the n polymorphic loci for each individual. The phenotype of an individual then is characterized by the set X of loci where it is a homozygote for allele number 1, the set Y of loci where it is a heterozygote, and the set $Z = N \backslash (X \cup Y)$ where it is a homozygote for allele number 0. We can characterize the phenotype by the two sets X and Y of loci, where the individual is a homozygote for allele number 1 and where it is a heterozygote, and we will address this phenotype as $\phi(X,Y)$, $X \in S(N)$ and $Y \in S(N \backslash X)$.

Phenotype $\phi(X,Y)$ may cover an array of genotypes AB. The requirement is that $X = A \cap B$ and $Y = A * B$, so any genotype AB with $A = X \cup Z$ and $B = X \cup (Y \backslash Z)$ produces phenotype $\phi(X,Y)$ when $Z \in S(Y)$. In the above considered population the frequency of the phenotype $\phi(X,Y)$, designated $P(X,Y)$, therefore is given by

$$(6.7) \qquad P(X,Y) = \sum_{Z \in S(Y)} Q(X \cup (Y \backslash Z), X \cup Z).$$

The marginal phenotype with respect to the loci in A is characterized by two sets X and $Y \in S(A)$ as before, and we will address this phenotype as $\phi_A(X,Y)$. The frequency of this phenotype is designated by $P_A(X,Y)$.

Consider a sample of individuals where $T(X,Y)$ individuals of phenotype $\phi(X,Y)$, $X \in S(N)$ and $Y \in S(N \backslash X)$, are observed. Then the marginals of this table is $T_A(X,Y)$, where

$$(6.8) \qquad T_A(X,Y) = \sum_{B \in S(N \backslash A)} \sum_{C \in S(N \backslash (A \cup B))} T(X \cup B, Y \cup C)$$

is the observed number of individuals of phenotype $\phi_A(X,Y)$, $X \in S(A)$ and $Y \in S(A \backslash X)$, and the total sample size is $T = T_\emptyset(\emptyset,\emptyset)$. The table $T(X,Y)$ has 3^n cells, and it is assumed to be multinomially distributed with probability parameters $P(X,Y)$ and number parameter T. The logarithm of the likelihood function of this sample is given by

$$(6.9) \qquad L = \sum_{X \in S(N)} \sum_{Y \in S(N \backslash X)} T(X,Y) \log(P(X,Y)) .$$

Now let us assume that the multi-locus genotypic frequencies are the Hardy-Weinberg proportions, i.e.,

$$(6.10) \qquad Q(A,B) = \pi(A)\pi(B), \ A \text{ and } B \in S(N) .$$

Under this hypothesis, using the gene-counting algorithm the maximum-likelihood estimates of the gamete frequencies are limits of the recurrence equations:

$$(6.11) \ \pi'(A) = \frac{1}{2T} \left[2T(A,\emptyset) + \sum_{a \in N} T(A \backslash \{a\}, \{a\}) + \sum_{\substack{Y \in S(N) \\ \#Y > 1}} T(A \backslash Y, Y) \frac{2\pi(A)\pi(A*Y)}{P(A/Y,Y)} \right]$$

where

$$(6.12) \qquad P(A \backslash Y, Y) = \sum_{Z \in S(Y)} \pi(Z \cup (A \backslash Y)) \pi((Y \backslash Z) \cup (A \backslash Y)) .$$

Equation (6.11) is constructed as follows. Gamete A is observed twice in every homogametic type that is homozygote 11 at the loci in A and homozygote 00 at the loci in $N \backslash A$ (first term). Gamete A is observed once in every heterogamete type that is heterozygote in exactly one locus a and that is homozygote 11 at the loci in $A \backslash \{a\}$ and homozygote 00 at the loci in $N \backslash (A \cup \{a\})$ (second term). Finally gamete A may have been observed in a heterogamete type that is heterozygote at two or more loci, that is homozygote 11 or heterozygote 01 at all the loci in A, and that is homozygote 00 or heterozygote 01 at all loci in $N \backslash A$. If this heterogamete type is a heterozygote at the loci in Y, then it must be a homozygote 11 at the loci in $A \backslash Y$ and a homozygote 00 at the loci in $N \backslash (A \cup Y)$. A genotype AZ is heterozygote at the loci in Y if and only if $Y = A * Z$, so $Z = A * Y$. Thus, in the phenotypic class $\phi(A \backslash Y, Y)$ the genotype made of an A gamete is $A(A*Y)$ and we expect a fraction $\pi(A)\pi(A*Y)/P(A*Y,Y)$ of the observed $\phi(A \backslash Y, Y)$ phenotypes to be of the genotype $A(A*Y)$, so this is twice the probability that in a $\phi(A \backslash Y, Y)$ individual we observe one A gamete. By counting these fractional observations we get the third term in Equation (6.12). These gene counting equations for the maximum-likelihood estimates are generalizations of the equations for two and three loci (Hill, 1974; Brown, 1975; Weir, 1979).

The maximum-likelihood estimates of the gene frequencies are identical to the one-locus estimates. To see this observe that $p_a = \pi_{\{a\}}(\emptyset)$ and make the proper summation in equation (6.11) to get

$$2Tp'_a = \sum_{A \in S(N\setminus\{a\})} 2T\pi'(A)$$

$$(6.13) \qquad = \sum_{Y \in S(N\setminus\{a\})} \sum_{A \in S(N\setminus\{a\})} \left[T(A\setminus Y, Y) \frac{2\pi(A)\pi(A*Y)}{P(A\setminus Y, Y)} \right.$$

$$\left. + T(A\setminus Y, Y \cup \{a\}) \frac{2\pi(A)\pi(A*Y*\{a\})}{P(A\setminus Y, Y \cup \{a\})} \right] .$$

Decomposing A into $A = B \cup C$ where $B \in S(Y)$ and $C \in S(N\setminus(Y \cup \{a\}))$ and summation over B provides from (6.12) that

$$(6.14) \qquad 2Tp'_a = \sum_{Y \in S(N\setminus\{a\})} \sum_{C \in S(N\setminus(Y \cup \{a\}))} [2T(C,Y) + T(C, Y \cup \{a\})] .$$

This is first of all an explicit expression for the maximum-likelihood equation in terms of the $\{a\}$-marginal table of observations, i.e., the maximum-likelihood estimate of the gene frequency at locus a is given by

$$(6.15) \qquad \hat{p}_a = [2T_{\{a\}}(\emptyset, \emptyset) + T_{\{a\}}(\emptyset, \{a\})]/(2T) ,$$

which is the ordinary estimate from the marginal distribution (Equation 1.2). Thus, the gene frequencies corresponding to the maximum likelihood estimate of the n-locus gamete frequencies are equal to the gene frequency estimates from the one locus marginal tables. Thus, the expected Robbins proportions stay the same for any marginal table of observations whether calculated as marginals in the full table of Robbins proportions or from the marginal table of interest.

6.3. PHENOTYPIC MOTHER-OFFSPRING COMBINATIONS

The lumping of genotypic classes in a sample of the adult population causes some difficulties in estimation, but the full information is present if Hardy-Weinberg proportions may be assumed. As we shall see this nice property does not carry to a population sample of mother-offspring combinations where only the genotype at each locus is determined.

Let us first consider the offspring phenotypic frequencies for given mother genotype. Focus on the loci where the offspring in the mother-offspring combination (AB,CD) is a heterozygote, and split these loci into the set $X \in S(A \cap B)$ where the mother is homozygote 11, the set $Y \in S(A*B)$ where the mother is heterozygote 01, and set

$Z \in S(N\backslash(A\cup B))$ where the mother is homozygote 00. Thus, $C*D = X\cup Y\cup Z$ and the transmission type of the offspring genotype is ambiguous if and only if $X = Z = \emptyset$. The offspring phenotype is $\phi([(A\cap B)\backslash X]\cup E, X\cup Y\cup Z)$ with $E = (C\cap D)\backslash(A\cap B)$ $\subseteq (A\cup B)$ being the loci where the offspring is homozygote 11 and the mother is heterozygote 01. This phenotype of offspring can be formed by the union of any male gamete of the form $[(A\cap B)\backslash X]\cup E \cup M \cup Z$, with $M \in S(Y)$, with the female gamete $(A\cap B) \cup E \cup (Y\backslash M)$. The mother-offspring genotypic combination (AB,CD) therefore is observed in an offspring phenotypic class with $2^{\#Y}$, $Y = (A*B) \cap (C*D)$, genotypes.

As an example let us consider a mother which is homozygote at a single locus. For simplicity, let her genotype be $\{a\}0$, so she segregates two gamete types, $\{a\}$ and $\emptyset$. Transmission ambiguity occurs for the genotype $\{a\}0$ where either the mother or the father may have transmitted the gamete $\{a\}$. Offspring phenotypes are $\phi(\emptyset,A)$, $\phi(\emptyset,A\cup\{a\})$ and $\phi(\{a\},A)$ for $A \in S(N\backslash\{a\})$. The phenotypic classes $\phi(\emptyset,A)$ and $\phi(\{a\},A)$, $A \in S(N\backslash\{a\})$, all contain only one genotype, and $A = \emptyset$ corresponds to the two possible homogametic classes. The phenotypic class $\phi(\emptyset,\{a\})$ is the class with transmission ambiguity, but it contain only one genotype. Finally, the phenotypic classes $\phi(\emptyset,A\cup\{a\})$, $A \in S(N\backslash\{a\})\backslash\{\emptyset\}$, each contain two genotypes formed either by union of mother gamete $\emptyset$ with $A \cup \{a\}$ or by union of mother gamete $\{a\}$ with A. In terms of the product structure of the basic statistical model all the phenotypic classes $\phi(\emptyset,A\cup\{a\})$, $A \in S(N\backslash\{a\})$, are mixed, in that their frequency is

$$(6.16) \qquad \beta(\emptyset|\{a\}\emptyset)\mu(A\cup\{a\}|\{a\}\emptyset)) + \beta(\{a\}|\{a\}\emptyset)\mu(A|\{a\}\emptyset).$$

This pattern of lumping of classes gives rise to pronounced symmetries between the β's and some aspect of the μ's. If the male gamete frequencies are in Robbins proportions then (6.16) may be written as

$$(6.17) \qquad [\beta(\emptyset|\{a\}\emptyset)q_a + \beta(\{a\}|\{a\}\emptyset)p_a]\mu_{N\backslash\{a\}}(A|\{a\}\emptyset)$$

which is exactly the structure giving symmetry problems in the one locus analysis (Equation 2.3). Østergaard and Christiansen (1981) discuss this further and provide an example in a two locus data from the *Zoarces viviparus* where the likelihood function has two local maxima giving almost symmetric estimates. Of these two maxima the lowest seems to be the more reasonable, in that the β's are close to Mendelian segregation and p_a and q_a are close to the values observed in the adult male population.

The more serious problem in the analysis of phenotypic mother-offspring combinations for multiple loci is the lumping of mother genotypes into phenotypic classes. This weakens the information on the segregation and transmission of female gametes to an extent where we must realize that most qualitative information on the process of genetic recombination is lost. For two loci the recombination frequency $r = 2R_{(1,2)}(\{1\})$, assuming Mendelian segregation, is buried in the statistical parameter

$$(6.18) \qquad (1/2-r)[Q(\emptyset,\{1,2\})-Q(\{1\},\{2\})]$$

where the quantity in the bracket, the *gametic phase disequilibrium*, is a property of the genotypic frequencies among mothers and therefore not observed (Østergaard and Christiansen, 1981). The only way out of this dilemma is to seek information on the recombination frequency by other means, e.g., from mother-offspring combinations with several offspring (Hjorth, 1971; Simonsen and Frydenberg, 1972; Simonsen and Christiansen, 1981, 1984).

This lack of qualitative information on the process of segregation is unfortunate, but it does not interfere with the basic property of the population sample of mother-offspring combinations that is used in the selection component analysis. The estimation of the phenotypic proportions among zygotes is still possible from the population sample of mother-offspring phenotypic combinations. Thus, the selection component analysis may proceed virtually unaltered if one is willing to make the assumption that natural selection refers to the phenotypic classes, in the sense we use here, which amounts to assuming that there is no position effect in selection. For two locus data this analysis has been worked out by Østergaard and Christiansen (1981).

References

Bennett, J.H. (1954), On the theory of random mating, *Ann. Eugenics* **18**, 311-317.

Bodmer, W.F. (1968), Demographic approaches to the measurement of differential selection in human populations, *Proc. Nat. Acad. Sci. USA* **59**, 690-699.

Brown, A.D.H. (1970), The estimation of Wright's fixation index from genotypic frequencies, *Genetica* **41**, 399-406.

Brown, A.D.H. (1975), Sample sizes required to detect linkage disequilibrium between two or three loci, *Theoret. Population Biol.* **8**, 184-201.

Bundgaard, J., and Christiansen, F.B. (1972), Dynamics of polymorphisms I. Selection components in an experimental population of *Drosophila melanogaster*, *Genetics* **71**, 439-460.

Christiansen, F.B. (1980), Studies on selection components in natural populations using population samples of mother-offspring combinations, *Hereditas* **92**, 199-203.

Christiansen, F.B. (1984), The definition and measurement of fitness, in *Evolutionary Ecology* (B.Shorrocks, ed.), British Ecol. Soc. Symp. Vol. 23, pp. 65-71. Blackwell Sci. Publ., Oxford, London, Edinburgh, Boston, Palo Alto, Melbourne.

Christiansen, F.B. (1987), The deviation from linkage equilibrium with multiple loci varying in a stepping-stone cline, *J. Genet.* **66**, 45-67.

Christiansen, F.B. (1988), Epistasis in the multiple locus symmetric viability model, *J. Math. Biol.* **26**, 595-618.

Christiansen, F.B. (1989), The effect of population subdivision on multiple loci without selection, to appear in *Mathematical Evolutionary Theory* (M. W. Feldman, ed.), Princeton University Press, Princeton.

Christiansen, F.B., Bundgaard, J., and Barker, J.S.F. (1977), On the structure of fitness estimates under post-observational selection, *Evolution* 31, 843-853.

Christiansen, F.B., and Feldman, M.W. (1985), *Population Genetics*, Blackwell Sci. Publ., Oxford, London, Edinburgh, Boston, Palo Alto, Melbourne.

Christiansen, F.B., and Frydenberg, O. (1973), Selection component analysis of natural polymorphisms using population samples including mother-offspring combinations, *Theoret. Population Biol.* 4, 425-445.

Christiansen, F.B., and Frydenberg, O. (1974), Geographical patterns of four polymorphisms in *Zoarces viviparus* as evidence of selection, *Genetics* 77, 765-770.

Christiansen, F.B., and Frydenberg, O. (1976), Selection component analysis of natural polymorphisms using mother-offspring samples of successive cohorts, in *Population Genetics and Ecology* (S. Karlin and E. Nevo, eds.), pp. 277-301. Academic Press, New York, San Francisco, London.

Christiansen, F.B., Frydenberg, O., and Simonsen, V. (1977), Genetics of *Zoarces* populations X. Selection component analysis of the *EstIII* polymorphism using samples of successive cohorts, *Hereditas* 87, 129-150.

Christiansen, F.B., Frydenberg, O., and Simonsen, V. (1984), Genetics of *Zoarces* populations XII. Variations at the polymorphic loci *PgmI, PgmII, HbI* and *EstIII* in fjords, *Hereditas* 101, 37-48.

Christiansen, F.B., Nielsen, B.H., and Simonsen, V. (1981), Genetical and morphological variation in the eelpout *Zoarces viviparus*, *Canadian J. Genet. Cytol.* 23, 163-172.

Christiansen, F.B., Nielsen, B.H., and Simonsen, V. (1988), Genetics of Zoarces populations XV. Genetic and morphological variation in Mariager Fjord, *Hereditas* 109, 99-112.

Clark, A.G., Feldman, M.W., and Christiansen, F.B. (1981), The estimation of epistasis in components of *Drosophila melanogaster* I. Fitness in experimental populations of a two stage maximum likelihood model, *Heredity* 46, 321-346.

Cooper, D.W. (1968), The use of incomplete family data in the study of selection and population structure in marsupials and domestic animals, *Genetics* 60, 147-156.

Feldman, M.W., and Christiansen, F.B. (1975), The effect of population subdivision on two loci without selection, *Genet. Res. (Camb.)* 24, 151-162.

Feldman, M.W., Christiansen, F.B., and Liberman, U. (1983), On some models of fertility selection, *Genetics* **105**, 1003-1010.

Feldman, M.W., Franklin, I.R., and Thomson, G.J. (1974), Selection in complex genetic systems I. The symmestric equilibria of the three-locus symmetric viability model, *Genetics* **76**, 135-162.

Frydenberg, O. (1956), On the observation of heterosis in natural populations, in *"Tercerira Semana de Genetica no Brasil"*, pp. 25-26, Escola Superior de Agricultura "Luiz de Queiroz", Piracicaba, Brazil.

Frydenberg, O., Gyldenholm, A.O., Hjorth, J.P., and Simonsen, V. (1973), Genetics of *Zoarces* populations III. Geographic variations in the esterase polymorphism ESTIII, *Hereditas* **73**, 233-238.

Geiringer, H. (1944), On the probability theory of linkage in Mendelian heredity, *Ann. Math. Statist.* **15**, 25-57.

Hill, W.G. (1974), Estimation of linkage disequilibrium in randomly mating populations, *Heredity* **33**, 229-239.

Hjorth, J.P. (1971), Genetics of *Zoarces* populations I. Three loci determining the phosphoglucomutase isoenzymes in brain tissue, *Hereditas* **69**, 233-242.

Holgate, P. (1976), Direct products of genetic algebras and Markov chains, *J. Math. Biol.* **6**, 289-295.

Karlin, S. (1977), Selection with many loci and possible relations to quantitative genetics, in *Proceedings of the International Conference on Quantitative Genetics* (E. Pollak, O. Kempthorne and T.B. Bailey, eds.), pp. 207-226. Iowa State University Press, Ames, Iowa.

Karlin, S., and Avni, H. (1981), Analysis of central equilibria in multilocus systems: A generalized symmetric viability regime, *Theoret. Population Biol.* **20**, 241-280.

Karlin, S., and Feldman, M.W. (1970), Linkage and selection. Two-locus symmetric viability model, *Theoret. Population Biol.* **1**, 39-71.

Karlin, S., and Liberman, U. (1978), Classification and comparisons of multilocus recombination distributions, *Proc. Nat. Acad. Sci. USA* **75**, 6332-6336.

Karlin, S., and Liberman, U. (1979a), Central equilibria in multilocus systems I. Generalized nonepistatic selection regimes, *Genetics* **91**, 777-798.

Karlin, S., and Liberman, U. (1979b), Central equilibria in multilocus systems II. Bisexual generalized nonepistatic selection models, *Genetics* **91**, 799-816.

Karlin, S., and Liberman, U. (1979c), A natural class of multilocus recombinations processes and related measures of crossover interference, *Adv. in Appl. Probab.* **11**, 479-501.

Kimura, M. (1953), "Stepping stone" model of population, *Ann. Rept. Nat. Inst. Japan* **3**, 62-63.

Lewontin, R.C., and Cockerham, C.C. (1959), The goodness-of-fit test for detecting natural selection in random mating populations, *Evolution* **13**, 561-564.

Nadeau, J.H., Dietz, K., and Tamarin, R.H. (1981), Gametic selection and the selection component analysis, *Genet. Res. (Camb.)* **37**, 275-284.

Nei, M. (1977), F-statistics and the analysis of gene diversity in subdivided populations, *Ann. Hum.Genet.* **41**, 225-233.

Nei, M., and Li, W. (1973), Linkage disequlibrium in subdivided populations, *Genetics* **75**, 213-219.

O'Donald, P. (1980), *Genetic Models of Sexual Selection*, Cambridge University Press, Cambridge.

Østergaard, H., and Christiansen, F.B. (1981), Selection component analysis of natural polymorphisms using population samples including mother-offspring combinations, II, *Theoret. Population Biol.* **19**, 378-419.

Prout, T. (1965), The estimation of fitnesses from genotypic frequencies, *Evolution* **19**, 546-551.

Prout, T. (1973), *Appendix to:* Population genetics of marine pelecypods. III. Epistasis between functionally related isoenzymes in *Mytilus* edulis (by J.B. Mitten and R.C. Koehn), *Genetics* **73**, 487-496.

Robbins, R.B. (1918), Some applications of mathematics to breeding problems, III, *Genetics* **3**, 375-389.

Schmidt, J. (1917), Racial investigations I. *Zoarces viviparus* L. and the local races of the same, *C. R. Trav. Lab. Carlsberg* **13**(3), 277-397.

Schnell, F.W. (1961), Some general formulations of linkage effects in inbreeding, *Genetics* **46**, 947-957.

Sick, K. (1965), Haemoglobin polymorphism in cod in the Baltic and the Danish Sea, *Hereditas* **54**, 19-48.

Siegismund, H.R. (1985), Genetic studies of *Gammarus* IV. Selection component analysis of the *Gpi* and the *Mpi* loci in *Gammarus* oceanicus, *Hereditas* **102**, 241-250.

Siegismund, H.R., and Christiansen, F.B. (1985), Selection component analysis of natural polymorphisms using population samples including mother-offspring combinations, III, *Theoret. Population Biol.* **27**, 268-297.

Simonsen, V., and Christiansen, F.B. (1981), Genetics of *Zoarces* populations XI. Inheritance of electrophoretic variants of the enzyme adenosine deaminase, *Hereditas* **95**, 289-294.

Simonsen, V., and Christiansen, F.B. (1984), Genetics of *Zoarces* populations XIII. Three loci determing isoenzymes of glutamic-oxaloacsetic transaminase, *Hereditas* **101**, 129-136.

Simonsen, V., and Frydenberg, O. (1972), Genetics of *Zoarces* populations II. Three loci determining esterase isozymes in eye and brain tissue, *Hereditas* **70**, 235-242.

Slatkin, M. (1972), On treating the chromosome as the unit of selection, *Genetics* **72**, 157-168.

Wahlund, S. (1928), Zusammensetzung von Populationen und Korrelationserscheinungen vom Standpunkt der Vererbungslehre aus betrachtet, *Hereditas* **11**, 65-106.

Wallace, B. (1958), The comparison of observed and calculated zygotic distributions, *Evolution* **12**, 113-115.

Watt, W.B., Carter, P.A. and Blower, S.M. (1985), Adaptation at specific loci. IV. Differential mating success among glycolytic allozyme genotypes of *Colias* butterflies. *Genetics* **109**, 157-175.

Weir, B.S. (1979), Inferences about linkage disequilibrium, *Biometrics* **35**, 235-254.

Wright, S. (1931), Evolution in Mendelian populations, *Genetics* **16**, 97-159.

Wright, S. (1943), Isolation by distance, *Genetics* **16**, 114-138.

Wright, S. (1969), *Evolution and the Genetics of Populations.* Vol. 2 *The Theory of Gene Frequencies,* Univ. Chicago Press, Chicago, London.

POPULATION GENETICS THEORY – THE PAST AND THE FUTURE

W.J. Ewens
Department of Mathematics
Monash University
Clayton, Victoria 3168
Australia
and
Department of Biology
University of Pennsylvania
Philadelphia, PA 19104
U.S.A.

ABSTRACT. Classical population genetics theory was largely directed towards processes relating to the future. Present theory, by contrast, focuses on the past, and in particular is motivated by the desire to make inferences about the evolutionary processes which have led to the presently observed patterns and nature of genetic variation. There are many connections between the classical prospective theory and the new retrospective theory. However, the retrospective theory introduces ideas not appearing in the classical theory, particularly those concerning the ancestry of the genes in a sample or in the entire population. It also introduces two important new distributions into the scientific literature, namely the Poisson-Dirichlet and the GEM: these are important not only in population genetics, but also in a very wide range in science and mathematics. Some of these are discussed. Population genetics theory has been greatly enriched by the introduction of many new concepts relating to the past evolution of biological populations.

1. Introduction

These notes are based on lectures given to an audience consisting of biologists, statisticians and mathematicians, and they reflect the breadth of interests of the participants. I have preferred to seek out connections and analogies between these disciplines rather than to pursue any topic in depth, to show how questions of interest to geneticists have led to mathematical developments in areas quite different from biology, and how in turn various mathematical developments lead to a more complete understanding of the evolutionary process.

The title does not imply an ambitious attempt to give an overview of population genetics theory. Rather, it can be interpreted in two different and more specific ways. First, it is intended to suggest the view that even the most recent research has its origins in, and often borrows results from, the classical theory. Secondly, it reflects the view (the closing theme in Ewens (1979)) that the direction of interest in population genetics theory is changing from the prospective to the retrospective. The classical theory, aiming to prove the validity of Darwinian evolution as a Mendelian process, was prospective and

S. Lessard (ed.), Mathematical and Statistical Developments of Evolutionary Theory, 177–227.

looked forward in time. But insofar as a proof is possible, the neo-Darwinian theory is "proven". Our present research should now look backwards in time and focus on retrospective questions, using the great volume of data now becoming available on the genetic make-up of populations, to assess the course which evolution happens to have followed, and to find the reasons for this course (even if, as claimed by the non-Darwinian theory, much of the course of genetic evolution is directionless and without any selective basis). This change of focus towards the past can borrow much, so far as the mathematical theory is concerned, from the theory which looked to the future. In 1979 I suggested that the concept of the time reversibility of a stochastic process, increasingly exploited in the 1970's, would be the main vehicle for this borrowing. Since that time, theory directed specifically for looking towards the past has been developed, the most important being that associated with the Kingman coalescent process. These notes have as a core various aspects of the relation between going forwards to the future, and backwards to the past, in stochastic processes describing genetic evolution.

2. Classical discrete evolutionary models and their diffusion approximations

One of the great scientific achievements of the first half of this century was the establishment of the neo-Darwinian synthesis, that is the rewriting of the Darwinian evolutionary theory using the Mendelian mechanism to describe the transmission of hereditary material from parent to offspring. Evolutionary processes in this new paradigm do not follow a strict deterministic pattern: the fact that one of two genes at a locus is randomly chosen to be passed on from parent to offspring, together with the fact that two genetically identical individuals will not necessarily have the same number of offspring, imply a stochastic element to the neo-Darwinian procedure. In order to gain insight into the properties of this stochastic process, Sewall Wright (1931) and R.A. Fisher (1930) independently introduced and analysed what we recognize as a Markov chain model of the changes of the frequencies of alleles A_1 and A_2 at some gene locus A. Their analysis assumes a diploid population of size N ($2N$ genes at this locus in the population), but it is in all essential respects identical to the analysis of a haploid population of size $M = 2N$, and thus for simplicity we assume throughout a haploid population (in which we can identify a gene with an individual) of size M.

The model assumes that the probability p_{ij} that there will be j A_1 genes in generation $t + 1$, given i A_1 genes in generation t, is

$$(1) \qquad p_{ij} = \binom{M}{j}(i/M)^j(1-i/M)^{M-j}.$$

This model allows no mutation or selection, and shows that the genes in generation $t + 1$ can be thought of as being derived by random sampling, with replacement, from the genes of generation t.

Clearly (1) is an idealization; no doubt there is no real population evolving exactly as prescribed by this formula. A most important class of models extending (1), the generalized exchangeable models, was introduced by Cannings (1974). Suppose the genes in any generation are labelled $1,2,...,M$, and that these have $v_1, v_2,...,v_M$ offspring genes, where clearly $\sum v_j = M$. In these models the sole further requirement on the v_i is that the joint probability distribution of any subset $v_a, v_b,...,v_c$ be independent of the labels $a,b,...,c$. This is clearly the case for the model (1), but there are many further models beyond (1) having this property. Clearly $E\{v_i\} = 1$, and we write

$$(2) \qquad\qquad \mathrm{Var}\{v_i\} = \sigma_v^2 ,$$

where the right-hand side is necessarily independent of i .

A further model, introduced by Moran (1958), abandons the non-overlapping generations assumption implicit in (1), and allows individuals to die and reproduce according to a birth and death process. At unit time points an individual is chosen at random to die and at the same time an individual (possibly the dying one) is chosen at random to reproduce. Here the number of A_1 genes follows a Markov chain process with

$$(3) \qquad p_{ij} = \begin{cases} i\,(M-i)/M^2 & \text{for } j = i-1, i+1 , \\ 1 - 2i(M-i)/M^2 & \text{for } j = i , \\ 0 & \text{otherwise.} \end{cases}$$

Indeed many further models, plausibly describing evolutionary genetic behavior, could be introduced. Despite this, the most frequently analysed model, perhaps for historical reasons, is (1). Unfortunately, this model does not yield to a detailed analysis, and essentially the only known exact results arising from (1) are that the probability that the allele A_1 eventually fixes in the population is its current frequency, and that the eigenvalues of the matrix $P = \{p_{ij}\}$ are

$$(4) \qquad \lambda_0 = \lambda_1 = 1 , \lambda_j = \{1-M^{-1}\}\{1-2M^{-1}\}...\{1-(j-1)M^{-1}\} , j = 2,3,...,M .$$

Thus the leading non-unit eigenvalue, describing in a sense the rate of approach to homozygosity, is

$$(5) \qquad\qquad \lambda_2 = 1 - M^{-1} .$$

Cannings (1974) has shown that in the generalized exchangeable models, (4) is replaced by

$$\lambda_0 = \lambda_1 = 1 , \lambda_j = E\{v_1 v_2 ... v_j\} , j = 2,3,...,M ,$$

and this leads to a leading non-unit eigenvalue of

$$(6) \qquad \lambda_2 = 1 - \sigma_v^2 (M-1)^{-1} .$$

The eigenvalues of the transition matrix of the Moran model (3) are

$$(7) \qquad \lambda_0 = \lambda_1 = 1 , \lambda_j = 1 - j(j-1)M^{-2} , j = 2,3,...,M ,$$

with a leading non-unit eigenvalue

$$(8) \qquad 1 - 2M^{-2} .$$

Because no exact information beyond fixation probabilities and eigenvalues appeared available for the model (1), both Fisher and Wright considered the diffusion approximation for this model. Here the population size M is allowed to increase without limit, time is rescaled so that unit time is M generations, and attention focusses on the fraction, x, of A_1 genes, rather than absolute number. In the limit $M \to \infty$, x evolves as a continuous space continuous time random variable whose density function $f(x;t)$ at times t satisfies the forward Kolmogorov equation

$$(9) \qquad \frac{\partial}{\partial t} f(x;t) = -\frac{\partial}{\partial x} \{m(x)f(x;t)\} + \frac{1}{2} \frac{\partial^2}{\partial x^2} \{v(x)f(x;t)\} ,$$

where $m(x)\delta t$ and $v(x)\delta t$ are respectively the mean change in x, and the variance of the change, during the time interval $(t,t+\delta t)$, given the value x at time t. (Here and throughout we ignore terms of order $(\delta t)^2$.) For the model (1),

$$(10) \qquad m(x) = 0 , v(x) = x(1-x) ,$$

the latter deriving from the binomial variance formula. The diffusion equation (9) also describes the evolution of the frequency of A_1 in a generalized exchangeable model, provided that σ_v^2 remains finite as $M \to \infty$ and we replace $v(x)$ by

$$(11) \qquad v(x) = \sigma_v^2 x(1-x) .$$

This was proved formally by Donnelly (1986a).
Equation (9) also describes the evolution in the Moran model, provided that unit time corresponds to M^2 birth-death events and that $v(x)$ is replaced by

$$(12) \qquad v(x) = 2x(1-x) .$$

A more useful convention for the Moran model, which we follow, is to use (10) rather than (12) and measure time in units of $\frac{1}{2} M^2$ birth-death events. The complete solution of (9), with $m(x)$ and $v(x)$ defined by (10) (or more generally (11)), was achieved by

Kimura (1955). Many results are available for this process, in particular that if p is the initial frequency of A_1, the mean time for loss of either A_1 or A_2 is

(13) $$-2\{p \log p + (1-p)\log(1-p)\}$$

time units. Thus in the Wright-Fisher model (1) the mean time for loss of A_1 or A_2 is approximately $-2M\{p \log p + (1-p)\log(1-p)\}$ generations, for the Cannings models is approximately $-2M\{p \log p + (1-p)\log(1-p)\}/\sigma_v^2$ generations, and in the Moran model is approximately $-M^2\{p \log p + (1-p)\log(1-p)\}$ birth-death events.

Since the transition matrix of the Moran model is a continuant, many exact results are known for this model, in particular the exact mean time for loss of A_1 or A_2. Writing $k = Mp$ for the initial number of A_1 genes, this is

(14) $$Mk \sum_{j=k+1}^{M-1} j^{-1} + M(M-k) \sum_{j=1}^{k} (M-j)^{-1} .$$

This is clearly very close to the diffusion approximation in the previous paragraph.

We now consider the effect of mutation. Suppose that A_1 mutates to A_2 at rate u, with no reverse mutation. Then A_1 must eventually be lost from the population, and while an exact expression can be found for the density function of the time taken until A_1 is lost (Crow and Kimura (1970), 8.5.19), this is not particularly informative and interest centers on the mean time for loss to occur. Again, this is not known exactly for the Wright-Fisher model, but for the diffusion approximation, where we must let $u \to 0$ as $M \to \infty$ with

(15) $$\theta = 2Mu$$

held fixed, we find $m(x) = -\frac{1}{2}\theta x$, and that the mean time for loss of A_1 is $T(p)$, given by

(16) $$T(p) = \int_0^1 t(x)dx ,$$

where

(17a)
(17b)
$$t(x) = \begin{cases} 2x^{-1}(1-\theta)^{-1}\{(1-x)^{\theta-1}-1\} , & 0 \le x \le p , \\ 2x^{-1}(1-\theta)^{-1}(1-x)^{\theta-1}\{1-(1-p)^{1-\theta}\} , & p \le x \le 1 , \end{cases}$$

p being the initial value of x. An obvious limiting procedure gives the form of $t(x)$ when $\theta = 1$. The expression (16) is a particular case of a more general result, namely that the mean time that the frequency of A_1 takes values in any arbitrary interval (x_1, x_2), where $0 \le x_1 < x_2 \le 1$, before being absorbed at 0, is

(18)
$$\int_{x_1}^{x_2} t(x)dx \; ,$$

with $t(x)$ defined by (17).

We approximate the mean number of generations until loss of A_1 in the Wright-Fisher model by calculating (16) and multiplying by M . More generally, the mean number of generations that the frequency of A_1 takes a value in (x_1,x_2) before A_1 is lost is approximated by calculating (18) and multiplying by M . For the Cannings models, we replace (15) by

(19)
$$\theta = 2Mu/\sigma_v^2$$

and then use (16) or (18), the mean times (in generations) being found by multiplying by M/σ_v^2 .

For the Moran model, two approaches are possible. The first uses the diffusion approximation: here we put $\theta = Mu$ and then use (16), multiplying by $\frac{1}{2} M^2$ to convert to birth-death events. The second approach is exact. With one-way mutation, the transition probabilities (3) must be replaced by

(20)
$$p_{ij} = \begin{cases} \{i(M-i)+ui^2\}/M^2 & \text{for } j = i-1 \\ \{i(M-i)(1-u)\}/M^2 & \text{for } j = i+1 \\ 1 - \{2i(M-i)+ui^2-ui(M-i)\}/M^2 & \text{for } j = i \\ 0 & \text{otherwise} \end{cases}$$

Again, the transition matrix $\{p_{ij}\}$ is a continuant, so standard results give an exact expression for the mean number of birth-events until loss of A_1 , given initially k A_1 genes. This is best calculated in the form

(21)
$$T(k) = \sum_{j=1}^{M} t_{kj} \; ,$$

where t_{kj} is the mean number of times, before A_1 is lost, that there are exactly j A_1 genes. The expression (21) is analogous to (16) when the interpretation of $t(x)$ implicit in (18) is taken into account. While an explicit expression for t_{kj} is possible, we will calculate this only in two cases of particular interest, which we now discuss.

The first case is when initially all genes in the population are A_1 . In the diffusion process, the mean time until loss of A_1 is found from (16) and (17) by putting $p = 1$, namely

$$(22) \qquad T(1) = 2(1-\theta)^{-1} \int_0^1 x^{-1}\{(1-x)^{\theta-1}-1\}dx$$

$$(23) \qquad = 2 \sum_{j=1}^{\infty} \{j(j+\theta-1)\}^{-1}.$$

The mean number of generations for loss of A_1 is found, for the Wright-Fisher process and the Cannings models, by multiplying (22) (or (23)) by M and M/σ_v^2 respectively, while the mean number of birth-death events for the Moran model is found by multiplying (22) by $\frac{1}{2} M^2$.

In the Moran model, however, an exact expression is possible by exact calculation of (21) for the case $k = M$. We find (Watterson, 1976a)

$$(24) \qquad t_{Mj} = M(M+\theta)(1-\theta)^{-1}j^{-1}\left[\binom{M}{j}\binom{M+\theta-1}{j}^{-1}-1\right]$$

exactly, where we must now make the definition

$$(25) \qquad \theta = Mu/(1-u).$$

In the particular case $\theta = 1$, (24) must be replaced by

$$(26) \qquad t_{Mj} = M(M+1)j^{-1}\{M^{-1}+(M-1)^{-1}+...+(M-j+1)^{-1}\}.$$

As a check on the accuracy of (22), we find that for $\theta = 2$,

$$(27) \qquad t_{Mj} = M(M+2)/(M+1),$$

$$(28) \qquad t(M) = M^2(M+2)/(M+1),$$

while (22) gives

$$(29) \qquad t(x) = 2, \quad T(1) = 2.$$

Conversion to birth-death events by multiplying by $\frac{1}{2} M^2$ gives $T(1) = M^2$, in very close agreement with the exact result in (28). More generally, by using standard asymptotic formulae for the gamma function, we find

$$(30) \qquad t_{Mj} \sim M^2(1-\theta)^{-1}j^{-1}\{(1-j/M)^{\theta-1}-1\},$$

and in this form the parallel with (22) is immediate.

The second case of particular interest arises when there is initially only one A_1 gene. Thus $p = M^{-1}$ and if we ignore (17a) as leading to a negligible contribution, (16) and (17b) give

$$(31) \qquad T(M^{-1}) \approx 2 \int_{M^{-1}}^{1} x^{-1}(1-x)^{\theta-1} dx$$

generations for the Wright-Fisher process. For the Cannings models we divide $T(M^{-1})$, given by (31), by σ_v^2. For the Moran model, the diffusion approximation is

$$(32) \qquad T(M^{-1}) \approx M \int_{M^{-1}}^{1} x^{-1}(1-x)^{\theta-1} dx$$

birth-death events. Again, for this model, an exact expression is possible for the mean number of birth-death events until loss of A_1, namely

$$(33) \qquad \sum_{j=1}^{M} (M+\theta)j^{-1} \binom{M}{j} \binom{M+\theta-1}{j}^{-1},$$

θ being defined by (25). As a check on the accuracy of (32), when $\theta = 1$ (32) gives

$$(34) \qquad T(M^{-1}) \approx M \log M$$

birth-death events, which is close to the exact expression

$$(M+1)\{1^{-1}+2^{-1}+3^{-1}+\ldots+M^{-1}\}$$

found from (33). Asymptotic expressions for the gamma functions in (33) confirm the accuracy of (32) for all θ.

We turn finally to the case of two-way mutation, and suppose that A_1 mutates to A_2, and A_2 to A_1, at equal rates u. Here there exists a stationary distribution, in all models, for the frequency of A_1. The exact form of this distribution is unknown in the Wright-Fisher model, and while it is known, exactly, in the Moran model, we shall be content here with the diffusion approximation. Denoting the frequency of A_1 by x, it is found that the diffusion process stationary distribution is

$$(35) \qquad f(x) = \frac{\Gamma(2\theta)}{\{\Gamma(\theta)\}^2} x^{\theta-1}(1-x)^{\theta-1},$$

and this approximates the stationary distributions in the Wright-Fisher model, the Cannings model and the Moran model if we define θ by (15), (19) and (25) respectively for these models.

3.   The Dirichlet distribution

In all models considered in the previous section, two alleles only (A_1 and A_2) were allowed at the locus A . More general models allow some fixed finite number K of possible alleles, $A_1,...,A_K$. Our interest focusses on the symmetric mutation case where, for any i,j (i≠j) , an A_i gene mutates to A_j with probability u/(K-1) . Here there will exist a stationary distribution for the frequencies $x_1,...,x_{K-1}$ of $A_1,...,A_{K-1}$, generalizing (34). Again we focus on the diffusion process stationary distribution, which is

$$(36) \qquad f(x_1,...,x_{K-1}) = \frac{\Gamma(K\varepsilon)}{\{\Gamma(\varepsilon)\}^K} \{x_1...x_K\}^{\varepsilon-1} ,$$

where $x_K = 1 - x_1 - ... - x_{K-1}$ and $\varepsilon = \theta/(K-1)$. This is of course the well-known Dirichlet distribution, often used in statistics (and fortunately, as we note below) as a reasonable and flexible approximate joint distribution for frequencies $x_1,...,x_{K-1}$, but here arising quite naturally in the genetic context. In the application of (36) to finite models, we define θ , and hence ε , according to (15), (19) or (25), as appropriate.

Suppose now we consider the order statistics

$$x_{(1)} = \max\{x_1,...,x_K\}$$

$$(37) \qquad ....$$

$$x_{(K)} = \min\{x_1,...,x_K\} .$$

Clearly

$$(38) \qquad f(x_{(1)},...,x_{(K-1)}) = \frac{K!\Gamma(K\varepsilon)}{[\Gamma(\varepsilon)]^K} \{x_{(1)}...x_{(K)}\}^{\varepsilon-1} ,$$

and from this we can find the joint density function of $x_{(1)},...,x_{(r)}$, for any r (r = 1,2,...,K-2) by integration. We find (Watterson (1976b))

$$(39) \qquad f(x_{(1)},...,x_{(r)}) = \frac{K!\Gamma(K\varepsilon)}{(K-r)![\Gamma(\varepsilon)]^K} \{x_{(1)}...x_{(r)}\}^{\varepsilon-1} x_{(r)}^{(K-r)\varepsilon-1} I ,$$

where I is a complicated integral whose precise form we do not reproduce here.

Another representation of the random variables $x_1,...,x_{K-1}$ occurring in (36) is well known, namely

$$x_1 = z_1$$

$$x_2 = z_2 (1-z_1)$$

(40)

$$\cdots\cdots\cdots\cdots\cdots$$

$$x_{K-1} = z_{K-1}(1-z_1)(1-z_2)\ldots(1-z_{K-2}) ,$$

where $z_1, z_2, \ldots, z_{K-1}$ are independent random variables with

(41)
$$f(z_i) = \frac{\Gamma[(K-i+1)\varepsilon]}{\Gamma(\varepsilon)\Gamma[(K-i)\varepsilon]} \; z_i^{\varepsilon-1}(1-z_i)^{(K-i)\varepsilon-1} .$$

Both this representation, and the order statistics, will later turn out to be of the greatest importance to us.

4.   The infinitely many alleles model

The models described in the previous sections, which may be described as "classical" models, allow a gene to be one of only a finite collection $A_1, \ldots, A_K$ of allelic types ($K = 2$ in Section 2). The recognition of the molecular nature of the gene as a DNA sequence allowing an effective infinity of different allelic types (i.e. DNA sequences) led to a model of a new type, namely the infinitely many alleles model (Malécot (1948), Wright (1949), Kimura and Crow (1964)), in which alleles from the infinite collection $A_1, A_2, \ldots,$ may occur at the locus A . It is assumed in this model that each mutant gene is of an allelic type never previously seen in the population. If we assume a mutation rate u (to new allelic types), and that all allelic types are selectively equivalent, a natural generalization of (1) is a model under which, if alleles $A_1, A_2, \ldots$ occur in generation t with respective numbers $n_1, n_2, \ldots$, then the probability that in generation $t + 1$ they will occur with numbers $m_1, m_2, \ldots$, and that there will be m_0 new mutant genes (all necessarily of different and novel allelic types), is

(42)
$$M![\textstyle\prod m_i!]^{-1} u^{m_0} \prod_{i>0} \{n_i(1-u)/M\}^{m_i} .$$

Clearly, in this model, any specific allele eventually becomes lost from the population, never to reappear, and there is no stationary distribution for the frequency of any allelic type.

Often, however, we are interested not in particular alleles but in allelic patterns. Specifically, in any generation the alleles present will describe some (random) partition

$$\beta = \{\beta_1, \beta_2, \ldots, \beta_M\} ,$$

where β_i is the (random) number of allelic types which are each represented by exactly i genes. Clearly $\sum i\beta_i = M$, while $K = \sum \beta_i$ is the (random) number of allelic types present in that generation.

We focus attention on the Markov chain process, admitting a stationary distribution for the vector $\boldsymbol{\beta}$. As with the model (1), very little exact information is known about this process other than its eigenvalues. There are $P(M)$ states for $\boldsymbol{\beta}$, where $P(M)$ is the number of partitions of the number M, and a (very complex) set of transition probabilities from one state to the next, implied by (42). The eigenvalues of the transition matrix of $\boldsymbol{\beta}$ are (Ewens and Kirby (1975))

$$(43) \qquad \lambda_1 = 1 \, , \lambda_j = (1-u)^j \{1-M^{-1}\}\{1-2M^{-1}\}...\{1-(j-1)M^{-1}\} \, ,$$

with λ_j occurring with multiplicity $P(j) - P(j-1)$, $j = 2,3,...,M$.

Very little is known exactly about the stationary distribution of $\boldsymbol{\beta}$. However, we are able to use the classical results outlined above to find diffusion approximations to several quantities of considerable interest by using the following intuitive arguments (which can be made more rigorous by using ergodicity results). First, since on average Mu new allelic types enter the population each generation, and since the mean number of generations that any allele remains in the population, once it enters it, is given by (31), the (stationary) mean of K for the model (42) is approximated by

$$(44) \qquad E(K) = \int_{M^{-1}}^{1} g(x)dx \, ,$$

where

$$(45) \qquad g(x) = \theta x^{-1}(1-x)^{\theta-1} \, ,$$

θ being defined by (15). More generally, the stationary mean number of alleles present in the population having a frequency in the interval (x_1,x_2), is, from (18), approximately

$$(46) \qquad E\{K(x_1,x_2)\} = \int_{x_1}^{x_2} g(x)dx \, .$$

The function $g(x)$ clearly gives information about the likely allelic frequencies to be observed in the population, and for this reason has become known as the "frequency spectrum". It will be a central feature of our subsequent investigations, and may be used immediately to give further information. Suppose, for example, that a gene is taken at random from the population. The probability that the frequency of the allelic type of the gene drawn is in $(x,x+\delta x)$ is clearly $xg(x)\delta x$. In other words, the probability density of the frequency of the allelic type of a randomly drawn gene is

$$(47) \qquad p(x) = \theta(1-x)^{\theta-1} \, , 0 < x < 1 \, .$$

The probability that two genes drawn at random from the population are of the same allelic type is

(48) $$\int_0^1 x^2 g(x)dx = (1+\theta)^{-1},$$

agreeing with a standard result for this model (Malécot (1948), Kimura and Crow (1964)).

As with the infinitely many alleles Wright-Fisher model, the infinitely many alleles Cannings models do not, apart from eigenvalues, allow many exact results. Nevertheless, all the diffusion formulae given above may be used as approximations, with θ defined by (19). The same is true for the infinitely many alleles Moran model, with θ now defined by (25).

However, as might be expected, the Moran model allows an exact formula for the probability distribution of β to be calculated. Consider the possible transitions of β to some new vector β' after one birth-death event. Thus for $j > 1$, $i \neq j$, $j - 1$, $j - 2$, and given β,

(49)

$$\mathrm{Prob}\{\beta' = (\beta_1,\ldots,\beta_i-1,\beta_{i+1}+1,\ldots,\beta_{j-1}+1,\beta_j-1,\ldots,\beta_M)\}$$

$$= (1-u)ij\beta_i\beta_j M^{-2}.$$

This arises because the transition from β to β', as defined in (49), requires the death of one of the $j\beta_j$ genes of one or other allelic type represented by j genes and a birth to one of the $i\beta_i$ genes of one or other of the allelic types represented by i genes, the probability of which is as given in (49). Trajstman (1974) and Watterson (1976a) list the probabilities of ten essentially different transitions, of which (49) is one, of β to some vector β'. From a detailed check of the stationary distribution requirements, they are able to find the exact stationary distribution of β : this is given in (69) below. From this, many specific results follow. We can already, however, use the classical results of the previous sections to write down some of these. For example, since on average u new alleles enter the population at each birth-death event, we can use an analogue of the argument which led to (44) to state, using (33), that if K is the random number of allelic types present at any time,

(50) $$E(K) = \theta \sum_{j=1}^M j^{-1} \binom{M}{j} \binom{M+\theta-1}{j}^{-1},$$

which may be written more conveniently

(51) $$E(K) = \frac{\theta}{\theta} + \frac{\theta}{1+\theta} + \frac{\theta}{2+\theta} + \ldots + \frac{\theta}{M-1+\theta}.$$

The (discrete) analogue of the frequency spectrum $g(x)$ is

$$(52) \qquad g(j) \;=\; \theta j^{-1} \binom{M}{j} \binom{M+\theta-1}{j}^{-1} \,,$$

while the probability that a gene drawn at random is of an allelic type having exactly j genes representing it is

$$(53) \qquad (j/M)g(j) \;=\; \theta M^{-1} \binom{M}{j} \binom{M+\theta-1}{j}^{-1} \,.$$

We now consider practical applications of these results. The development of the rudiments of the (Wright-Fisher) infinitely many alleles model, in particular the derivation of (44)-(48), coincided, in the 1960's, with two other events. The first was the arrival of substantial data on allele numbers and frequencies in natural populations, and the second was the resulting hypothesis (Kimura (1968)) that much of the genetic variation observed in these populations was not determined by selection but arose by purely random processes acting on selectively neutral, or equivalent, alleles.

It became natural to ask whether the infinitely many alleles model, which seemed appropriate for the data then existing, or soon to appear, could be used to test the neutral theory. Since samples of genes from a natural population, rather than the entire collection of genes in the population, would provide the data for any test, and since the null hypothesis (i.e. neutral theory) distribution of the number and pattern of genes in this sample would be necessary as the basis for such a test, the need arose to develop theory concerning a sample of selectively neutral alleles. This theory is taken up in the next section.

5.   The Ewens sampling formula

We sketch the development of the sampling theory in a way which will be useful for our discussion below (and which also, by coincidence, follows closely the original derivation (Ewens (1972), Karlin and McGregor (1972)). Imagine that a sample of n genes is taken at random at stationary from a population evolving according to (42). Imagine these genes arrayed in a line, and that we examine the genes one by one, starting at one end of the line. (We describe this as sampling "through space".)

Using the diffusion theory approximation results developed in the classical theory, a generalization of the argument which led to (48) shows that the stationary probability that the first i genes examined are all of the same allelic type is

$$(54) \qquad \int_0^1 x^i g(x)dx \;=\; (i-1)!\,\theta/S_i(\theta) \,,$$

where $S_i(\theta) = \theta(1+\theta)(2+\theta)...(i-1+\theta)$. From this, the probability that gene i is of a novel allelic type, given that the previous $i-1$ are of the same allelic type, is

(55) $\theta/(i-1+\theta)$.

Further, the mean number of allelic types observed at the times when i genes have been examined is

(56) $\int_0^1 \{1-(1-x)^i\}g(x)dx = \frac{\theta}{\theta} + \frac{\theta}{1+\theta} + \frac{\theta}{2+\theta} + \dots + \frac{\theta}{i-1+\theta}$.

From (56), the unconditional probability that the ith gene sampled is of a novel allelic type is also given by (55). This raises the conjecture that (55) provides the probability that the ith gene sampled is of a novel type, irrespective of the number of alleles seen, and their frequencies, in the first i - 1 genes. Assuming this conjecture to be true, it is easy to show that if we describe the allelic composition of the sample by b , where

(57) $b = \{b_1, b_2, \dots, b_n\}$

and b_j is the number of alleles in the sample each having j genes (so that $\sum jb_j = n$) , then

(58) $Prob(b) = \frac{n!}{S_n(\theta)} \prod_j \frac{\theta^{b_j}}{j^{b_j} b_j!}$.

The truth of (58) was established by Karlin and McGregor (1972) by induction, following its conjecture by Ewens (1972).

We may regard (58), known as the Ewens sampling formula, in two ways. The first is as an approximation to the (very complicated, and completely unknown) distribution of b arising for the infinitely many alleles Wright-Fisher model (42), and for the analogous Cannings model. Secondly, we will see that (58) is an *exact* formula in three different models. First, Ethier and Kurtz (1981) have introduced an infinitely many alleles generalization of the diffusion process (9), describing the evolution of allele frequencies in a hypothetically infinite population. This process is complicated by the fact that at any time there is an infinite number of alleles present in the population, with no natural labelling of alleles. We consider this process later, but note now that its stationary distribution is known, and if a (multinomial) sample of size n is taken form this stationary distribution, the distribution of the allelic partition of the sample is given exactly by (58). Second, Donnelly and Tavaré (1987) have introduced a genealogical model for an infinitely large population for which (58) is again an exact sampling formula. (In fact the Ethier and Kurtz formulation, and that of Donnelly and Tavaré, are two different ways of describing the same process, a fact we return to later.) Finally, (58) is an exact formula for the infinitely many alleles version of the Moran model (20), as we note later, with θ defined by (25).

There are several consequences of (58). First, the marginal distribution of $k = \sum b_j$ is

(59)
$$\text{Prob(k)} = |S_n^k| \theta^k / S_n(\theta) ,$$

where S_n^k is a Stirling number of the first kind. The conditional distribution of b, given k, is then

(60)
$$\text{Prob}\{b|k\} = n! / [|S_n^k| \prod_j j^{b_j} b_j!] ,$$

showing that k is a sufficient statistic for θ. The consequences of this for the estimation of θ have been explored by Guess and Ewens (1972). However, the most important consequence of the sufficiency of k for θ is that (60), being free of all unknown parameters, can be used to provide objective tests of neutrality. We consider these tests in a later section.

Second, (59) shows that the only functions of θ which admit unbiased estimation are linear combinations of the form

(61)
$$\{(a+\theta)(b+\theta)...(c+\theta)\}^{-1} ,$$

where $a,b,...,c$ are integers such that $1 \leq a < b < ... < c \leq n-1$. It is of some interest to note why this is so. It one whose value, when $k = 2$, is may be shown from (59) (see Ewens (1972)) that an unbiased estimator of $(1+\theta)^{-1}$ is

(62)
$$\{\frac{1}{2} + \frac{1}{3} + ... + \frac{1}{n-1}\} / \{\frac{1}{1} + \frac{1}{2} + \frac{1}{3} + ... + \frac{1}{n-1}\} .$$

This may be regarded as an empirical estimator, in the following sense. Equation (48) shows that in the Ethier-Kurtz process, $(1+\theta)^{-1}$ is the probability that two genes, taken at random, are of the same allelic type. From (60) the probability, given $k = 2$, that of the two alleles sampled, n_1 are of one allelic type and $n - n_1$ of the other, is

(63)
$$n! / [|S_n^2| n_1(n-n_1)] .$$

Given n_1, the probability that two genes sampled are of the same allelic type is

$$\left\{ \binom{n_1}{2} + \binom{n-n_1}{2} \right\} \binom{n}{2}^{-1} .$$

Multiplying this expression by (63) and summing over all n_1 gives (62).

Similarly, the probability that the final gene drawn in the sample is of a novel allelic type is, from (55),

(64)
$$\theta/(n-1+\theta) .$$

This may be written as $1 - (n-1)/(n-1+\theta)$ and hence admits unbiased estimation. The unbiased estimator, when $k = 2$, is

$$(65) \qquad [(n-1)\{\frac{1}{1} + \frac{1}{2} + \frac{1}{3} + \ldots + \frac{1}{n-1}\}]^{-1} ,$$

and for general k is

$$(66) \qquad \text{coeff } \theta^{k-2} \text{ in } (\theta+1)(\theta+2)\ldots(\theta+n-2)/|S_n^k| .$$

The expression (65) may also be regarded as an empirical estimator in the following sense. If $k = 2$, the last gene drawn is different from all preceding genes only if, first, the allelic partition is $\beta_1 = 1$, $\beta_{n-1} = 1$, and, second, the singleton is drawn last. The probability of this is, from (60), exactly (65). The more general expression (66) can be arrived at in a similar way.

These observations suggest why there is no unbiased estimator of

$$(67) \qquad \theta/(n+\theta) ,$$

since (67) is the probability that a further $((n+1)^{th})$ gene sampled is of a new allelic type, and empirical estimation of this from a sample of n genes is impossible. The expression (67) will subsequently be of some interest to us and we will later regard the conclusion of the previous sentence from a different perspective.

A further remark concerning (58) is the following. The factor

$$(68) \qquad n!/\prod_j j^{b_j} b_j!$$

appearing in (58) is the number of permutations of the integers $(1,2,\ldots,n)$ which fall into $k = \sum b_j$ cycles, where b_j of the cycles are of length j. We will, later, find two reasons for the occurrence of this term in (58), the first being given by Joyce and Tavaré (1987) as a result of examining the ancestry of the genes in the sample, and the second arising non-genetically through an examination of standard properties of permutations.

We have noted that (58) is an exact result in the Ethier-Kurtz diffusion process and an approximate result in the Wright-Fisher and Cannings finite Markov chain models. We have also noted that exact results are available for the Moran infinitely many alleles model. It was shown by Trajstman (1974), following a conjecture by Watterson (1974), that (58) holds exactly at stationary for a sample of n genes from a Moran model population, with θ defined by (25). Further, an analogous conclusion holds for the entire Moran population: at stationarity, the probability distribution of the population vector β is

$$(69) \qquad \text{Prob}(\beta) = \frac{M!}{S_M(\theta)} \prod_j \frac{\theta^{\beta_j}}{j^{\beta_j} \beta_j!} ,$$

θ again being defined by (25).

All the models discussed above, which share (58) as an exact or approximate sampling formula, are infinitely many alleles models. It is therefore remarkable that (58) can be obtained by a limiting argument from the "finite number of alleles" distribution (36), by allowing K to increase indefinitely (Watterson, (1976b)). If we are given the frequencies $x_1,...,x_K$ in the diffusion distribution (36), the probability that a sample of n genes gives n_i genes of allelic type A_i ($i = 1,2,...,K$) is a standard multinomial. Averaging this multinomial over the density function (36) yields an absolute probability

$$(70) \qquad n!\Gamma(K\varepsilon) \prod_j [\{\Gamma(\varepsilon+n_j)\}/\{\Gamma(\varepsilon)\}^K \Gamma(K\varepsilon+n) \prod_j n_j!] \,.$$

Denoting by k the number of non-zero n_i values, which we arrange in decreasing order $n_{(1)} \geq n_{(2)} \geq ... \geq n_{(k)}$, we find from (70)

$$(71) \quad \text{Prob}\{k;n_{(1)},n_{(2)},...,n_{(k)}\} = \frac{n!K!\Gamma(K\varepsilon) \prod_{j=1}^{k} \{\Gamma(\varepsilon+n_{(j)})\}}{\{\Gamma(\varepsilon)\}^K \{\Gamma(K\varepsilon+n)\} \{K-k\}!b_1!...b_n! \prod_j n_{(j)}!}$$

where $b_1,...,b_n$ are as defined in (57). If we now let $K \rightarrow \infty$, $\varepsilon \rightarrow 0$ with $K\varepsilon = \theta$, we find easily that (71) converges to (58).

This is a remarkable conclusion, since (36) has no non-trivial limit as $K \rightarrow \infty$, which is precisely the limiting operation we have just used. To resolve this curious point, we turn to what initially appears to be a topic quite unconnected with genetics, namely heaps and search procedures.

6.   Heaps and the Poisson-Dirichlet distribution

Suppose K items are stored in a linear order, say in a heap. Items are requested for use one by one, and independently, the probability that item i is requested being x_i. Each item is restored to the heap before the next one requested is withdrawn. We consider properties of two strategies for the storage of items in the heap. The first is the "ordered" strategy: the heap is always maintained so that the most popular item is on top, the next most popular is immediately below it, and so on. The second is the "self-regulating" strategy: any item once used is returned to the top of the heap. Here the most popular items will tend to be at or near the top. If the order statistics of the popularities of the items are denoted $x_{(1)},x_{(2)},...,x_{(K)}$, with $x_{(i)} \geq x_{(i+1)}$, then when an item is requested, and the heap is searched from the top down, the mean search time until the item requested is found is

$$(72) \qquad \mu_1 = \sum (i-1)x_{(i)}$$

under the "ordered" strategy, and

(73)
$$\mu_2 = \sum_{i \neq j} \sum x_i x_j / (x_i + x_j)$$

under the "self-regulating" policy (Burville (1974), Kingman (1975)); for further information on this model see Hendricks (1972), Burville and Kingman (1973). These simple observations turn out to have remarkable consequences, in particular for our genetic models.

Suppose now, with Kingman (1975), that $x_1, \ldots, x_K$ are random variables drawn from the Dirichlet distribution (36). The expected values of μ_1 and μ_2, defined in (72) and (73), become

(74)
$$E(\mu_1) = (K-1)[\frac{1}{2} - \frac{\Gamma(1+2\varepsilon)}{\{\Gamma(1+\varepsilon)\}^2 2^{1+2\varepsilon}}] ,$$

(75)
$$E(\mu_2) = \theta/(1+2\varepsilon) .$$

Kingman noted a conclusion parallel to that discussed at the end of the previous section, that while there is no non-trivial limit for (36) as $K \to \infty$, both (74) and (75) do have non-trivial limits, namely

(76)
$$E(\mu_1) \to \theta \log 2 ,$$

(77)
$$E(\mu_2) \to \theta .$$

We may alternatively say that in the limit $K \to \infty$,

(78)
$$E(\sum i x_{(i)}) = 1 + \theta \log 2 ,$$

(79)
$$E(\sum i \bar{x}_i) = 1 + \theta ,$$

where by $\bar{x}_i$ we mean the stationary probability that the item in position i in the heap is the next item requested, assuming the self-regulating policy.

Note that if θ is small, (78) and (79) imply that

(80)
$$E(x_{(1)}) \approx 1 - \theta \log 2 ,$$

(81)
$$E(\bar{x}_1) \approx 1 - \theta ,$$

equations which we will return to later.

The existence of the non-trivial limits (76) and (77) as $K \rightarrow \infty$, together with the convergence of (71) to (58) under the same limiting operation, lead to the investigation of non-degenerate limits of (36) as $K \rightarrow \infty$. The key observation was made by Kingman (1975), namely that there exists a non-degenerate limit for the joint density function (38) of the order statistics $x_{(1)}, \ldots, x_{(K)}$. Let ∇ be the space of all vectors $x = (x_{(1)}, x_{(2)}, \ldots)$ for which $x_{(i)} \geq x_{(i+1)}$, $x_{(i)} \geq 0$, $\Sigma x_{(i)} = 1$. Then the limiting order statistics vector for the Dirichlet distribution is defined on ∇ and has non-degenerate probability measure on this space. This measure has been called by Kingman the Poisson-Dirichlet distribution, and turns out to be of the greatest importance in population genetics theory. (We note in passing the good fortune, for population genetics theory, that Kingman chose the joint density function (36) – more or less arbitrarily, and without reference to population genetics theory, where (36) arises naturally – in his development of the theory of heaps.)

The actual form of the Poisson-Dirichlet distribution is implied by its finite dimensional marginals, and Watterson (1976b) shows that the finite r-dimensional marginal density $f(x_{(1)}, x_{(2)}, \ldots, x_{(r)})$ is

$$(82) \qquad f(x_{(1)}, \ldots, x_{(r)}) = \theta^r [x_{(1)}, \ldots, x_{(r-1)}]^{-1} x_{(r)}^{\theta-2} h\{(1-x_{(1)}-\cdots-x_{(r)})/x_{(r)}\} ,$$

where the function $h(\cdot)$ is most conveniently defined by the Laplace transform equation

$$(83) \qquad \int_0^\infty h(z)\exp(-tz)dz = \Gamma(\theta)t^{-\theta}\exp\{-\theta G(t)\} ,$$

with

$$(84) \qquad G(t) = \int_t^\infty y^{-1}\exp(-y)dy .$$

In particular,

$$(85) \qquad f(x_{(1)}) = \theta x_{(1)}^{\theta-2} h\{(1-x_{(1)})/x_{(1)}\} .$$

The expression (82) simplifies considerably, when $x_{(1)} + x_{(2)} + \ldots + x_{(r-1)} + 2x_{(r)} \geq 1$, to

$$(86) \qquad f(x_{(1)}, \ldots, x_{(r)}) = \theta^r \{x_{(1)} \ldots x_{(r)}\}^{-1} \{1-x_{(1)}-\cdots-x_{(r)}\}^{\theta-1} .$$

In particular, when $x_{(1)} \geq \frac{1}{2}$,

$$(87) \qquad f(x_{(1)}) = \theta x_{(1)}^{-1}(1-x_{(1)})^{\theta-1} .$$

The identity of (87) and (45) when $x_{(1)} \geq \frac{1}{2}$ is not, of course, coincidental, since at most one allele can have a frequency exceeding $\frac{1}{2}$ and if it has, it must be the most frequent allele.

We return to the Poisson-Dirichlet distribution, and the heaps process in which it originated, later. For the moment we make four comments about it. First, it is arrived at as a limiting $(K \to \infty)$ distribution from the Dirichlet, focussing on the order statistics. Other limiting operations as $K \to \infty$ are also possible and provide us with in some ways more useful, and far simpler, formulae. Second, the sampling formula (58) can be derived directly from the Poisson-Dirichlet (Kingman (1977), Hoppe (1987)), without going via the K-dimensional Dirichlet: indeed, there is an intimate relation between the Poisson-Dirichlet distribution and the Ewens sampling formula, deriving from the concept of partition structures. Third, the Poisson-Dirichlet, focussing as it does on order statistics, derives from the "ordered" heap strategy: an equally important distribution arises from the "self-regulated" heap. Finally, the Poisson-Dirichlet distribution is, like (58), a stationary distribution, in this case of the Ethier-Kurtz diffusion process. All of these comments will be returned to, and expanded upon, later.

7. The coalescent

Having made an incursion, albeit serendipitously, into population genetics theory through the Poisson-Dirichlet distribution, Kingman made two further decisive contributions through the concepts of the coalescent and of partition structures. We discuss these in this and the next section.

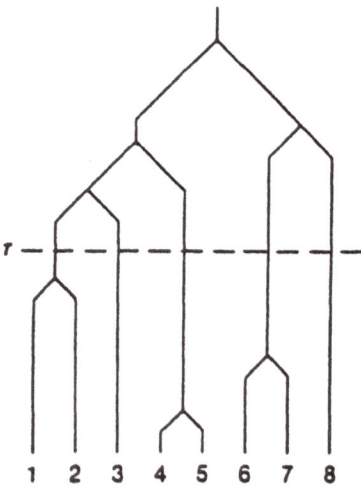

Figure 1

Suppose we take a sample of n genes (n ≪ M) from a population of M genes, and trace back their ancestry. At some time two of these genes will have a (most recent) common ancestor. We may say that their ancestral lines have coalesced at this ancestor. Further back in time a second coalescence will occur, and so on until the entire ancestry coalesces at a single individual, the most recent common ancestor of all genes in the sample. Figure 1 shows an example for a sample of n = 8 genes. Kingman (1982 a,b,c) observed that properties of a sample of genes are most fruitfully examined through considering their ancestry, and we now outline his analysis of the coalescent process that this ancestry describes.

We consider first a specific model, the Kingman coalescent process, describing the tracing back of the ancestry of n genes. In doing this we move backwards in time, and it is convenient to introduce a notation acknowledging this. We denote by τ a time τ in the past before the sample is taken: thus if $\tau_1 > \tau_2$, time τ_1 is further in the past, in real time, than τ_2 .

At any time τ we may describe the composition of the ancestry by an equivalence relation on the integers $\{1,2,...,n\}$. For example, if the equivalence classes at time τ are

$$(88) \qquad\qquad (1,2),(3),(4,5),(6,7),(8) \quad ,$$

then at this time genes 1 and 2 have a common ancestor, as do 4 and 5 and also 6 and 7 . Such at time τ is shown on Figure 1. We write E as the set of all such equivalence relations, and if ξ and η are members of E , we write $\xi \to \eta$ if η can be obtained from ξ by amalgamation of two of its equivalence classes. Thus as we trace back the ancestry of our n genes, we can next move from ξ only to some η for which $\xi \to \eta$. For some equivalence relation ξ , we define

$$(89) \qquad\qquad k = |\xi| \, , m = \frac{1}{2} |\xi|\{|\xi|-1\} \, ,$$

where $|\xi|$ is the number of equivalence classes in ξ .

E possesses two particular equivalence relations, the "finest" and "coarsest" members of E , namely

$$(90) \qquad\qquad \phi_n = (1),(2),(3),....,(n) \, ,$$

and

$$(91) \qquad\qquad \phi_1 = (1,2,3,...,n) \, .$$

The Kingman coalescent process is a Markov chain in continuous time, moving from ϕ_n to ϕ_1 through a random sequence of members of E , in such a way that if the process is in state ξ at time τ , the probability of being in state η at time $\tau + \delta\tau$, for each η for which $\xi \to \eta$, is

$$(92) \qquad\qquad \text{Prob}\{\eta \text{ at } \tau+\delta\tau | \xi \text{ at } \tau\} = \delta\tau + O(\delta\tau)^2 \, ,$$

(93) $\text{Prob}\{\xi \text{ at } \tau+\delta\tau|\xi \text{ at } \tau\} = 1-m\delta\tau + O(\delta\tau)^2,$

with the probability of other transitions being of order $(\delta\tau)^2$.

It is sometimes convenient to ignore the times at which transitions from one equivalence relation to another occur, and to focus on the random sequence $\{\phi_n, \phi_{n-1}, \ldots, \phi_2, \phi_1\}$ through which the process moves from ϕ_n to ϕ_1. Clearly ϕ_k is an equivalence relation admitting exactly k equivalence classes, and Kingman (1982b) shows that for any equivalence relation ξ having k classes,

(94) $\text{Prob}\{\phi_k=\xi\} = \dfrac{(n-k)!k!(k-1)!}{n!(n-1)!} \lambda_1!\lambda_2!\ldots\lambda_k!,$

where $\lambda_1, \ldots, \lambda_k$ are the sizes of the equivalence classes in ξ.

Suppose that we now allow mutation in such a way that for any gene present in the ancestry at time τ, the probability of a mutation in $(\tau, \tau+\delta\tau)$ is

(95) $\text{Prob}\{\text{mutation in } (\tau, \tau+\delta\tau)\} = \dfrac{1}{2}\theta\,\delta\tau,$

where here and below we ignore terms of order $(\delta t)^2$. Thus in the Kingman coalescent process a mutation, if it occurs, occurs simultaneously to all members of an equivalence class. We follow the infinitely many alleles model and assume that all mutations are to new allelic types. If we now group the genes in the sample according to their allelic type, (so that two genes are of the same allelic type if there are no mutations along the lines of ascent from each gene to their most recent common ancestor), it may be shown (Kingman (1982a)) that (94) implies that the partition probability of allelic types at time 0 is given by (58). In other words, (58) derives immediately from, and is in many ways best regarded as a consequence of, the coalescent properties of the ancestry of the sample.

Much more can be said than this. For example, suppose we trace back the ancestry of a sample of n genes. From time to time we will encounter a "defining event", taken either as a coalescence or a mutation. Figure 2 describes such an ancestry, (identical to that of Figure 1), but with crosses marked to indicate mutations. We now exclude from any further tracing back any line in which a mutation occurs: these lines may be thought of as stopping at the mutation, as indicated in Figure 3 (describing the same ancestry as Figure 2). Thus coalescence can only occur between lines in which no mutation, as we trace back the lines, has so far occurred. If at time τ there are k individuals remaining as we trace back according to this convention, the probability of a defining event in $(\tau, \tau+\delta\tau)$ is

(96) $\dfrac{1}{2}k(k+\theta-1)\delta\tau.$

The mean time, given k, between the previous and the next defining event is thus

(97) $2\{k(k+\theta-1)\}^{-1}$

and continuing back up the ancestry, we find that the mean age of the oldest allele in the sample is simply the mean time until the last (going backwards in time) defining event, namely

(98)
$$2 \sum_{j=1}^{n} \{j(j+\theta-1)\}^{-1} .$$

We will see later that the identity of the limiting $(n \to \infty)$ value

(99)
$$2 \sum_{j=1}^{\infty} \{j(j+\theta-1)\}^{-1}$$

with (23) is no coincidence.

Note that the tracing back in the ancestry, as just described, does not necessarily go back as far as the nearest common ancestor of all n genes in the sample, and will do so only if his allelic type is represented in the sample.

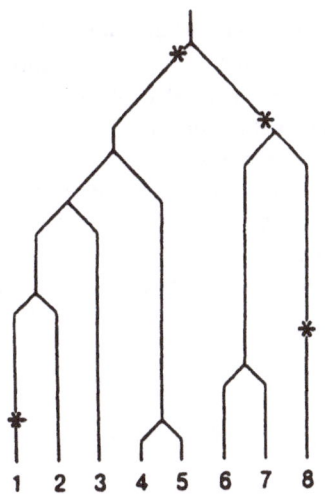

Figure 2

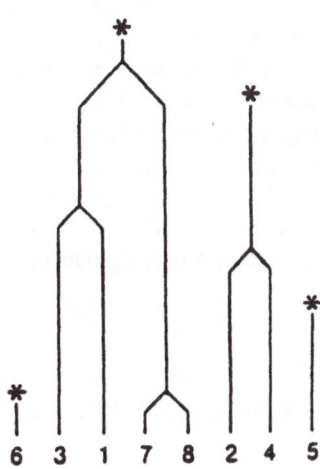

Figure 3

Another important conclusion which we reach from the Kingman coalescent is found by now moving forward in real time, (rather than backwards), down the tree of descent as described in Figure 3. Just as, in reverse time, we regard a defining event as the coalescence of two lines of ascent or the elimination of a line through a mutation (see Figure 3), moving forward in real time allows us to regard a defining event either as the addition of a new gene as the offspring of a currently existing gene (corresponding to a coalescence in backwards time), or the addition of a new mutant gene (corresponding to the initiation of a new line of descent). If at time t there are i-1 genes in the ancestry, then the coalescent process shows that the probabilities of these two events, in $(t,t+\delta t)$, are respectively

(100) $$\frac{1}{2} i(i-1)\delta t \text{ and } \frac{1}{2} i\theta\delta t \, .$$

Thus given that a defining event occurs, the probability that it introduces a new allelic type into the ancestry is

(101) $$\theta/(i-1+\theta) \, ,$$

whatever is the allelic composition of the i-1 genes in the ancestry immediately before this event occurs. But (101) is simply (55) which was assumed, in the argument preceding (58), to be the probability that the ith gene sampled "through space" is of a novel allelic type, whatever the allelic composition of the first i-1 genes sampled might be. Thus deriving the sample "through space" and "through time" are identical concepts. This is a powerful idea, due originally to Donnelly (1986b) and later Hoppe (1987), to which we return at some length later.

How accurate is the Kingman coalescent process as a description of sample genealogies in the Wright-Fisher model? Clearly it is at best an approximation (for example in the ancestry of a sample of genes in this process, several equivalence classes can coalesce simultaneously). Nevertheless, Kingman has shown that this model, and the Cannings generalized exchangeable models for which σ_v^2 remains finite as $M \to \infty$, lie within the "domain of the coalescent". Thus consider the ancestry of n genes taken, in generation 0 , from a Wright-Fisher process. In some ancestral generation these genes will have ancestry described by some equivalence relation ξ . For each η for which $\xi \to \eta$, the probability that in the previous generation the ancestry relation is η , given ξ , is

(102) $$M^{-1} + O(M^{-2}) \, ,$$

while the probability that the ancestry relation in the previous generation is still ξ is

(103) $$1 - mM^{-1} + O(M^{-2}) \, ,$$

m being defined by (89). All other possible equivalence relations for the previous generation have probability $O(M^{-2})$ or less. If P_M is the matrix of transition probabilities of equivalence relations as we trace the ancestry back, (102) and (103) show that

(104) $$\lim_{M \to \infty} P_M^{[Mt]} = \exp(Pt) \, ,$$

where P is the infinitesimal generator of the Kingman coalescent defined in (92) and (93). Thus in the limit $M \to \infty$ all conclusions we have reached concerning samples, in particular (58) and (101), asymptotically apply. Similarly, any Cannings model for which σ_v^2 remains finite as $M \to \infty$ is in the domain of the coalescent. The same conclusion is reached for the Moran model, using, of course, a different time scale. Thus (58) is robust to the particular model used to describe the evolutionary process.

Again, an exact conclusion is possible for the Moran model. We can construct an ancestry of a sample of n genes, as in Figure 3. Moving down the ancestry in real time through a sequence of birth-death events, a defining event will occur, when there are i-1 individuals in the ancestry, only if one of the existing individuals produces a non-mutant offspring (probability $(1-u)(i-1)M^{-1}$) or a new mutant joins the ancestry (probability u) . Given that a defining event does occur, it is the addition of a new allelic type with probability

$$(105) \qquad u/[u+(1-u)(i-1)M^{-1}] = \theta/(\theta+i-1) ,$$

where θ is defined by (25). The identity of this with (101) leads directly to the exact truth of (58), (69) and (98) in the Moran process. Further, equation (98), with n replaced by M , gives the mean age of the oldest allele in the entire (Moran) population.

We conclude this section by making several observations about the Kingman coalescent. The first observation is obvious: the coalescent provides the most natural vehicle for investigating the ancestry of a sample and thus, for example, for finding properties of the ages of alleles. These properties can often be obtained by using processes moving forward in time, and then exploiting time reversibility arguments, but this is clearly less natural than an approach using the coalescent. Second, this approach can be used in reverse: results concerning the future can be found from the coalescent, possibly in conjunction with time reversibility arguments. Third, the coalescent is increasingly providing the natural vehicle for deriving practical results of a surprisingly wide nature. In this connection one should mention in particular the work of Hudson and Kaplan (1986) on transposable elements and the evolutionary properties of recombination (1985), and of Kaplan and Hudson (1987) on multigene families and on highly repeated interspersed DNA sequences (1988).

Our fourth comment explains the occurrence of the combinatorial term (68) in the sampling formula (58). Joyce and Tavaré (1987) note that the lines of descent in the (equivalent) Figures 1, 2 and 3 may be viewed as describing a pure birth and immigration process, the introduction of a new allelic type (see Figure 3) being viewed as an immigration. From time to time some family size is increased by one (through a birth) or an immigrant ancestor of a new family arrives. We relabel the n individuals in the sample so that the first individual in the process is labelled 1, the next (either an offspring of the first, or a new immigrant) 2, and so on: a possible relabelling is shown in Figure 3. The sequence of birth and immigration events leading to the present collection of n individuals can be encapsulated in a permutation of {1,2,...,n} according to the rule that when a birth occurs, the number for the new individual is written immediately to the left of that of his parent, while the number for a new immigrant is written to the right of all numbers present at the time of the immigration. Thus a possible history of the process shown in Figure 3 is encapsulated in the permutation $\{3, 8, 7, 1, 4, 2, 5, 6\}$. Any such permutation can be written as a collection of cycles (here $\{(3871)\ (42)\ (5)\ (6)\}$) which describe family membership (or, in the genetic context, allelic type). The coalescent shows that for a sample of n individuals, falling into $\Sigma\ b_j$ families with b_j families of size j , the probability of any such permutation, that is the probability of a specified line of descent, is

$$(106) \qquad \theta^{\Sigma b_j}/S_n(\theta) .$$

Multiplication of this by the factor (68) leads to (58). Joyce and Tavaré (1987) and Joyce (1988), to whom this argument is due, investigate a variety of further consequences of this approach.

Finally, we note that (58) and (94) are particular cases of a more general formula, due to Watterson (1984). The n genes in the sample are descended, for some given time τ in the past, from a set of ancestor genes existing at time τ . It is possible that in some of the lines of descent no mutations occurred between τ and the present. Let there be k such lines, with λ_i genes in the sample being in the ith such line. Other genes may be descended from mutations occurring since τ . These are partitioned into ℓ classes according to the rule that all genes of the same allelic type go into the same class, the classes are then labelled, and we write μ_j for the number of genes in the class labelled j . Watterson then shows that

$$(107) \qquad \text{Prob}\{\ell;\lambda_1,...,\lambda_k;\mu_1,...,\mu_\ell|k\} = \frac{(n-k)!k!\theta^\ell\Pi\lambda_i!\Pi(\mu_j-1)!}{n!(k+\theta)(k+\theta+1)...(n+\theta-1)}$$

exactly for the Moran model (θ being given by (25)), and Donnelly and Tavaré (1986) show, in the spirit of the arguments leading to (104), that (107) is asymptotically valid for effectively the same range of models as is (58).

If there is no mutation there are no "new" classes, so that $\theta = \ell = 0$ and (107) reduces to (94) . When there is mutation, then in the limit $\tau \to \infty$ there will be no "old" classes, so that $k = 0$ and (107), when multiplied by an appropriate combinatorial term to remove the effect of the labelling process, reduces to (58).

So far we have considered only the genealogy of a sample of genes of fixed size n , and the proofs of our conclusions rely on the fixed nature of the size of this sample. It is more difficult to make statements about the genealogy of the entire population. Donnelly and Tavaré (1987) introduce an infinite population model, with transitions occurring in continuous time, which is the natural population analogue of the (sample) coalescent process discussed above. Of the various results they obtain for this model (which we do not describe in detail here), we note only the following. Suppose that in the population there are k "old" alleles present at time 0 which were also present at some previous time τ , and write their frequencies $x_1, x_2, ..., x_k$. The remaining "new" alleles present at time 0 were not present at time τ : we write their frequencies, ordered by the times at which these alleles arose in the population, as $x_{k+1}, x_{k+2}, \ldots$. Then, given k, we may represent $x = (x_1, x_2, \ldots)$ as

$$x = (v_k u_1, \ldots, v_k u_k, (1-v_k)z_1, (1-v_k)z_2(1-z_1), \ldots) ,$$

where $u_1, \ldots, u_k$ are uniformly distributed on $u_i \geq 0$, $\sum u_i = 1$, v_k is a random variable having density function

$$f_k(v) = \frac{\Gamma(k+\theta)v^{k-1}(1-v)^{\theta-1}}{\Gamma(\theta)(k-1)!}$$

and $z_1, z_2, \ldots$ are i.i.d. random variables with density function

$$f(z) = \theta(1-z)^{\theta-1} , 0 < z < 1 .$$

The interpretation of these random variables is immediate: given k, v_k is the random sum of the frequencies of the old alleles (see also Griffiths (1980), equation (18)). Given v_k, the frequencies of the "old" alleles are uniformly distributed with sum v_k. The relative frequencies of the new alleles can be represented as $z_1, z_2(1-z_1)$, $z_3(1-z_1)(1-z_2), \ldots$. As $\tau \to \infty$, $k \to 0$ almost surely and then the absolute frequencies of all the alleles, indexed by age, can be represented as $z_1, z_2(1-z_1)$, $z_3(1-z_1)(1-z_2), \ldots$, where $z_1, z_2, z_3, \ldots$ are as defined above. We will meet this representation later, in (143)-(145).

It is difficult to prove convergence results for the infinitely many alleles Wright-Fisher model, or more generally the generalized exchangeable model, to conclusions for the Donnelly-Tavaré population genealogy process. For substantial partial results, see Donnelly (1989) and Donnelly and Joyce (1989). In particular, (99) provides the limiting $(N \to \infty , u \to 0 , 4Nu = \theta)$ mean age of the oldest allele in a Wright-Fisher population, a conclusion which typifies the nature of the convergence results in that their conclusions are as expected, but their proofs are very difficult.

8.   Partition structures

The concept of a partition structure derives from the observation (Kingman, 1978a) that for any sample of n genes, there is no particular significance to be attached to the sample size. A sample of n genes could have arisen from a sample of $n+1$ genes, one of which was accidentally lost, and the requirement that the same family of partition probabilities should describe samples of both sizes imposes a consistency structure on probabilities of allelic partitions with respect to sample size, which we now explore.

Suppose then, with Kingman (1978a,b), that the partition (57) of a sample of n genes was indeed obtained from a partition of $n+1$ genes, a random one of which was lost. Considering all the possible allelic partitions of the $n+1$ genes, it is clear that a consistent family of probabilities must satisfy

$$P_n(b_1,b_2,\ldots) \;=\; \frac{b_1+1}{n+1}\, P_{n+1}(b_1+1,b_2,\ldots)$$

(108)

$$+ \sum_{r=2}^{n+1} \frac{r(b_r+1)}{n+1} P_{n+1}(b_1,\ldots,b_{r-1}-1,b_r+1,\ldots) \;.$$

It is easy to see that (58) satisfies (108), but Kingman raised a much more general question: how may one characterize "partition structures", that is probability structures satisfying (108)? For partitions of relevance in biology, the answer is the following. Let ∇ be the set of all $\mathbf{x} = (x_{(1)}, x_{(2)}, \ldots)$ such that $x_{(i)} \geq 0$, $x_{(i)} \geq x_{(i+1)}$, $\sum x_{(i)} = 1$. (The set ∇ was introduced earlier as the domain of the Poisson-Dirichlet distribution.) Let $P_n(b_1, b_2, \ldots, |\mathbf{x})$ be the probability of the partition $\mathbf{b} = (b_1, b_2, \ldots)$ under multinomial sampling with probability vector $\mathbf{x}$ (i.e. each individual in the sample is independently of type "i" with probability $x_{(i)}$). Then

(109)
$$P_n(b_1,b_2,\ldots|\mathbf{x}) = n! \prod_j (j!)^{-b_j} \{\textstyle\sum x_{(1)}^{\gamma_1} x_{(2)}^{\gamma_2}\ldots\} \;,$$

where the inner sum is taken over all sets $(\gamma_1, \gamma_2, \ldots)$ of integers $0 \leq \gamma_i \leq n$ such that exactly b_j of the γ_i are equal to j. If $\mathbf{x}$ is now itself random, with probability measure μ on ∇, the absolute probability of $\mathbf{b}$ is

(110)
$$P_n(b_1,b_2,\ldots) = \int P_n(b_1,b_n,\ldots|\mathbf{x})\mu(d\mathbf{x}) \;.$$

The central result of Kingman is that all "representable" partition structures (which include all cases of biological interest) can be represented as in (110). This beautiful result, bearing a close analogy to the famous theorem of de Finetti for exchangeable random variables, has applications outside genetics, as we note shortly.

We call μ the representing measure of $P_n(b_1, b_2, \ldots)$. Kingman showed that the representing measure for the partition probabilities (58) is the Poisson-Dirichlet distribution. We may also regard this conclusion from the following point of view. Imagine that we take successively larger samples (i.e., $n \to \infty$), where for each n the allelic partition has distribution (58). Then as $n \to \infty$,

(111)
$$n^{-1}(n_1,n_2,\ldots) \to (x_{(1)},x_{(2)},\ldots) \;,$$

where "$\to$" implies convergence in distribution, and $(x_{(1)}, x_{(2)}, \ldots)$ is a random (infinite) vector having the Poisson-Dirichlet distribution.

Kingman also took up the question of "non-interference", defined by the requirement that if a gene is taken at random from the sample, and all r genes of its allelic type removed, the probability structure of the remaining n-r genes should be that of an original sample of n-r genes. Non-interference requires that

$$(112) \qquad \frac{rb_r}{n}\, P_n(b_1,b_2,...,b_r,...) \;=\; c(n,r)P_{n-r}(b_1,b_2,...,b_r-1,...) \;,$$

where $c(n,r)$ does not depend on $b_1,b_2,...$ Kingman showed that the only representable partition structure which also satisfies (112) is (58). Correspondingly, for representable partitions the only representing measure for which (112) holds is the Poisson-Dirichlet.

We now give two applications of these conclusions, taken from the diverse fields of ecology and pure mathematics. Ecologists have long being interested in the non-interference of the species inhabiting a given area, so that for them (112) may be taken, with proper reinterpretation of the notation, as the defining requirement for species non-interference. If we make the further, reasonable, consistency requirement (108), then the above arguments show that the "species" partition must be (58). Indeed, the first use made of (58) was by Caswell (1976) in his study of the "non-interference" of species occupying a common area. Further ecological studies using this approach are made by Platt and Lambshead (1985), Lambshead and Platt (1985) and Lambshead (1986). We return to the ecological context, and non-interference, later.

The second application of (108), (110) and (112) arises in probability theory in connection with the concept of exchangeability (Aldous (1985), pp. 85-97, Kingman (1980)). Let B be a partition of the integers $\{1,2,...,n\}$, and write $B = \{B_1,B_2,...\}$, where $B_1,B_2,...$ are the "components" of B . Thus if $n = 8$ we might have $B = \{(4627), (813), (5)\}$.   If π is any permutation of $\{1,2,...,n\}$,   then $\pi B = \{(\pi B_1),(\pi B_2),...\}$ is also a partition of $\{1,2,...,n\}$. Suppose some probability distribution is imposed upon the partitions of $\{1,2,...,n\}$: this probability structure is exchangeable if B and πB have the same probability, for each B and π .

Clearly B and πB have the same sized components. We can express the component sizes of B in two ways, either by the vector $L(B) = (n_1,n_2,...)$, where $n_i \geq n_{i+1}$, $\Sigma\, n_i = n$, where $n_1,n_2,...$ are the sizes of the components of B , or by the partition (57), where b_j is the number of components of B whose size is j . The most general exchangeable probability structure is that for which

$$(113) \qquad\qquad P(B) = P(B*)$$

whenever B and B* have the same partition (57), and otherwise allows P(B) to be arbitrary, subject of course to an appropriate normalizing requirement over all partitions.

We now wish to consider exchangeable random partitions of $\{1,2,3,...\}$. To do this, it is useful to rewrite (113) as

$$(114) \qquad\qquad P_n(B_n) = P_n(B_n^*) \;,$$

to emphasize the dependence of both P and B on n , and to rewrite L(B) , defined above, as

(115) $L_n(B_n) = (n_1, n_2, \ldots)$.

Exchangeable random partitions on $\{1,2,3,\ldots\}$ are best defined through their restrictions to $\{1,2,3,\ldots,n\}$, for various n. If B_n and B_n^* are the restrictions to $\{1,2,3,\ldots,n\}$ of exchangeable partitions R and R* of $\{1,2,3,\ldots\}$, and if the partitions (57) of B and B* are equal, then we require (114) to hold, where P_n is the restriction of the probabilities of R and R* to $\{1,2,3,\ldots,n\}$. Further these restrictions must satisfy the consistency condition (108) (which will then guarantee, by the Kolmogorov extension theorem, the existence of probabilities of exchangeable partitions of $\{1,2,3,\ldots\}$). One way of stating Kingman's theorem for partitions, analogous to (109), (110), and (111), is as follows. Suppose that R is an exchangeable partition of $\{1,2,3,\ldots\}$ whose restriction to $\{1,2,3,\ldots,n\}$ we denote by B_n. Then as $n \rightarrow \infty$,

(116) $n^{-1}L_n(B_n) = n^{-1}(n_1, n_2, \ldots) \rightarrow (p_1, p_2, \ldots)$,

where $(p_1, p_2, \ldots)$ is a random element of the space $\bar{\nabla}$ (that is, $p_i \geq p_{i+1} \geq 0$, $\sum p_i \leq 1$), whose distribution we denote by μ_p. Further, conditional on $(p_1, p_2, \ldots)$, the composition of the partitions is arrived at by, independently for each integer, placing that integer into a unique partition (with probability $1 - \sum p_i$), or into labelled partition i with probability p_i.

More structure is added if we make the "non-interference" requirement (112), to be interpreted for partitions of $\{1,2,\ldots,n\}$ as the requirement that if we choose a number from this set at random, and then delete the component B_j containing this number, that the remaining partitions have the same distribution as partitions of $n-r$, where r is the number of integers in B_j. Of course there are many probability structures for which (108) and (112) do not both hold, but if they do, we are entitled to use (58) for finite n, and the Poisson-Dirichlet distribution asymptotically, to draw immediately a variety of interesting conclusions.

Aldous presents two examples. The first concerns the $n!$ permutations of $\{1,2,3,\ldots,n\}$, where the partitions are the cycles formed under each permutation. It is immediate that, if we give each permutation probability $(n!)^{-1}$, (108) and (112) hold. Thus (58) holds and it is easy to see that $\theta = 1$. The finite n results are immediate but the asymptotics are less obvious. Our approach shows that if a permutation is taken at random and its cycle lengths are arranged in order $n_1 \geq n_2 \geq n_3 \geq \ldots$, the asymptotic distribution of $n^{-1}\{n_1, n_2, n_3, \ldots\}$ is Poisson-Dirichlet with $\theta = 1$. Thus, for example, the asymptotic distribution of $x_{(1)} = n_1/n$ is (85) with $\theta = 1$, and Table III in Watterson and Guess (1977) shows that asymptotically

(117) $E\{n_1/n\} = .624 \ldots$.

The asymptotic probability that n_1/n exceeds $\frac{1}{2}$ is log 2 (from (87)): this result can also be found directly as the limit of the exact finite n formula $1 - \frac{1}{2} + \frac{1}{3} - \ldots \pm 1/n$.

These and other asymptotic results have been found, by methods other than using the Poisson-Dirichlet, by Goncharov (1962), Golomb (1964) and Shepp and Lloyd (1966).

In particular, Golomb and Shepp and Lloyd give the numerical value (117). (Indeed, equation (3.2.7) of Watterson and Guess (1977), from which (117) is calculated, reduces for $\theta = 1$ to equation (14) of Shepp and Lloyd (1966).) Golomb (1964) shows that $E(n_1/n)$ is a decreasing function of n ; the numerical result in Watterson and Guess (1977) confirm this and show that the limit (117) is approached rapidly (so that, for example, $E(n_1/n) = .627...$ when n = 100). Further results of this nature are given by Vershik and Shmidt (1977), Ignatov (1982) and Kerov and Vershik (1986).

The second example concerns random functions from $\{1,2,...,n\}$ to $\{1,2,...,n\}$. There are n^n such functions, and for any fixed one of these we put the integers i and j in the same "component" if some functional iterate of i under this function is equal to some functional iterate of j. If each function is given probability n^{-n}, it is easy to see that (112) is satisfied. However (108) does not hold exactly, but only in an asymptotic sense defined by Aldous. This asymptotic result implies that, while (58) cannot hold exactly for any finite n , the normalized component sizes have, asymptotically, a Poisson-Dirichlet distribution. Aldous shows that asymptotically two randomly chosen numbers are in the same component with probability 2/3 , and (48) then immediately identifies $\theta = \frac{1}{2}$. Many more asymptotic results are now possible. For example, (54) shows that the probability, for any fixed i , that i numbers taken at random are in the same component is asymptotically

(118) $\qquad \{2\times4\times6\times8\times...\times(2i-2)]/[3\times5\times7\times9\times...\times(2i-1)]$,

generalizing Aldous' result. Next, the density function (87) shows that the probability that the normalized size of the largest partitions exceeds $\frac{1}{2}$ is asymptotically

(119) $\qquad \frac{1}{2} \int_{1/2}^{1} x^{-1}(1-x)^{-\frac{1}{2}} dx = \log(1+\sqrt{2}) = .881374...$

Further, Table III of Watterson and Guess (1977) shows that the asymptotic normalized mean size of this largest partition is .758... , which may be compared to the value .624 for random permutations given in (117), and it is easy to see that the asymptotic median size is .786... . Many further elegant limiting $(n \to \infty)$ results can be obtained immediately, confirming and extending easily those found, often at great effort, in the literature (see, for example, Pittel (1983), Theorem 2, and Kolchin (1976), Theorem 9).

For finite n we have noted that results calculated from (56) or (58) are not exactly correct. For example, neither (118) nor (119) is correct for finite n , although numerical results indicating the rate and mode of convergence of finite n values are given by Ewens and Padmadisastra (1988). Similarly, the approximate value for the mean number of components (found from (56) with $\theta = \frac{1}{2}$) , namely

(120) $\qquad 1 + \frac{1}{3} + \frac{1}{5} + \frac{1}{7} + ... + \frac{1}{2n-1}$,

diverges as $n \to \infty$ and thus does not admit a non-trivial convergence result, although it may be noted that (120) differs asymptotically from the true mean number (calculated by Kruskal (1954) and Ross (1981)) by only $\frac{1}{2} \log 2$.

A another application of the Poisson-Dirichlet distribution arises with prime numbers. Let n be an integer drawn randomly from the values $1,2,...,N$, allvalues being equally likely. Write $n = p_1 p_2 p_3...$, where $p_1, p_2, p_3,...$ are the prime factors of n, with $p_1 \geq p_2 \geq p_3 \geq...$ Put $x_i = \log p_i / \log n$. Then asymptotically, as $N \to \infty$, the vector $(x_1, x_2,...)$ has the Poisson-Dirichlet distribution with $\theta = 1$. This and similar remarkable results is given by Vershik (1986).

We may regard the calculation of probabilities using (108) and (112) as following an axiomatic approach: if the "axioms" (108) and (112) are satisfied, exactly or asymptotically, the conclusions implicit in (58) (for exact cases) or the Poisson-Dirichlet distribution may be applied, once θ is identified. A recent axiomatic approach to the partition or particles into "cells", or energy levels, in statistical mechanics (Constantini (1987)) has some interesting similarities with, and differences from, partitions in genetics arrived at, as (58) can be, by the "axioms" (108) and (112). Constantini presents three axioms which imply, in the "symmetric" case, that if particles are placed sequentially into cells, the probability that particle $n+1$ is placed into cell j, given that n_j of the first n particles are placed in cell j, is

(121)
$$(n_j + \varepsilon)/(n + K\varepsilon) ,$$

where ε is a constant. The Maxwell-Boltzmann, Bose-Einstein and Fermi-Dirac statistics follow when $\varepsilon \to \infty$, $\varepsilon = 1$, $\varepsilon = -1$, respectively.

But, as is well known, (121) follows immediately from (70) (see Blackwell and MacQueen (1973), Hoppe (1987)) when $\theta > 0$, as the probability when sampling from the Dirichlet distribution that the $(n+1)^{\text{th}}$ gene drawn is of allelic type A_j, given that n_j of the first n genes drawn of this type. The well-known Bose-Einstein probability

(122)
$$P\{n_1,...,n_k\} = 1 \bigg/ \binom{K+n-1}{K-1}$$

is, for example, simply (70) with $\varepsilon = 1$. We might thus hope, since (58) and the Poisson-Dirichlet distribution can be obtained from (70), that connections can be made between the "physical" and the "biological" partition probabilities. This hope is not realized in the sense that, for example, (122) does not satisfy the "axioms" (108) and (112). One reason for this is that (122) allows empty cells $(n_j = 0)$ whereas (58) considers only observed alleles. Even the well-known probability

(123)
$$P\{n_1,...,n_K\} = 1 \bigg/ \binom{n-1}{K-1}$$

analogous to (122), but requiring each cell to be occupied, satisfies neither (108) nor (112). The fundamental axioms of the physical partitions differ from those of the biological partitions. Despite this comment, Keener et al (1987) have been able to derive (58), the Maxwell-Boltzmann and the Bose-Einstein distributions as particular cases of a more general exchangeable probability distribution, differing only according to the value of a certain "mixing" parameter.

9. The Watterson testing theory

The robustness of (58) to the details of the evolutionary model, demonstrated by the coalescent, makes it reasonable for us to use (58), and hence (60), as the null hypothesis distribution of the partition b in (57) in a test of selective neutrality. Our preference for using (60) rather than (58) derives from the fact that (60) is free of the (unknown) parameter θ , so that once a test statistic is chosen, exact significance levels for tests using (60) can be found.

The choice of a test statistic, however, is not immediate. Under the Neyman-Pearson testing theory, a test statistic emerges from a likelihood ratio, and to find this ratio we would have to calculate the probability of the partition (57) under some selective model. Unfortunately the great variety of selective models possible makes this a hopeless task, and so we focus on one selective scheme of particular interest to geneticists, namely that in which in a diploid population all heterozygotes have fitness $1+s$ $(s>0)$ and all homozygotes have fitness 1. It is convenient to consider first the K allele model, and then let $K \rightarrow \infty$.

In the selective case the probability distribution corresponding to (36) is

$$(124) \qquad f(x_1,...,x_{K-1}) = \text{const}\{x_1...x_K\}^{\varepsilon-1}\exp\{-S\Sigma\ x_i^2\}$$

where $S = Ms = 2Ns$. Likelihood ratio arguments suggest the use of $F = \Sigma\ x_i^2$ as test statistic, but further arguments are required since the tests are based on samples, whereas (36) and (124) apply to populations. This implies that we need to find the selective analogue of (58) and (60). After extensive calculations, Watterson (1977) showed that the selective analogue of (60) differs from (60) itself by a factor whose leading term, for small s , is a function of

$$(125) \qquad \hat{F} = \Sigma\ j^2 b_j/n^2 .$$

We therefore use $\hat{F}$ as test statistic, with sufficiently small values of $\hat{F}$ rejecting the null hypothesis (of selective neutrality). Significance points are found in principle by finding the null hypothesis distribution of $\hat{F}$, using (60). Unfortunately this distribution does not follow any well-known standard statistical form, and empirical percentage points for $\hat{F}$ are used, found by large-scale computer simulation. A selection of these points, for various (n,k) combinations, was given by Anderson (1978) and is reproduced in Ewens (1979, Appendix C).

Our preceding considerations make us confident about the robustness of this testing procedure to the actual details of the population model. There is however one further important consideration: (58) and (60) are stationary distributions, and we must investigate the robustness of the testing procedure to non-stationarity. This problem is taken up in the next section.

10.   The Ethier-Kurtz diffusion process

Reference has been made several times above to an infinite-dimensional diffusion process, analogous to the finite process (42), describing the evolution of the (ordered) gene frequencies in an infinitely large population. So far the main interest in this process has been simply that it is well defined, so that, for example, equations such as (47) are exactly true, at stationarity, for it. There has therefore been no necessity to introduce this process, and in particular to consider its time-dependent properties, until now. However, our investigation of the robustness of the Watterson testing theory to non-stationarity now requires us to examine the properties of this diffusion process, which we may think of as the $M \to \infty$ limit of (42).

To obtain the diffusion analogue of (42), Ethier and Kurtz (1981, 1986) start with the K-allele generalization of the diffusion process (9), whose stationary distribution (for ordered gene frequencies) is (39). They then show the existence of a unique diffusion process on the space

$$\nabla = \{x : x_{(1)} \geq x_{(2)} \geq x_{(3)} \geq \ldots ; x_{(i)} \geq 0, \Sigma x_{(i)} = 1\}$$

(on which the Poisson-Dirichlet distribution is defined), towards which the K-dimensional diffusion process converges, in an appropriate way, as $K \to \infty$. This is the existence theorem which we need.

Griffiths (1979a,b, 1980) has shown that the non-stationary analogue of (58) is, for this process, of the form

(126) $\text{Prob}\{b|t,z\} = \text{Prob}\{b|t=\infty\}[1+\rho_2(t)G(\theta,n,\hat{F},z)+O(\rho_3(t))]$,

where

(127) $G(\theta,n,\hat{F},z) = \dfrac{(\theta+1)^2(\theta+2)(\theta+3)n^2}{2\theta(\theta+n)(\theta+n+1)} [\hat{F}-E(\hat{F})][\Sigma z_i^2-(1+\theta)^{-1}]$,

$\hat{F}$ is given in (125), z is the initial vector of gene frequencies, $\text{Prob}\{b|t=\infty\}$ is the stationary probability (58) and the eigenvalues $\rho_j(t)$ are given by

(128) $\rho_j(t) = \exp\{-\tfrac{1}{2} j(j+\theta-1)t\}$,

the continuous-time analogues of (43). A corresponding transient expression for $E(k)$ may be found. The outcome of the analysis is that non-stationarity is a potentially serious problem for the Watterson testing procedure, since Griffiths shows that the effects of non-stationarity are almost identical to the effects of heterozygote selective advantage. (Kirby (1974) attempted to overcome this problem by using test statistics with a fast convergence to their stationary distributions.) The warnings of Griffiths (1979b) are most important and timely so far as testing for neutrality using the Watterson, or any other, method is concerned.

Expressions such as (126) and (127) can also be derived, and are indeed more naturally derived, by using the genealogy of the sample, and this parallels the perhaps more natural genealogical approach of Donnelly and Tavaré (1987) in describing the evolution of the entire population. The eigenvalues (128) describe the rate of occurrence of "defining events" (described above (96)), which are central events in describing the genealogy of the sample. At stationarity all defining events for the sample have almost surely occurred, so that there is no eigenvalue correction in (126) and (58) is recovered as the stationary distribution of b.

11. Reversibility and the ages of alleles

A discrete Markov chain with transition matrix $\{p_{ij}\}$ and admitting a stationary distribution $\{\pi_i\}$ is said to be (time) reversible if

$$(129) \qquad\qquad \pi_i p_{ij} = \pi_j p_{ji} \text{ for all } i,j .$$

Reversibility implies

$$(130) \quad \text{Prob}\{x_t = a, x_{t+1} = b, \ldots, x_{t+s} = c\} = \text{Prob}\{x_t = a, x_{t-1} = b, \ldots, x_{t-s} = c\} .$$

Thus the probability structure for the random variable in reversed time is the same as in normal forward time.

Some processes in population genetics are reversible, others are not. The process defining the transitions of the allelic configuration vector β is reversible (Watterson (1976a)) for the Moran infinitely many alleles model (described in part in (49)), but is not reversible for the Wright-Fisher analogue (42). The Ethier-Kurtz diffusion process is reversible. We thus limit attention in this section to the finite Moran process (for which θ is defined by (25)) and the Ethier-Kurtz process (where θ is defined by an equation identical to that given immediately below (15)).

It was mentioned in the Introduction that there exists much classical theory concerning the probable future evolution of a population, and one hopes to be able to use reversibility arguments, together with this theory, to derive information about the probable past history of the population, in particular about the ages of the currently existing alleles. Unfortunately this latter aim cannot immediately be satisfed for the Moran process described earlier, since this process considers only the evolution of a vector β describing allele configurations and contains no information about ages. A parallel comment applies for the Ethier-Kurtz process. However, for the Moran process, Watterson (1976a) and Kelly (1979) introduce more detailed variables incorporating age information as well as configuration information and show that the processes describing the evolution of these variables are reversible in the sense that, at stationarity in the Moran process, the joint distribution of the ages of the alleles present at any time is identical to the joint distribution of their (future) extinction times. Ethier (1989) has recently shown that a corresponding procedure can be carried out for the Ethier-Kurtz diffusion, implying the same conclusion. We thus assume in our future discussion that the Ethier-Kurtz diffusion is reversible in this sense.

We can therefore say, for example, that the mean age of an allele in the Ethier-Kurtz diffusion whose current frequency is p is given by the mean extinction time (16), while in the Moran process the mean age of an allele represented by k genes is given by (21). (Watterson (1976a) gives the complete age distribution in each case.) Further results are immediately possible. For example, the age of any allele must be independent of the relative frequencies of the remaining alleles (since its extinction time is). Further (Kelly, (1979)), the age of the oldest allele is independent of all allele frequencies, including its own. This follows because the time to extinction of the longest surviving allele is identical to the time to extinction of all currently existing alleles. This mean time is thus (23) for the diffusion process and $t_{M1}+...+t_{MM}$, with t_{Mj} given by (24), for the Moran process. Note the identity of (23) with (99): (23) is found by looking forward in time, (99) was found, using the coalescent, by looking backwards, and reversibility arguments verify their identity. A similar observation is possible for the Moran model.

Kelly exploits reversibility to show that, in the Moran model, the probability distribution of the number i of genes of the oldest allele in the population is

$$(131) \qquad P(i) = \frac{\theta}{M}\binom{M}{i}\binom{M+\theta-1}{i}^{-1}, \qquad i = 1,2,...,M .$$

Further, an allele represented by j genes in a sample of n is the oldest in a Moran population with probability

$$(132) \qquad j(\theta+M)/\{M(\theta+n)\} ,$$

and the probability Q that the oldest allele in the population is represented at all in the sample is

$$(133) \qquad Q = n(\theta+M)/\{M(\theta+n)\} .$$

The diffusion process analogues of (131) and (133), which may be found directly or by letting $M \to \infty$ in (131) and (133), are

$$(134) \qquad f(x) = \theta(1-x)^{\theta-1}, \ 0 < x < 1 ,$$

(135)
$$Q = n/(n+\theta) .$$

We note that the density function (134) of the frequency of the oldest allele in the population is identical to that of the frequency of the allelic type of a randomly selected gene, given in (47). We also note that the probability (135) that the oldest allele in the population is represented in a sample of n genes is identical, from (67), to the probability that the allelic type of a further ((n+1)[th]) gene is represented in the sample. We show later why these conclusions are true.

Watterson and Guess (1977) take up in great detail the question of whether the most frequent allele in a population is the oldest in the population, and whether the most frequent allele in a sample is the oldest in the sample (or even the entire population). The probability that the most frequent allele in the population is the oldest is simply the probability that it will last longest, by reversibility. Given its current frequency p , this is just p (by symmetry). Thus the probability in question is simply the mean frequency of the most frequent allele, which is, in the diffusion process,

(136)
$$E\{x_{(1)}\} = \int_0^1 x_{(1)} f(x_{(1)}) dx_{(1)} ,$$

where $f(x_{(1)})$ is defined by (85). This expression must be evaluated numerically. Various quick bounds are available: (87) shows that

(137)
$$E\{x_{(1)}\} \geq (\tfrac{1}{2})^\theta ,$$

and Watterson and Guess (1977) show also that

(138)
$$E\{x_{(1)}\} \leq 1 - \theta(1-\theta)\log 2 .$$

Combining this with (137), we get for θ small

(139)
$$E\{x_{(1)}\} \simeq 1 - \theta \log 2 .$$

This agrees, as it must, with (80), found from using the theory of heaps. Note that for $\theta = \tfrac{1}{2}$, the lower and upper bounds are .707 and .827, and these bracket the true mean .758 given below (119) in the context of random functions. Watterson and Guess (1977) give many further numerical values and a large variety of further results, not only for the diffusion process but also for the Moran model, and not only for the entire population but also for samples from the population.

12. Hoppe's urn

Hoppe (1984, 1987) (see also Watterson (1984)) introduced an urn model which gives a colorful approach to confirming many of the results given above, as well as to providing new ones. We imagine an urn containing one black ball, of mass θ, as well as, at time n, a further n colored, or equivalently labelled, balls, each of mass unity. At unit time points a ball is drawn at random from the urn with probability proportional to its mass (the ball chosen at the initial time zero must be the black ball, since at that time that is the only ball in the urn). If at any time the black ball is drawn, a ball having a new label, not so far used for any of the labelled balls currently in the urn, is added. If a labelled ball is drawn, a new ball carrying the same label is added to the urn and the ball drawn is also replaced. The number of balls in the urn clearly increases by unity after each draw, and it is convenient to use as labels the numbers 1,2,3... as needed. Clearly

(140) $\text{Prob}\{i^{th} \text{ ball has a novel label}\} = \theta/(i\text{-}1\text{+}\theta)$,

(141) $\text{Prob}\{i^{th} \text{ ball has a label } j\} = \gamma_j/(i\text{-}1\text{+}\theta)$,

where γ_j is the number of balls having label j at times i-1.

These equations show that the urn process may be regarded as a colorful description of sampling "through space" , which led to the sampling formula (58); (compare (55) and (140)). Thus (58) applies for the urn process.

But (100) shows that the urn process can be thought of as tracing the coalescent of the sample forward in time, and thus deriving the sample "through time" (as in effect we noted in the discussion following (100) and (101)), with the labelling of the balls in the urn corresponding to the labelling of alleles by ages. This is an exact statement, in genetics, for the Moran model and (equivalently) the Donnelly-Tavaré population genealogical process and the Ethier-Kurtz diffusion. We note later an equivalent interpretation in connection with size-biased sampling. This implies (Donnelly, 1986) the following conclusion, proved directly by Hoppe (1987).

Suppose we continue the urn process indefinitely and let $S_j(n)$ be the number of balls, at times n, having label j. Then as $n \to \infty$,

(142) $S_j(n)/n \overset{a.s.}{\to} x_j$,

where $x_1, x_2,...$ are random variables which can be represented simply in the form

(143) $x_1 = z_1$,

(144) $x_j = z_j(1\text{-}z_1)(1\text{-}z_2)...(1\text{-}z_{j\text{-}1})$,

$j = 2,3,...$, where $z_1, z_2,...$ are i.i.d. random variables with common density function

(145) $f(z) = \theta(1\text{-}z)^{\theta\text{-}1}$, $0 < z < 1$.

We have met (143)-(145) before. They arose at the end of Section 7 as describing the population frequencies indexed by age order: since the labelling in the urn process corresponds to age ordering, this is no surprise.

A representation corresponding to (143)-(145) for a Moran population is as follows (Hoppe, (1987)).   Let $M_1, M_2, \ldots$ be the numbers of genes of the oldest, second oldest, ... alleles in a Moran population. Then we can represent $M_1, M_2, \ldots$ by

$$(146) \qquad M_i = 1 + Bin(M-M_1-\ldots-M_{i-1}-1, p_i) \ ,$$

where $p_1, p_2, \ldots$ are i.i.d. random prior probabilities, each having density function

$$(147) \qquad f(p) = \theta(1-p)^{\theta-1} \ .$$

The number K of alleles in the population is the (random) first integer K such that $M_1 + M_2 + \ldots M_K = M$. This representation is a natural analogue of (143)-(145), but we will later give a different, and perhaps more useful, representation for $M_1, M_2, \ldots, M_K$.

The representation (143)-(145) can also be found by a limiting procedure from (40) and (41). Suppose we take a gene at random from a population where the (fixed number K of) alleles present have joint frequency distribution (36). Allele A_i will be chosen with probability x_i , its current frequency. Thus the frequency z_1 of the allele of the gene drawn can be found from the *size-biased* version of (41) for $i = 1$, namely

$$(148) \qquad f(z_1) = \frac{\Gamma[(K-i+1)\epsilon+1]}{\Gamma[1+\epsilon]\Gamma[(K-i)\epsilon]} \ z_1^{\epsilon} \ (1-z_1)^{(K-i)\epsilon-1} \ .$$

If we allow K to increase without limit and ϵ to decrease to 0 with $K\epsilon = \theta$, (148) converges to (145).   If we continue sampling until a second allele is found, its frequency x_2 can be represented as in (40), and under the limiting operations just described, $f(z_2)$ is also given by (145). This continues for all alleles and gives the same representation as (143)-(145). The convergence behavior, as $K \rightarrow \infty$, of the order statistics of the Dirichlet distribution to the Poisson-Dirichlet has been established above, so that we reach the important conclusion that the representation (143)-(145) gives not only the age-labelled, but also the size-biased, representation of the alleles in the Poisson-Dirichlet distribution. This was first noted by Patil and Taillie (1977), who analysed ecological rather than genetic processes: the formal proof making the arguments in this paragraph precise is given by Donnelly and Joyce (1989).

The identity in distribution of age labelled and size biased alleles can be verified in another way. The argument that led to (136) shows that alleles when labelled by size biasing have the same distribution as when labelled by the length of their future persistence in the population, and reversibility then shows that this is the same as the distribution when labelled by age.

These observations immediately explain the identity of the density function of the frequency of the allelic type of a randomly chosen gene, given by (47), and that of the oldest allelic type in the population, given by (134), since by its derivation (47) is a size-biased allele frequency distribution. It is interesting to note that Sawyer (1977) obtains the size-biased distribution (47) directly from a genealogical argument.

The relation between (67) and (135) can be explained in a similar way (Donnelly (1986b)). Having taken a sample of n genes, we may regard the allelic frequency of a further $((n+1)^{th})$ gene as deriving from size-biased sampling. Thus the probability (67) that this further gene is of an allelic type not observed in the sample of n genes must be identical to the probability that the oldest allele in the population is not observed in the sample. This verifies the complementary probability (135).

We noted, in the discussion following (67), that we cannot use the information in a sample of n genes to obtain an unbiased estimate of the probability that a further gene is of an allelic type not represented in the sample. The conclusion of the previous paragraph implies immediately that we cannot use the sample information to obtain an unbiased estimator of the probability that the oldest allele in the population is not represented in the sample. On the other hand, the inequality

$$\frac{n-1}{n} \cdot \frac{\theta}{n-1+\theta} < \frac{\theta}{n+\theta} < \frac{\theta}{n-1+\theta} \, ,$$

together with the unbiased estimator (66) of $\theta/(n-1+\theta)$, implies that we can get very tight bounds on a unbiased estimator.

We now turn to other properties of size-biased sampling. The Poisson-Dirichlet distribution was found as a limiting $(K \to \infty)$ distribution from the Dirichlet, by focussing on order statistics. We have just found another limiting $(K \to \infty)$ distribution by the operation of size-biased sampling. This limiting distribution has two advantages over the Poisson-Dirichlet, the first being the simple representation (143)-(145), and the second being the fact that the frequencies $x_1, x_2...$ in (143) and (144) have the important interpretation as those of age-ordered alleles.

We may connect (143)-(145) with the heaps process of Section 6. In a self-regulating heap, the item of the top is item i with probability equal to the popularity of this item. Thus labelling the items from the top down 1,2,3... is a size-biased labelling, and, arguing informally, the asymptotic $(K \to \infty)$ distribution of the popularities, (assuming $x_1,...x_K$ are random variables drawn from (36)), is (143)-(145). (A rigorous derivation of this conclusion is given by Donnelly (1988).) In particular, if $\bar{x}_1$ is the popularity of the item, at some random time, at the top of the heap,

$$(149) \qquad E(\bar{x}_1) = \int_0^1 \theta x(1-x)^{\theta-1} dx = 1/(1+\theta) .$$

This gives a more accurate statement than (81). There is, however, no immediate way to use the representation (143)-(145) to give a more accurate statement than that provided by (80) (and (139)).

We make two concluding remarks concerning the urn process. First, (140) is a particular case of models for which

(150)   Prob$\{i^{th}$ ball has a novel label$\}$ = f_i ,

where f_i is independent of the number of labels used, and their frequencies, for the first
i - 1 balls. Donnelly (1986) proves the remarkable result that the choice (140) for f_i is
the only one for which the consistency relation (108), defining partition structures, holds.

Second, we consider properties of the urn process in reverse time. Hoppe (1987)
shows in effect that if $P_n(b)$ is the probability of the label partition b at times n , then
the $P_n(b)$ satisfy (108), the defining relations of a partition structure. From this point of
view, we may put a different interpretation on the argument at the beginning of Section 9
for the consistency relation between samples of n and n + 1 genes. Instead of regarding
a sample of n genes as deriving from a sample of n + 1 , one of which was lost, we can
trace back the ancestry of the n + 1 genes until there are n genes in the ancestry.
Consistency requires that the same probability structure holds for this "ancestor" sample as
for the "descendent". This again leads to (108) under the assumptions of the coalescent,
suggesting why (108) describes the reverse-time structure of Hoppe's urn.

13.   The GEM distribution

We devote this section to the properties of the distribution of x = $(x_1, x_2, ...)$ defined by
(143)-(145). This distribution was introduced, and many of its significant properties
developed, by McCloskey (1965) and Engen (1975) in the context of ecology. Its
relevance in genetics as describing age-ordered allele frequency distributions was first
established by Griffiths (unpublished notes). I thus call it here the GEM distribution – in
view of its beautiful properties, an appropriate acronym. Further properties of the GEM
distribution, in particular for the heaps process and for ecological processes, were found
by Patil and Taillie (1977).

The GEM distribution is a particular case of a residual allocation model (RAM). These
were introduced by Halmos (1944) who discussed the problem of distributing gold dust
among a countably infinite sequence of beggars. The first beggar receives a fraction z_1 of
the gold dust, the second beggar a fraction z_2 of the remainder, and so on, so that in the
notation of (143)-(145) the ith beggar receives a fraction x_i of the original amount.
Halmos found many properties of residual allocation models but unfortunately did not use
(145) for f(z) and thus did not arrive at any results of genetical relevance.

We now consider various properties of the GEM distribution. We have seen that the
GEM distribution is the size-biased Poisson-Dirichlet, and it is then immediate that the
Poisson-Dirichlet is the distribution of the order statistics of the GEM distribution. Both
must then have the same frequency spectrum, namely (45). This may be found by
integration using the GEM representation (143)-(145), but is not easily found directly from
the Poisson-Dirichlet.

The GEM distribution is invariant under size biasing (McCloskey (1965), Engen
(1975), Patil and Taillie (1977)). Thus if x has the GEM distribution and if y is defined
by y_1 = x_i with probability x_i , and given y_1 = x_i , y_2 = x_j with probability
$x_j/(1-x_i)$, $(i \neq j)$, and so on, then x and y have the same distribution. The GEM

distribution is the only infinite RAM having i.i.d. residual fractions $z_1, z_2,...$ enjoying this property.

This invariance property has several important consequences. Thus while the frequency spectrum (45) can be found from (143)-(145) by integration, as noted above, its most elegant derivation (Hoppe (1987)) used invariance. Let x have the GEM distribution and let $h(x_1)$ be some function of x_1. From invariance under size-biased sampling,

$$(151) \qquad E\{h(x_1)|x\} = \sum h(x_i)x_i$$

and taking unconditional expectations,

$$(152) \qquad E\{h(x_1)\} = E \sum h(x_i)x_i .$$

But, from (143),

$$(153) \qquad E\{h(x_1)\} = \int_0^1 h(x)\theta(1-x)^{\theta-1}dx ,$$

so that

$$(154) \qquad E \sum h(x_i)x_i = \int_0^1 h(x)\theta(1-x)^{\theta-1}dx .$$

Now put $h(x_i) = 1/x_i$ if $x_i \geq w$, 0 otherwise. Then

$$(155) \qquad E\{\#x_i \geq w\} = \int_w^1 \theta x^{-1}(1-x)^{\theta-1}dx ,$$

leading to (45).

Invariance also leads to an important result concerning random deletions. Using genetic terminology, choose a gene at random and delete all genes of its allelic type. If the initial distribution of allelic frequencies has the GEM distribution, so do the relative frequencies of the remaining alleles. Thus the Poisson-Dirichlet distribution has the same property, because of its relation with the GEM distribution. Hoppe (1986) observed this and showed that the non-interference property (112) of the Ewens sampling formula is inherited from the corresponding property, just noted, of its representing measure, the Poisson-Dirichlet. This conclusion is of interest in ecology, as noted earlier, in the assessment of the non-interference of species.

Another connection between the Poisson-Dirichlet and the GEM distribution is the following. We have seen that the evolution of a genetic population can be modelled equivalently as a diffusion process (Ethier-Kurtz) or a genealogy process (Donnelly-Tavaré). In the former, alleles are ordered by frequency (as in the Poisson-Dirichlet) and in the latter, by age order (as in the GEM distribution). The comparative tractability of the genealogical approach is reflected in the greater tractability of the GEM distribution compared to the Poisson-Dirichlet.

The GEM distribution gives us immediately properties of random permutations and random functions, as discussed towards the end of Section 8. The results so far as permutations are concerned are obvious in the sense that the finite permutation results are trivial. The length of the cycle containing 1 in a random permutation of $\{1,2,...,n\}$ is uniformly distributed on $(1,2,...,n)$. If the length of this cycle is m , and if j is the lowest number not in the first cycle, the length of the cycle containing j is uniformly distributed on $(1,2,...,n-m)$. This continues until all numbers have been accounted for (in a finite RAM). Since $\theta = 1$ for a permutation, the uniform distribution (145) with $\theta = 1$ provides the natural limit as $n \to \infty$ for the finite uniform distributions just noted. For random functions the value of θ is $\frac{1}{2}$, and asymptotically the proportional size of the equivalence class containing 1 has density function, from (145),

$$(156) \qquad f(x) = \frac{1}{2} (1-x)^{-\frac{1}{2}}.$$

Thus, for example, the probability that the proportional size of this equivalence class exceeds $\frac{1}{2}$ is asymptotically $1/\sqrt{2}$, a result which may be compared with the value $\log 2$ for the corresponding result for random permutations. The mean proportional size of this equivalence class is $\frac{2}{3}$, a result which may be compared with the corresponding value $\frac{1}{2}$ for random permutations, and the mean proportional size .758 for the largest equivalence class for random functions. Further results, including in particular numerical estimates of the rate of convergence, as n increases, to those limiting values, are given by Ewens and Padmadisastra (1988).

14.   Age-ordered alleles

We are now in a position to exploit the results of the preceding sections, particularly as they concern age-ordered alleles. We consider first the Ethier-Kurtz diffusion process (or, equivalently, the Donnelly-Tavaré genealogy process) and later the Moran model.

We have already found (in (135)) the probability that the oldest allele in the population is included in a sample of n genes. The GEM distribution shows more generally that the probability that the oldest allele in the population has i representing genes in the sample is

$$(157) \qquad P(i) = \frac{\theta}{n+\theta} \binom{n}{i} \binom{n+\theta-1}{i}^{-1}, \quad i = 0,1,2,...,n,$$

(Donnelly, (1986b)). The probability that the oldest allele in the sample has i representing genes is, immediately,

(158)
$$\frac{\theta}{n} \binom{n}{i} \binom{n+\theta-1}{i}^{-1} , \; i = 1,2,...,n$$

which is identical in form to (131). The probability that the oldest allele in the sample is the ith oldest in the population is, immediately from the GEM distribution formulation, (Donnelly, (1986b)),

(159)
$$[\theta/(n+\theta)]^{i-1}[n/(n+\theta)] , \; i = 1,2,3... .$$

This result was found, using another approach, by Saunders et al. (1984).

We now turn to ages of alleles. We have established the mean age of the oldest allele in the population (see (99) and (23)). Suppose next that a gene is taken at random from the population: what is the mean age of its allelic type? We can use (145) (as the probability density of the frequency of the allelic type of this gene), together with the mean time formula (16), to calculate the mean time that this allelic type persists as

(160)
$$\int_0^1 T(x)\theta(1-x)^{\theta-1}dx ,$$

and then use reversibility arguments. But (160) reduces to $2/\theta$, and this result can be found more immediately by looking backwards in time in the coalescent tracing back the line of descent of this gene, since mutations occur at Poisson rate $\theta/2$ along this line.

Many further beautiful results, analogous to (157)-(160), are given by Donnelly (1986b), Donnelly and Tavaré (1986), (1987) and Hoppe (1987), the GEM distribution often playing a key part in their derivations. We will return to the most important of these later. We can also use (157)-(160) to get results for heaps, cycles and random functions, although here the concept of taking a sample, and the analogues of moving backwards and forwards in time, are not of natural interest.

For the Moran model, equations (157), (158) and (159) continue to hold, but if we wish to find the mean age of the allelic type of a gene chosen at random, (160) must clearly be replaced by a new formula involving a summation. Here we use (131), interpreted now as the probability distribution of the number of genes of the allelic type of a randomly chosen gene, together with the mean time (21), and calculate

(161)
$$\sum_{k=1}^{M} T(k)P(k) .$$

The coalescent, however, shows immediately that the required expression is simply M/u.

We turn finally to the most important formula deriving from the GEM distribution, namely the age-labelled analogue of (58). We suppose that a sample of n genes taken from a population described by the Ethier-Kurtz diffusion, or equivalently the Donnelly-Tavaré genealogical process, is labelled by age order. Denoting by n_j the number of genes of allele j (i.e., the jth oldest allele in the sample), we seek

(162) $$\text{Prob}\{k;n_1,\ldots,n_k\}$$

where k is the (random) number of alleles in the sample. We can arrive at this probability immediately by calculating the size-biased version of (58), which is

(163) $$\frac{\theta^k(n-1)!}{S_n(\theta)n_k(n_k+n_{k-1})\ldots(n_k+n_{k-1}+\ldots+n_2)} .$$

This expression was first given explicitly by Donnelly and Tavaré (1986), although it was also known to Griffiths (unpublished notes), so I call it here the Donnelly-Tavaré-Griffiths formula. Tavaré (1987) has also shown that (163) also arises as a description of family sizes in a linear pure birth and immigration process, an interpretation associated with the work of Joyce and Tavaré (1987). The expression (163) bears the same relationship to the GEM distribution as the Ewens sampling formula does to the Poisson-Dirichlet. It is also an explicit formula for the RAM expression obtained immediately from the Hoppe representation (146) by replacing M by m , M_i by n_i and K by k . (This observation in turn shows that (163) describes the frequency distribution of the age-ordered alleles in a Moran population, making the converse notational changes. Kelly (1979, Ex. 7.2.5) derived this latter distribution, using time reversibility.) We observe that k remains a sufficient statistic for θ and that the conditional distribution of $n_1,\ldots,n_k$, given k and n , is

(164) $$(n-1)!/[\,|S_n^k|\,n_k(n_k+n_{k-1})\ldots(n_k+n_{k-1}+\ldots+n_2)]\, .$$

Suppose now that we wish to construct a test of selective neutrality more powerful than the Watterson test, by exploiting age-order information. The first step is to use (164) to find the probability distribution of any test statistic, analogous to (125), which exploits age-order information. There is no obvious test statistic to use, and we do not have any age-ordered theory in the selective case to guide us, using likelihood ratios. One possible test statistic is

(165) $$F^* = \sum_{i=1}^{k} (n_i-\mu_i)^2/\sigma_i^2 \, ,$$

where $\mu_i = E(n_i|k,n)$, $\sigma_i^2 = \text{var}(n_i|k,n)$. Although it is not hard to use (164) to find μ_i and σ_i^2 (see Tavaré et al. (1988)), there is no simple form for the probability distribution of F^* (or of any other reasonable statistic using age-order information). We therefore employ the same approach as was used with (125) for finding significance points, namely simulating many $(n_1,\ldots,n_k)$ vectors, followed by an empirical estimation of significance points.

Fortunately, it is not difficult to simulate a random vector from (164). We find

(166) $$P(n_1 = j \mid k,n) = (n-1)! \, |S_{n-j}^{k-1}| \, / [(n-j)! \, |S_n^k|] \, ,$$

and the form of (164) shows that the conditional distribution of $n_2, \ldots, n_k$, given n_1, k and n, is of the same form as (164) with n replaced by $n - n_1$ and k by $k - 1$. More precisely,

(167) $$P(n_2 = j \mid n_1, k, n) = (n - n_1 - 1)! \, |S_{n-n_1-j}^{k-2}| \, / [(n - n_1 - j)! \, |S_{n-n_1}^{k-1}|] \, .$$

Corresponding results hold for n_3, n_4 and so on. Thus we can simulate an n_1 value using (166), and given this an n_2 value using (167), and so on, thus rapidly finding a vector $(n_1, \ldots n_k)$. Many repetitions of this process lead to the desired empirical percentage points of F^*, or of any other test statistic.

Having found percentage points, it is necessary to simulate a selective model and check how often the observed value of F^*, of any similar statistic, reaches a significant value. At the same time the corresponding observations can be made for F, defined in (125), which we can think of as a test statistic which ignores age-order information. When this is done (Tavaré et al. (1988)) it is found, quite unexpectedly, that the most powerful test of neutrality is that which uses F as test statistic, that is, which ignores age-order information. The reason for this conclusion is presumably that the age-ordered likelihood ratio reduces, for small selective values, to a function of F (as does the non-age-ordered likelihood ratio). This is a useful conclusion, as well as an unexpected one, since it implies that we do not have to attempt to find age-order information to improve our testing methods. At this stage we can do no better than use the Watterson testing theory.

Acknowledgements

Very little of the above is original with me, and I thank those who have developed much of the theory, including in particular David Aldous, Peter Donnelly, Stewart Ethier, Bob Griffiths, Fred Hoppe, Dick Hudson, John Kingman, Simon Tavaré and Geoff Watterson for many discussions and much valuable advice.

References

Aldous, D.J. (1985), Exchangeability and related topics, in: *École d'été de probabilités de Saint-Flour XIII-1983* (P.L. Hennequin, éd.), Lecture Notes in Mathematics 1117, Springer-Verlag, Berlin, 2-198.

Anderson, R. (1978), Some stochastic models in population genetics, Unpublished M.Sc. thesis, Monash University.

Blackwell, D. and MacQueen, J.B. (1973).   Ferguson distributions via Polya urn schemes, *Ann. Statist.* **1**, 353-355.

Burville, P.J. (1974), Heaps: a concept in optimization, *J. Inst. Math. Appl.* **13**, 263-278.

Burville, P.J. and Kingman J.F.C. (1973), On a model for storage and search, *J. Appl. Probab.* **10**, 697-701.

Cannings, C. (1974), The latent roots of certain Markov chains arising in genetics: a new approach. 1. Haploid models, *Adv. in Appl. Probab.* **6**, 260-290.

Caswell, H. (1976), Community structure: a neutral model analysis. *Ecological Monographs* **46**, 327-353.

Constantini, D. (1987), Symmetry and the indistinguishability of classical particles, *Phys. Lett. A* **123**, 433-436.

Crow, J.F. and Kimura, M. (1970), *An Introduction to Population Genetics Theory*, Harper and Row, New York.

Donnelly, P. (1986a), Dual processes in population genetics, in: *Stochastic Spatial Processes* (P. Tautu, ed.), Lecture Notes in Mathematics 1212, Springer-Verlag, Berlin, 94-105.

Donnelly, P. (1986b), Partition structures, Polya urns, the Ewens sampling formula, and the ages of alleles. *Theoret. Population Biol.* **30**, 271-288.

Donnelly, P. (1988), Heaps processes and size-biased permutations. Submitted.

Donnelly, P. (1989), Weak convergence to a death process with an entrance boundary: ancestral processes in population genetics. To appear in *Ann. Probab.*

Donnelly, P. and Joyce, P. (1989), Consistent ordered sampling distributions: characterization and convergence. Submitted.

Donnelly, P. and Tavaré, S. (1986), The ages of alleles and a coalescent, *Adv. in Appl. Probab.* **18**, 1-19.

Donnelly, P. and Tavaré, S. (1987), The population genealogy of the infinitely-many neutral alleles model, *J. Math. Biol.* **251**, 381-391.

Engen, S. (1975), A note on the geometric series as a species frequency model, *Biometrika* **62**, 694-699.

Ethier, S.N. (1989), The infinitely-many-neutral-alleles diffusion model with ages. Preprint.

Ethier, S.N. and Kurtz, T.G. (1981), The infinitely-many-neutral- alleles diffusion model, *Adv. in Appl. Probab.* **13**, 429-452.

Ethier, S.N. and Kurtz, T.G. (1986), *Markov Processes: Characterization and Convergence*, Wiley, New York.

Ewens, W.J. (1972), The sampling theory of selectively neutral alleles, *Theoret. Population Biol.* **3**, 87-112.

Ewens, W.J. (1979), *Mathematical Population Genetics*, Springer-Verlag, Berlin.

Ewens, W.J.and Kirby, K (1975), The eigenvalues of the neutral alleles process, *Theoret. Population Biol.* **7**, 212-220.

Ewens, W.J. and Padmadisastra, S. (1989), Asymptotic and numerical results for random functions. In preparation.

Fisher, R.A. (1930), *The Genetical Theory of Natural Selection*, Clarendon Press, Oxford.

Golomb, S.W. (1964), Random permutations. *Bull. Amer. Math. Soc.* **70**, 747.

Goncharov, V. (1962), Du domaine d'analyse combinatoire, *Amer. Math. Soc. Transl.* (2) **19**, 1-46.

Griffiths, R.C. (1979a). A transition density expansion for a multi-allele diffusion model. *Adv. in Appl. Probab.* **11**, 310-325.

Griffiths, R.C. (1979b), Exact sampling distributions from the infinite neutral alleles model, *Adv. in Appl. Probab.* **11**, 326-354.

Griffiths, R.C. (1980), Lines of descent in the diffusion approximation of neutral Wright-Fisher models, *Theoret. Population Biol.* **17**, 37-50.

Guess, H.A. and Ewens, W.J. (1972), Theoretical and simulation results relating to the neutral allele theory, *Theoret. Population Biol.* **3**, 434-447.

Halmos, P.R. (1944), Random alms, *Ann. Math. Statist.* **15**, 182-189.

Hendricks, W.J. (1972), The stationary distribution of an interesting Markov chain, *J. Appl. Probab.* **9**, 231-233.

Hoppe, F.M. (1984), Polya-like urns and the Ewens sampling formula, *J. Math. Biol.* **20**, 91-99.

Hoppe, F.M. (1986), Size-biased filtering of Poisson-Dirichlet samples with an application to partition structures in genetics, *J. Appl. Probab.* **23**, 1008-1012.

Hoppe, F.M. (1987), The sampling theory of neutral alleles and an urn model in population genetics, *J. Math. Biol.* **25**, 123-159.

Hudson, R.R. and Kaplan, N.L. (1985), Statistical properties of the number of recombination events in the history of a sample of DNA sequences, *Genetics* **111**, 147-164.

Hudson, R.R. and Kaplan, N.L. (1986), On the divergence of alleles in nested subsamples from finite populations, *Genetics* **113**, 1057-1076.

Ignatov, T. (1982), On a constant arising in the theory of symmetric groups and on Poisson-Dirichlet measures, *Theory Probab. Appl.* **27**, 136-147.

Joyce, P. (1988), Age-ordered distributions associated with some neutral population genetics models. Unpublished Ph.D. thesis. University of Utah.

Joyce, P. and Tavaré, S. (1987), Cycles, permutations and the structures of the Yule process with immigration, *Stochastic Process. Appl.* **25**, 309-314.

Kaplan, N.L. and Hudson, R.R. (1987), On the divergence of genes in multigene families, *Theoret. Population Biol.* **31**, 178-194.

Kaplan, N.L. and Hudson, R.R. (1988), An evolutionary model for highly repeated interspersed DNA sequences, in: *Mathematical Evolutionary Theory* (M.W. Feldman, ed.), Princeton University Press, Princeton.

Karlin, S. and McGregor, J.L. (1972), Addendum to a paper of W. Ewens, *Theoret. Population Biol.* **3**, 113-116.

Keener, R., Rothman, E. and Starr, N. (1988), Distributions on partitions, *Ann. Statist.* **15**, 1466-1481.

Kelly, F.P. (1979), *Reversibility and Stochastic Networks*, Wiley, New York.

Kerov, S.V. and Vershik, A.M. (1986), Characters of infinite symmetric groups and probability properties of Robinson-Shenstead-Knuth's algorithm, *SIAM J. Algebraic Discrete Methods* **7**, 116-124.

Kimura, M. (1955), Solution of a process of random genetic drift with a continuous model, *Proc. Nat. Acad. Sci. U.S.A.* **41**, 144-150.

Kimura, M. (1968), Evolutionary rate at the molecular level, *Nature* **217**, 624-626.

Kimura, M. and Crow, J.F. (1964), The number of alleles that can be maintained in a finite population, *Genetics* **49**, 725-738.

Kingman, J.F.C. (1975), Random discrete distributions, *J. Roy. Statist. Soc. Ser. B.* **37**, 1-22.

Kingman, J.F.C. (1977), The population structure associated with the Ewens sampling formula, *Theoret. Population Biol.* **11**, 274-284.

Kingman, J.F.C. (1978a), Random partitions in population genetics, *Proc. Roy. Soc. London Ser.* A **361**, 1-20.

Kingman, J.F.C. (1978b), The representaton of partition structures, *J. Lond. Math. Soc.* **18**, 374-380.

Kingman, J.F.C. (1980), *Mathematics of Genetic Diversity*, SIAM, Philadelphia.

226   EWENS

Kingman, J.F.C. (1982a), On the genealogy of large populations, *J. Appl. Probab.* **19A**, 27-43.

Kingman, J.F.C. (1982b), The coalescent, *Stochastic Process. Appl.* **13**, 235-248.

Kingman, J.F.C. (1982c), Exchangeability and the evolution of large populations, in: *Exchangeability in Probability and Statistics* (G. Koch and F. Spizzichino, eds.), North-Holland, Amsterdam, 97-112.

Kirby, K. (1974), Unpublished Ph.D. thesis. Princeton University.

Kolchin, V.F. (1976), A problem of the allocation of particles into cells and random mappings, *Theory Probab. Appl.* **21**, 48-63.

Kruskal, M.D. (1954), The expected number of components under a random mapping function, *Amer. Math. Monthly* **61**, 392-397.

Lambshead, P.J.D. (1986), Sub-catastrophic sewage and industrial waste contamination as revealed by marine neatode faunal analysis, *Marine Ecology Progess Series* **29**, 247-260.

Lambshead, P.J.D. and Platt, H.M. (1985), Structural patterns of marine benthic assemblages and their relationship with empirical statistical models, in: *Nineteenth European Marine Biology Symposium*, Cambridge University Press, Cambridge, 371-380.

Malécot, G. (1984), *Les mathématiques de l'hérédité*, Masson, Paris.

McCloskey, J.W. (1965), A model for the distribution of individuals by species in an environment. Unpublished Ph.D. thesis, Michigan State University.

Moran, P.A.P. (1958), Random processes in genetics, *Proc. Camb. Phil. Soc.* **54**, 69-71.

Patil, G.P. and Taillie, C. (1977), Diversity as a concept and its implications for random communities, *Bull. Inst. Internat. Statist.* **47**, 497-515.

Pittel, B. (1983), On distributions relating to transitive closure of random finite mappings, *Ann. Probab.* **11**, 428-441.

Platt, H.M. and Lambshead, P.J.D. (1985), Neutral model analysis of patterns of marine benthic species diversity. *Marine Ecology Progress Series* **24**, 75-81.

Ross, S.M. (1981), A random graph. *J. Appl. Probab.* **18**, 309-315.

Saunders, I.W., Tavaré, S. and Watterson, G.A. (1984), On the genealogy of nested subsamples from a haploid population, *Adv. in Appl. Probab.* **16**, 471-491.

Sawyer, S. (1977), On the past history of an allele now known to have frequency p, *J. Appl. Probab.* **14**, 439-450.

Shepp, L.A. and Lloyd, S.P. (1966), Ordered cycle lengths in a random permutation, *Trans. Amer. Math. Soc.* **121**, 340-357.

Tavaré, S. (1984), Line-of-descent and genealogical processes and their applications in population genetics, *Theoret. Population Biol.* **26**, 119-164.

Tavaré, S. (1987), The birth process with immigration, and the genealogical structure of large populations, *J. Math. Biol.* **25**, 161-171.

Tavaré, S., Ewens, W.J. and Joyce, P. (1988), Is knowing the age-order of alleles useful in testing neutrality? Preprint.

Trajstman, A.C. (1974), On a conjecture of G.A. Watterson, *Adv. in Appl. Probab.* **6**, 489-493.

Vershik, A.M. (1986), The asymptotic distribution of factorizations of natural numbers into prime divisors, *Soviet Math. Dokl.* **34**, 57-61.

Vershik, A.M. and Shmidt, A.A. (1977), Limit measures arising in the theory of groups. I., *Theory Probab. Appl.* **22**, 79-85.

Watterson, G.A. (1974), The sampling theory of selectively neutral alleles, *Adv. in Appl. Probab.* **6**, 463-488.

Watterson, G.A. (1976a), Reversibility and the age of an allele. I. Moran's infinitely-many neutral alleles model, *Theoret. Population Biol.* **10**, 239-253.

Watterson, G.A. (1976b), The stationary distribution of the infinitely-many neutral alleles diffusion model, *J. Appl. Probab.* **13**, 639-651.

Watterson, G.A. (1977), Heterosis or neutrality?, *Genetics* **85**, 789-814.

Watterson, G.A. (1984), Lines of descent and the coalescent, *Theoret. Population Biol.* **10**, 239-253.

Watterson, G.A. and Guess, H.A. (1977), Is the most frequent allele the oldest?, *Theoret. Population Biol.* **11**, 141-160.

Wright, S. (1931), Evolution in Mendelian populations, *Genetics* **16**, 97-159.

Wright, S. (1949), Genetics of populations, *Encyclopaedia Britannica*, 14th ed., vol. 10, 111.

A DISCUSSION OF EVOLUTIONARILY STABLE STRATEGIES

W.G.S. Hines with D. Anfossi
Department of Mathematics
and Statistics
University of Guelph,
Guelph, Ontario   N1G 2W1
Canada

ABSTRACT. Consideration of the question "What behaviours can be expected in stable biological populations?" leads first to a description of the classical ESS population model and determination of the two ESS conditions, formulated explicitly for the constant payoff matrix case, with examples. After discussing an instructive method of summarizing populations (the mean-covariance approach), and after considering the dynamics of the evolution of populations with diversities of behaviour, we use the model to consider the effects of changes in payoff, the effects of sexual reproduction and of finite population sizes, and the implications of the models studied for the possibility of cooperative behaviour.

1.   Introduction

The theoretical study of behaviour involves modelling of roles, interrelations, interactions and decision processes on the part of one or more individuals, features which allow for an approach based on game theory. A pioneering step in this direction was the introduction of the concept of the evolutionarily stable strategy or ESS by Maynard Smith [1974]. Starting from his formulation of the concept, with consideration of behaviours as strategies, encounters as games and reproductive fitness as payoff, ESS theory has proven fruitful in its applications, the insights it provides and the numerous variants it suggests.

The perspective adopted in this presentation is to establish the relevance of ESS theory to "real" biological systems, by examining its apparent weaknesses and resolving some of them (if possible), so as to investigate its robustness and to extend its applicability. This brief survey of the topic is by no means exhaustive, and, with few exceptions, results are stated or paraphrased, but are not derived. For a more detailed treatment and a guide to the relevant literature, the reader is referred to texts and survey articles by Dawkins [1976], Barash [1982], Maynard Smith [1982], Riechert and Hammerstein [1983], Parker [1984] and Hines [1987a].

Section 2, which considers the mathematical formulation of the basic ESS argument, also discusses the relevant biological context, noting that simplifying assumptions made in the original ESS formulation are seriously at odds with important or essential aspects of real biological systems. With the constant-payoff model as a starting example, Section 3 introduces the use of statistical summaries as a useful alternative approach for population

S. Lessard (ed.), Mathematical and Statistical Developments of Evolutionary Theory, 229–267.

studies. That section focuses on a few relevant aspects of the behaviour of a system, using a mean-and-covariance analysis applicable to biology. ESS theory is then applied to polymorphic populations in Section 4, and extended to sexual populations in Section 5. The dynamics of the system are described both in terms of evolution of the population summary parameters and through results from complementary approaches. The methodology is then used, in the subsequent sections, in the study of biological population models which allow for a finite population size (Section 6), and for cooperative behaviour of the players (Section 7), in an attempt to determine whether the ESS formulation might hold when particular assumptions are relaxed.

2. The classic ESS population model

The initial question that ESS theory seeks to answer can be roughly expressed as:

What behaviours can be expected in stable biological populations?

A number of assumptions concerning the context, the population and the method at hand, first have to be made explicit. Although based on classical game theory, the ESS approach involves some relatively major changes, such as:
• the behaviour considered in ESS theory is not "rational" and is not affected by any available information (opponent's nature, various histories, and so on);
• the gains and losses resulting from individual contests (always between two individuals) are expressed in terms of biological fitness, usually interpreted as the number, or expected number, of eventual offsprings;
• players have no objectives which influence their actions or alter their strategies – rather, players whose strategies happen to be atypically advantageous at a given time have greater reproductive success and thereby increase the frequencies with which their particular strategies are used in succeeding generations.

As well, the initial model assumes an infinite population of asexual individuals, differing only in behavioural strategies which are transmitted exactly to the offsprings. We first investigate consequences of these elementary assumptions before considering the effects of introducing modifications to the model which are biologically plausible or essential.

Basic argument . With the working assumption that stable populations do exist, we expect that they should be composed of individuals who all use equally successful (but possibly distinct) strategies which cannot be bettered by any new one: the objective is to identify and characterize such strategies.

Assume first that the population is "monomorphic", with all individuals using some common strategy, say u . This strategy u is an ESS if it places any invading component using any other strategy v at a disadvantage in the population. Let E(u) denote the frequency dependent fitness of strategy u and E(u,v) the fitness of strategy u when used against strategy v. If ε is the fraction of the population, and so, by assumption, the fraction of encounters that will be with a v-user opponent, then we want:

$$E(u) > E(v) ,$$

with

(2.1) $$E(u) = (1-\varepsilon)E(u,u) + \varepsilon\ E(u,v) ,$$

and

(2.2) $$E(u) = (1-\varepsilon)E(v,u) + \varepsilon\ E(v,v) .$$

While we might want this condition to be met for all ε, it is not clear that this will always be possible, and we consider instead the case of initial invasions by v-users. For ε close to 0, we find that two necessary conditions for (2.1) and (2.2) to hold are that for all u different from v $(u \neq v)$,

(2.3) $$\text{I. } E(u,u) \geq E(v,u) ,$$

(2.4) $$\text{II. } E(u,v) > E(v,v) , \text{ when } E(u,u) = E(v,u) .$$

Condition I (analogous to that of the Nash equilibrium) is sufficient to ensure that u is an equilibrium strategy, while II is necessary to establish stability and, as we will see, does so even under more general situations that those considered here. These basic conditions now allow for a variety of paths of analysis. We restrict our attention to one simple and instructive case, the constant payoff matrix model.

3.   The constant-payoff model

This simple case is convenient for presenting the basis terminology introducing the general procedure of analysis, obtaining basic results, and for discussing extensions to the model which are of biological interest and theoretical importance.   It provides one means, therefore, of studying the effects from imposing convenient but biologically severe or implausible assumptions.

3.1. CONVENTIONS

This model considers a population satisfying the assumptions previously mentioned as well as the following:

• the generations are non-overlapping and the population size remains fixed;
• the number of contests an individual can enter is fixed;

• the number of possible tactics is finite, say n in number, and the probability that a given individual will choose any one of those is also fixed (this is biologically plausible in the case of non-continuous traits).

Recall that individuals differ only in their strategies, so that we might as well use the strategies to label and to distinguish among the various individuals. The strategies refer to the frequencies with which individuals use the various tactics available, so that a user of a strategy u , a u-user, will use tactic i with probability u_i , and a strategy u is then described by the n-dimensional column vector $(u_1,...,u_n)^T$. Necessarily, all components of u are non-negative (although some components might be 0), and the components sum to unity.

It is convenient to have a term for the situation in which various strategies all make use only of [some of] those tactics which are used by some particular strategy: we say that those strategies are *in the support* of that strategy. One basic result in ESS theory for the constant-payoff model is that any strategy in the support of the ESS has the same fitness as the ESS if the mean strategy of the population is the ESS, as occurs, for example, if the vast majority of the population are using the ESS as their individual strategies and if the few who depart form that strategy do so in a diversity of directions so as still to have the ESS as their *mean* strategy.

The set of mathematically possible strategies constitutes a probability simplex defined by:

$$\Delta = \{u = (u_i,...,u_n) \mid u_i \geq 0,\ h^T u = 1\} ,$$

where $h^T = (1,...,1)$. The simplex, which forms a triangle for n = 3 , a tetrahedron for n = 4 , etc., can further be divided into two sets. The interior of Δ , $\mathring{\Delta}$ contains the "interior" strategies which make use of all tactics available, i.e., no $u_i = 0$. The complement of the interior is the boundary of Δ , $\delta\Delta$. The boundary, where at least one of the u_i is 0 , consists of the edges and vertices of Δ , representing respectively, "boundary mixed" strategies (with no $u_i = 1$ and with some $u_i = 0$) , and pure strategies (with some $u_i = 1$ and, necessarily, all other $u_i = 0$) which use exclusively one of the tactics available. The term *mixed strategy* refers to both "interior" and "boundary mixed" strategies.

In Figure 1, the pure strategies of Δ correspond to the vertices r , s and p [for Rock, Scissors, Paper — a classic example of a three tactic game]; the boundary strategies include these pure strategies and the other strategies, such as w , along the edge of Δ ; the interior strategies such as u and v are on the "inside" of Δ ; and the mixed strategies are all points of Δ other than the three points r , s , and p . [Figure 1(a) shows how Δ might look if it were propped in a corner of the room, with each of the points r , s and p being one metre from the corner. Figure 1(b) dispenses with the axes of the corner of the room, for greater graphical clarity, although the coordinates of a given strategy u can be used relative to those axes to locate the strategy in the simplex Δ .]

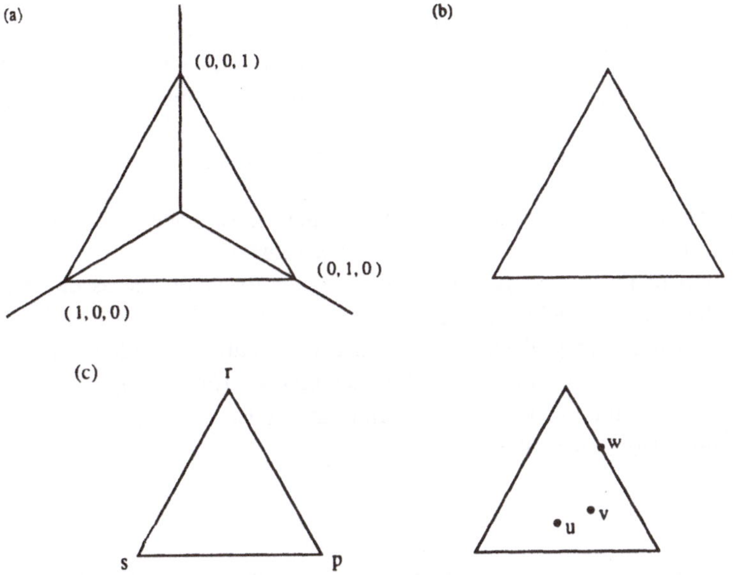

Figure 1

The contest is characterized by the payoff matrix $A = (a_{ij})$, where a_{ij} is the return of using tactic i against tactic j. For example, one particular case of a standard contest in ESS theory, the Hawk-Dove game with two tactics – the threatening and potentially expensive behaviour "Hawk", and the relatively unthreatening behaviour "Dove" – could have a payoff matrix of the following form, with columns and rows each with labels H and D indicating the tactic chosen by the individual (giving the row used of A) and by the opponent (determining the column) and with the entries showing the benefits received by the first player:

$$\begin{array}{cc} & \begin{array}{cc} H & D \end{array} \\ A = & \begin{pmatrix} 1 & 4 \\ 0 & 2 \end{pmatrix} \begin{array}{c} H \\ D \end{array} \end{array}$$

In this contest, any individual who decides to play "Hawk" against an opponent who chooses, independently, to play "Dove" will receive 4 units of benefit. The table also shows that the opponent in that case will receive 0 units.

Faced with an opponent who uses strategy v, the vector of payoffs resulting form using tactic $i = 1,\dots,n$ is then given by Av. Some payoff matrices possess special properties that reflect the nature of the game. For example, in a *zero-sum* game [your win is exactly my loss], $A + A^T = 0$, while in a *symmetric* game, $A = A^T$. [In a symmetric game, you and I win equally at all times, so that what improves your situation improves mine equally, making our interests effectively identical].

Given the possible behaviours, the type of game and the payoff structure, and the assumption that individuals choose their tactics with constant probabilities independently of their opponent's nature or possible choice, the gain to a u-user resulting from a contest with a v-user will be expressed as a weighted average of the returns of each tactic:

$$E(u,v) = \sum_i \sum_j a_{ij} u_i v_j = u^T A v ,$$

(where the subscript on the summations indicate the indices summed over, and summation is always understood to be over the set of all possible values).

While this form of the fitness function follows naturally given the assumed nature of the competition between individuals, it can be inappropriate when individuals' strategies are not described in terms of probabilities, or when contributions to fitnesses acquired by individuals over several contests do not combine additively. (For example, the cumulative fitness obtained by an individual over its lifetime might be the maximum, minimum or product of its returns from all contests.)

3.2. ESS CONDITIONS

To characterize an ESS under this model, the conditions (2.3) & (2.4) will be formulated in terms of various projection matrices, Π and $\Pi(u)$, which allow for considerable notational simplifications.

Matrix Notation. Recall the n-dimensional vector $h = (1,...,1)^T$. Denote the $n \times n$ identity matrix by Id, and define the projection matrix Π, by $\Pi = Id - n^{-1} h\, h^T$. Important properties of Π include the following:

(i) $\Pi = \Pi^T$ (Π is symmetric)
(ii) Π is a projection matrix. That is, $\Pi = \Pi^2$, so that $\Pi (\Pi\text{-Id}) = 0$, and the eigenvalues of Π are each either zero and one.
(iii) Trace $(\Pi) = n\text{-}1$. This and (ii) imply that the eigenvalue zero has multiplicity 1, and one has multiplicity $n - 1$.

Immediate consequences are that:
• $\Pi h = 0$ (the projection is orthogonal to h) ;
• if $h^T x = \sum x_i = 0$, then $\Pi x = x$ (vectors orthogonal to h are not altered by the projection operation).

Similarly, let $h(u) = (h_1(u),...,h_n(u))^T$, where $h_i(u) = 0$ if $u_i = 0$, and $= 1$ otherwise. Note that $h(u)$ locates the *support* of u, the set of tactics given non-zero probabilities by u. Simple calculation shows that there are $k(u) = h^T h(u) = h(u)^T h(u)$ tactics used by any particular (boundary or other) strategy u. Correspondingly, let $Id(u)$ be a diagonal matrix with an entry of unity at the i-th diagonal position if $u_i > 0$, and with all other entries equal to 0. If we define the related projection matrix $\Pi(u) = Id(u) - k^-$

$^1(u)h(u)h^T(u)$, then that projection matrix has a set of exactly $k - 1$ (orthogonal) non-zero eigenvectors x such that $Id(u)x = x$ and $h(u)^Tx = 0$.

The first ESS condition. Consider a strategy u^* satisfying equation (2.3). That equation requires that for all $v \in \Delta$ different from u^*,

$$u^{*T} Au^* \geq v^T Au^* ,$$

or

(3.1) $$(u^*-v)^TAu^* \geq 0$$

Suppose that there exist two particular strategies $u^* - x$ and $u^* + x$ in the support of u^* . For such x , (3.1) implies that $x^TAu^* = 0$. If u^* is interior, then for all (sufficiently small) x such that $\Pi x = x$ we have a set of $n - 1$ conditions on Au^* (i.e., $x^T \Pi^T Au^* = 0$)   and an additional, generically independent, condition $h^T\Pi^TAu^* = 0$, so that $\Pi A u^* = 0$ and u^* is a null right eigenvector of ΠA , which generically completely determines u^* as unique when combined with the requirement on u (which holds for any strategy vector) that its probabilities sum to unity ($h^Tu = 1$). If u^* is not an interior strategy, then the above discussion can be modified, using $\Pi(u)$ in place of Π and with the same conditions on x , leading to the requirements that $\Pi(u)A u^* = 0$, $\Pi(u)u^* = u^*$, and $h(u)^Tu^* = 1$.

Note: It can be verified that if u^* is an ESS, then $Id(u^*)Au^* = c h(u)$ for some c .

Aside: A property is generic it it holds except possibly for some exceptional cases, if, in addition, it holds for very minor modifications to these cases. For example, the property that "Any two straight lines intersect" is generic since any particular exception – a pair of parallel lines – can be modified by altering the slope of one or other of the lines by an arbitrarily small amount to make the statement true. Generic properties are "typically" true.

The second ESS condition. Suppose that u^* is an equilibrium strategy in the sense that it satisfies (3.1). Consider the two possible cases, that u^* is interior, and that it is not interior.

(i) u^* is an interior strategy with $\Pi Au^* = 0$, which implies that for all v in Δ , $u^{*T}Au^* = v^TAu^*$, then (2.4) applies and can be reformulated as

$$u^{*T} Av > v^T Av \text{ (for } u \neq v)$$

or

$$u^{*T} Av - v^T Av - (u^*-v)Au^* > 0$$

by including a term which is equal to zero. This expression simplifies to

$$-(u^*-v)^T A(u^*-v) > 0 \qquad \text{(again, for } u = v\text{)}$$

or with $u^* - v = x \neq 0$ (so that $h^T x = h^T(u^* - v) = 1 - 1 = 0$, implying that $\Pi x = x$)

(3.2) $$x^T(A+A^T)x < 0.$$

where we have used the fact that $x^T A x = x^T A^T x$ (since the transpose of a 1×1 matrix is itself). Note that $x^T(A+A^T) x = xT \ \Pi^T(A+A^T)\Pi \ x$.

(ii) If u^* is not an interior strategy, then, confining our attention to the support of u^*, we require that $\Pi(u^*)Au^* = 0$, and that for all $x = 0$ with $\Pi(u^*) x = x$, the stability condition is

(3.3) $$x^T(A+A^T) x < 0,$$

where we note that $x^T(A+A^T) x = x^T \ \Pi(u^*)^T(A+A^T)\Pi(u^*) x$.

(iii) Note that in the case where u^* is interior, (3.2) is required to hold for *all* x, while if u^* is not interior, the parallel condition (3.3) is only imposed on *some* x. We can combine the two conditions into one of "strong stability": for all $x \neq 0$ with $\Pi x = x$,

(3.4) $$x^T(A+A^T) x < 0.$$

or, say, for all y, $y^T\Pi(u^*)^T(A+A^T)\Pi(u^*)y \leq 0$, with equality only if all components of y are equal, so that for some constant c, $y = c\,h$.

Some standard games of considerable interest do not satisfy the stability condition (strong or otherwise), even though they possess equilibrium strategies satisfying (3.1). Examples include Dawkins' "Battle of the Sexes", the zero-sum game previously mentioned, games with multiple ESSs [Bishop and Cannings, 1976] and a 2-locus model with ESS and cycling [Akin and Hofbauer, 1982].

Digression. Finding the actual strategy satisfying conditions (3.1) and (3.3) using standard computer package capabilities.

Provided that the ESS is in the interior of Δ, $\mathring{\Delta}$, the procedure is in two stages: the first step involves finding that interior equilibrium, and the second step involves checking for stability.

(i) The first step requires the solving of n linear equations in n unknowns: $\Pi^T Au^* = 0$ and $h^T u^* = 1$. (The first [matrix] equation provides n somewhat redundant equations in n unknowns. One of these [and often an arbitrary one] can be deleted. Alternatively, u^* can be found as a right eigenvector, e_o, of $\Pi^T A$ corresponding to the eigenvalue 0, and then setting $u* = (h^T e_0)^{-1} e_0$, provided that all components of $u*$ are then non-negative. [The case where this non-negativity requirement is not met is more complicated, requiring deletion of various strategies.]

(ii) The verification of the strong stability condition involves:

• calculating $\Pi(A+A^T)$, or equivalently $\Pi(A+A^T)\Pi$.
• finding the eigenvalues of this matrix. (They will necessarily be real rather than imaginary or complex, since the matrix is symmetric.)
• checking that there is exactly one (real) eigenvalue of zero, and $n-1$ strictly negative real eigenvalues.

The existence of even one positive real eigenvalue indicates a "direction" of possible instability.

3.3. THREE EXAMPLES

A contest with a pure ESS, and the strong stability condition. This is a two-player two-tactic "Hawk-Dove" game, with the payoff matrix A:

$$A = \begin{pmatrix} 1 & 4 \\ 0 & 2 \end{pmatrix}.$$

• The ESS found is not interior: $u*^T = (1,0)^T$, so that $h(u*)^T = (1,0)^T$.
• $Id(u*) = \begin{pmatrix} 1 & 0 \\ 0 & 0 \end{pmatrix}$, $\Pi(u*) = \begin{pmatrix} 0 & 0 \\ 0 & 0 \end{pmatrix}$, $\Pi = 0.5 \times \begin{pmatrix} 1 & -1 \\ -1 & 1 \end{pmatrix}$.
• The stability condition (3.3) is satisfied trivially, given the nature of $\Pi(u*)$.
• The strong stability condition is also satisfied, since

$$\Pi(A+A^T)\Pi = 0.5^2 \begin{pmatrix} 1 & -1 \\ -1 & 1 \end{pmatrix} \left(\begin{pmatrix} 1 & 4 \\ 0 & 2 \end{pmatrix} + \begin{pmatrix} 1 & 0 \\ 4 & 2 \end{pmatrix} \right) \begin{pmatrix} 1 & -1 \\ -1 & 1 \end{pmatrix} = 0.5 \begin{pmatrix} -1 & 1 \\ 1 & -1 \end{pmatrix}$$

has two eigenvalues, 0 and -1, as required.

A contest with an interior ESS, and the strong stability condition. This is a more standard two-player two-tactic "Hawk-Dove" game, with the payoff matrix

$$A = \begin{pmatrix} -1 & 4 \\ 0 & 2 \end{pmatrix}.$$

• The ESS found is interior: $u*^T = 3^{-1}(2,1)^T$, so that $h(u*)^T = (1,1)^T$.
• $Id(u*) = \begin{pmatrix} 1 & 0 \\ 0 & 0 \end{pmatrix}$, $\Pi(u*) = \Pi = 0.5 \times \begin{pmatrix} 1 & -1 \\ -1 & 1 \end{pmatrix}$.

- The relevant stability condition is (3.2), equivalent to the strong stability condition.
- The strong stability condition is satisfied, since

$$\Pi(A+A^T)\,\Pi \;=\; 0.5^2 \begin{pmatrix} 1 & -1 \\ -1 & 1 \end{pmatrix}\left(\begin{pmatrix} -1 & 4 \\ 0 & 2 \end{pmatrix}+\begin{pmatrix} -1 & 0 \\ 4 & 2 \end{pmatrix}\right)\begin{pmatrix} 1 & -1 \\ -1 & 1 \end{pmatrix} \;=\; 0.5\begin{pmatrix} -3 & 3 \\ 3 & -3 \end{pmatrix}$$

has two eigenvalues, 0 and -3 , as required.

A contest with two pure ESSs, but without the strong stability condition. This an improper "Hawk-Dove" game, with the payoff matrix

$$A \;=\; \begin{pmatrix} 1 & 0 \\ 0 & 2 \end{pmatrix}.$$

- Two ESSs can be found; neither is interior. They are

$$u^{*(1)T} \;=\; (1,0)^T, \text{ with } h(u^{*(1)})^T \;=\; (1,0)^T,$$

and

$$u^{*(2)T} \;=\; (1,0)^T, \text{ with } h(u^{*(2)})^T \;=\; (1,0)^T.$$

- There are two corresponding sets of matrices

$$Id(u^{*^{(1)}}) = \begin{pmatrix} 1 & 0 \\ 0 & 0 \end{pmatrix},\ Id(u^{*^{(2)}}) = \begin{pmatrix} 0 & 0 \\ 0 & 1 \end{pmatrix},\ \Pi(u^{*^{(1)}}) = \Pi(u^{*^{(2)}}) = \begin{pmatrix} 0 & 0 \\ 0 & 0 \end{pmatrix}.$$

- In each case, the stability condition (3.3) is satisfied trivially, given the nature of $\Pi(u^*)$.
- The strong stability condition is *not* satisfied, since

$$\Pi(A+A^T)\,\Pi \;=\; 0.5^2 \begin{pmatrix} 1 & -1 \\ -1 & 1 \end{pmatrix}\left(\begin{pmatrix} 1 & 0 \\ 0 & 2 \end{pmatrix}+\begin{pmatrix} 1 & 0 \\ 0 & 2 \end{pmatrix}\right)\begin{pmatrix} 1 & -1 \\ -1 & 1 \end{pmatrix} \;=\; 0.5\begin{pmatrix} 3 & -3 \\ -3 & 3 \end{pmatrix}$$

has two eigenvalues, 0 (unavoidable, given the presence of Π in the expression) and +3 .

It can be shown that the strong stability condition, which is not satisfied in this last example, precludes the possibly of multiple ESSs.

3.4. EXTENSIONS

The previous outline of ESS theory applied to the basic monomorphic population case indicates some of the strengths and weaknesses of such an approach, but perhaps more importantly it opens the field to further questions and investigations. Some of these will be discussed here or in the following sections while others will only be briefly suggested.

Effects of Changes in Payoff. The payoff matrix can be modified to reflect deviations from the initial case, in the nature and structure of the game.
• If a certain amount of variation in payoff dependent only on the opponent's choice of tactic is introduced in the payoff matrix by adding constants to the columns of A, so that A becomes $A + hd^T$, the position and stability of the ESS, u^*, are not affected. [Zeeman, 1981, who notes that "If a constant is added to a column of A then the pay-off to all strategies is increased equally, and so the advantage of each strategy is unaltered; hence the dynamic is unaltered. Therefore, by subtracting a suitable constant from each column, we can simplify A by reducing its diagonal to zero, without altering the flow."] The result follows since we now require u^* to satisfy $\Pi(A+hd^T)u^* = 0$, which reduces to $\Pi Au^* = 0$ because $\Pi h = 0$, and since we now require that, for all y,

$$y^T \Pi(u^*)^T[(A+hd^T) + (A+hd^T)^T)]\Pi(u^*) y \leq 0,$$

which reduces to the previously imposed requirement that for all y,

$$y^T \Pi(u^*)^T(A+A^T)^T)\Pi(u^*) y \leq 0,$$

again since $\Pi h = 0$.
• If gains and losses depending only on the first player's strategies are arbitrarily added to or subtracted from the initial payoffs, by adding constants to the rows of A, so that A becomes $A + d h^T$, the locations of the equilibrium for the modified game is changed to $u^\sim$, say, if $d = -Au^\sim$ or if $d = -\Pi au^\sim$. (We can say that the payoff matrix is "tilted" to locate the ESS at $u^\sim$.)

The argument used to show this is little different from the argument just used for the previous case.
• If small random amounts are added to A, the solution to the (perturbed) equilibrium equation $\Pi Au^\sim = 0$ is relocated. Such modifications could correspond, for example, to the introduction of random environmental (or other) variation in the model, affecting the structure of the game. It can be shown [Hines, 1982] that strategies at the ESS for the mean payoff matrix have strictly greater than average fitness.

The ESS formulation allows for algebraic ways of analyzing basic constant-payoff games. It can be proved that none of the three cases above generically alters the strong stability condition of A if it is initially satisfied and if the changes are small enough. In the first two cases, the relevant eigenvalues are unaltered, while in the last, the relevant negative eigenvectors remain negative for sufficiently small changes in the payoff matrix, while the zero eigenvalue remains zero, by virtue of the presence of the Π's in the stability condition.

Other changes. Given the practical usefulness of ESS theory, a natural question, or set of questions, is whether or not the theory is indeed relevant to situations in which some other assumptions are introduced (or where some of the original assumptions are relaxed). Some of the questions that arise are:

• Will the theory apply to populations with a diversity of strategies? If not the case, this would be a major obstacle in the application of ESS theory. We examine the situation in Section 4.

• What can be said if assumptions are modified to improve the plausibility of the model in a biological context, specifically assumptions dealing with:

- the use of sexual rather than asexual reproductions? Some results are discussed in Section 5.

- the possibility of mutations altering strategies? Some results are discussed as part of Section 7, on cooperation.

- sexual populations, where strategy transmission is not exact, so that offspring strategies are not identical to those of the parents?

- the effect of spatial distribution on the interactions between strategies present in the population?

- the situation where generations overlap and hence notions of population size and ratios of sizes of strategy-user groups have to be redefined?

• What are the consequences of assuming a finite population size? Preliminary work in this area will be presented in Section 6.

Some of the above questions which are not discussed in this article are discussed, for example, in the review article by Hines. Others have not been explored deeply.

4.   ESSs and polymorphic populations

A monomorphic population, where all members use a strategy u^* characterized by the ESS conditions cannot be invaded by a monomorphic group of individuals using any differing strategy. What is true in the case of polymorphic populations, in which a diversity of strategies exists? While polymorphic populations do not immediately lend themselves to the same direct application of ESS theory, the concept of ESS still proves relevant, although for such populations there are various methods of analysis. The most common approach by far is the analysis of the interrelated evolutions of individual frequency components, often by the numerical or analytic solution of systems of differential equations. These methods are direct but not always tractable or intuitive in their conclusions.

An alternative approach is to study the evolution of summaries of population structure – in particular the evolution of the mean strategies as influenced by the diversity of strategies, as measured by covariance matrices. While perhaps lacking the definitiveness of conclusion which can be obtained with more detailed mathematical analyses, the approach is concise because of the relatively simple summaries used, attractive biologically, because it requires a limited description of actual biological populations, and flexible since it allows for the consideration of such phenomena as mutational and environmental fluctuations.

4.1. THE MEAN COVARIANCE SUMMARY

To this point, we have considered a special strategy, u* used by (almost) all members of a population. We now focus attention on the behaviour of the population's *mean* strategy μ, and will be interested in a particular mean strategy, $\mu*$, which is relevant to populations with stable compositions. We will find that the strategy $\mu*$ satisfies the same conditions as the previously encountered strategy $\mu*$, so that the two strategies will be numerically the same. We retain the difference in notation, however, to emphasize the difference in meaning. In order to index quantities changing over time, we introduce the subscript [t], as needed, to indicate a quantity at time t, so that, for example, $\mu_{[t]}$ indicates a mean at time t.

Notation. Consider first the question of describing a polymorphic population at some given time, say with strategies u,r,..., all lying in a corresponding simplex Δ. As one form of description of the population, denote the fraction of the population using strategy u by p(u) an so on.

The *mean strategy* μ of the population is defined by the weighted average

$$\mu \quad = \quad u\ p(u) + r\ p(r) + \dots$$

$$= \quad \Sigma_u\ u\ p(u) \qquad \text{[for finite or countable number of strategies]}$$

(4.1)

$$= \quad \int_u\ u\ p(u)\ du \qquad \text{[for a continuous distribution of strategies on } \Delta]$$

$$= \quad \int_u\ u\ dP(u) \qquad \text{[for a general distribution of strategies on } \Delta]\ ,$$

where, as before the subscript on Σ or on $\int$ indicates the variable which is summed or integrated over its range of possible values [here, over all u in Δ].

We will concentrate on the first of these situations, since it is notationally and conceptually simpler, and frequently more relevant to actual biological populations, but parallel discussions are straightforward. An interior mean strategy, $\mu*$ ($\mu* \in \overset{\circ}{\Delta}$), analogous to u*, has the property that $A\mu* = ch$, for some c.

For finite or countable numbers of employed strategies, the *covariance matrix* C, is defined as a weighted sum of matrices,

(4.2)
$$C = \Sigma_u\ (u\text{-}\mu)\ (u\text{-}\mu)^T\ p(u)\ ,$$

while, more generally,

$$C = \int_u (u-\mu)(u-\mu)^T \, dP(u).$$

This matrix gives a convenient indication of the various degrees and directions of strategy diversity. As an introduction to useful properties of the covariance matrix, let us consider

$$u - \mu = (u_1-\mu_1, u_2-\mu_2)^T,$$

the deviations of strategy u from μ in each of the possible tactics. For an arbitrary vector w, the vector inner product

$$w^T(u-\mu) = w_1(u_1-\mu_1)+\ldots+w_n(u_n-\mu_n)$$

gives a weighted combination of these deviations. For example, if $w^T = (1,0,0,\ldots,0)$, then $w^T(u-\mu) = (u_1-\mu_1)$ is the deviation of the frequency with which strategy u uses tactic 1 from the frequency with which strategy μ uses that same tactic. A detailed knowledge about all of these deviations would tell us a lot (too much?) about the diversity of strategies in the population. One aspect of these deviations is how large they typically are, as measured, say, by the average of the squared deviations.

The averaged square value of $w^T(u-\mu)$ is

(4.3)
$$\Sigma[w^T(u-\mu)]^2 \, p(u) = \Sigma \, w^T(u-\mu)(u-\mu)^T w \, p(u)$$

$$= w^T C w.$$

For example, if $C_{11} = 0$, it follows that the average squared value of $(u_1-\mu_1)$ is 0, (just take $w = (1,0,0,\ldots)^T$), and we can conclude that there is then no variability in the frequencies with which tactic #1 is used in the population.

In general, when

- $C = 0$,   there is no diversity in strategies,

- C is small,   all strategies tend to be similar,

- C is large,   there is large variation in strategies used.

Remarks.

- The expression (4.3), $w^T C w \geq 0$ (referred to as implying that C is *positive semi-definite*), immediately indicates that $w^T(Cw) \geq 0$. That is, the action of C on the vector w is to change its direction by at most 90°. In contrast to this, the payoff matrix A is negative definite, and the effect of A acting on (non-zero) x, ax, is to redirect x to lie strictly within 90° of $-x$, since $x^T A x < 0$.

- Recall the vector $h = (1,1,\ldots,1)^T$. Since the probabilities in every strategy sum to unity, $h^T u = 1 = h^T \mu$ for all $u \in \Delta$. Therefore $[h^T(u-\mu)]^2 = 0$ for *each* u and so the average of this square, $h^T Ch$, is also 0. This result is not surprising, since there is no diversity in the totals of the probabilities for the various strategies, regardless of whatever other diversity might exist.

In fact, $Ch = \Sigma (u-\mu)(u-\mu)^T hp(u) = 0$, and so C always has h as a null eigenvector.

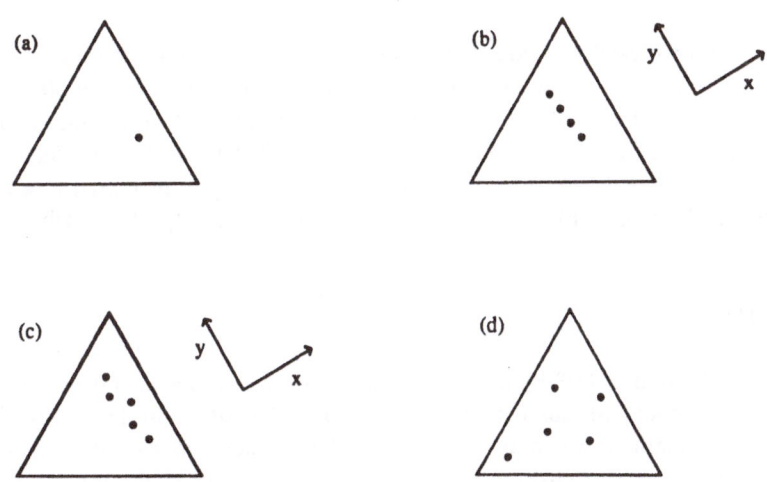

Figure 2

• Recall the projection matrix Π onto the surface in which the simplex lies. For all x such that $\Pi x = x$ (that is, for all x which are invariant under the projection Π), $x^T CX$ measures the strategy diversity in the direction of x. That is, as Figure 2 suggests, if x is a possible direction of differences in strategies (so that $hx = 0$), the population has
  - no variability of strategies in the direction x if $x^T Cx = 0$,
  - little variability of strategies in the direction x if $x^T Cx$ is small,
  - considerable variability of strategies in the direction x if $x^T Cx$ is large.
• Since all tactics used in a given population are in the support of the mean strategy μ, for any strategy employed in the population, $\Pi(\mu)(u-\mu) = u - \mu$. This implies, in (4.2), that

$$C = \Sigma_u \Pi(\mu)(u-\mu)(u-\mu)^T \Pi(\mu) p(u)$$

$$= \Pi(\mu)\Sigma_u (u-\mu)(u-\mu)^T p(u) \Pi(\mu)$$

(4.4)

$$= \Pi(\mu) C \Pi(\mu)$$

$$= C \Pi(\mu) = \Pi(\mu) C, \text{ (by similar calculations)}$$

• Since $\Pi(\mu^*)A\mu^* = c\, h(\mu^*)$ for some constant c, and since $\Pi(\mu)\, h(\mu^*) = 0$ for all mean strategies μ with support within the support of μ^*, the previous remark implies that

$$CA\mu^* = C\,\Pi(\mu^*)\, A\mu^* = C\,\Pi(\mu^*)^2\, A\mu^*$$

(4.5)

$$= C\,\Pi(\mu^*)\, c\, h(\mu^*) = 0,$$

provided that the diversity of strategies measured by C is within the support of μ^*.

• If the strategies u are thought of as points on the strategy simplex surface with associated weights corresponding to their probabilities $p(u)$, then the mean strategy is the center of gravity of such a distribution of weight. Changes in the current frequencies of use of existing strategies or the introduction of a new strategy then lead to a redistribution of weight, and so, possibly, to a relocating of the center of balance (the mean) of the weights.

4.2. DYNAMICS

As Taylor and Jonker [1978] recognized, the dynamic aspect is intrinsic to ESS theory since the outcomes of contests (or games) are described in terms of the eventual reproductive fitness of an individual. The differing successes of the various strategies used therefore determine changes in frequencies of strategies in the following generations. [The strategies employed are assumed to be perfectly transmitted from parent to offspring.] Let $p_{[0]}(u)$ and $p_{[1]}(u)$ be the frequencies of u in successive (non-overlapping) generations, and recall that $\mu_{[0]}$ and $\mu_{[1]}$ denote the corresponding mean strategies. We can conclude that:

(i) For generation 0, the fitness of a u-user playing an opponent with any strategy v in the simplex is:

$$E(u) = \sum_v u^T A v\, p_{[0]}(v)$$

(4.6)

$$= u^T A \sum_v v\, p_{[0]}(v)$$

$$= u^T A\, \mu_{[0]}$$

Note that this result does not depend on the diversity of strategies employed in the populations, but just on the mean strategy: it is the same whether all individuals in the population play $\mu_{[0]}$, all play pure strategies (with the successive components of $\mu_{[0]}$ giving the proportions playing tactic #1, tactic #2, and so on), or all play arbitrarily selected strategies in Δ which average out to $\mu_{[0]}$.

(ii) The (expected) fraction of the next generation that uses strategy u equals the product of the number of u-users in generation 0 and their fitnesses, divided by the total size of generation 1. That is,

$$p_{[1]}(u) = \alpha_{[0]} E(u) p_{[0]}(u) = \alpha_{[0]} u^T A \mu_{[0]} p_{[0]}(u) ,$$

where $\alpha_{[0]}$ is a scaling factor holding the total strategy frequency equal to 1:

$$\sum_u p_{[1]}(u) = \sum_u \alpha_{[0]} u^T A \mu_{[0]} p_{[0]}(u) = 1$$

which gives

$$\alpha_{[0]}\mu_{[0]}^T A \mu_{[0]} = 1 ,$$

so that

(4.7)
$$\alpha_{[0]} = (\mu_{[0]}^T A \mu_{[0]})^{-1} .$$

(iii) The mean strategy of the following generation is then

$$\mu_{[1]} = \sum_u u\, p_{[1]}(u)$$

$$= \sum_u \alpha_{[0]} u\, u^T A \mu_{[0]} p_{[0]}(u)$$

$$= \alpha_{[0]} \sum_u (u-\mu_{[0]}+\mu_{[0]})(u-\mu_{[0]}+\mu_{[0]})^T p_{[0]}(u) A \mu_{[0]}$$

(4.8)
$$= \alpha_{[0]} \{\sum_u(u-\mu_{[0]})(u-\mu_{[0]})^T p_{[0]}(u) A \mu_{[0]}$$

$$+ 0 + 0 + \alpha_{[0]} \sum_u \mu_{[0]}\mu_{[0]}^T p_{[0]}(u) A \mu_{[0]}\}$$

$$= \alpha_{[0]} \{C_{[0]}+\mu_{[0]}\mu_{[0]}^T A \mu_{[0]}\}$$

$$= \alpha_{[0]} \{C_{[0]}^T A \mu_{[0]}+\mu_{[0]}\} ,$$

recalling the previously cited condition on $\alpha_{[0]}$, and where the two "0" terms follow since, for example,

$$\sum_u(u-\mu_{[0]})\mu_{[0]}^T p_{[0]}(u) A \mu_{[0]} = \{\sum_u(u-\mu_{[0]}) p_{[0]}(u)\}\mu_{[0]}^T A \mu_{[0]}$$

$$= \{\mu_{[0]}-\mu_{[0]}\}\mu_{[0]}^T A \mu_{[0]} = 0 .$$

Changes in Mean. As before, let $\mu_{[0]}$ and $\mu_{[1]}$ be the mean strategy in successive generations. The change in mean strategy of the population for generation 0 to generation 1 is

$$\mu_{[1]} - \mu_{[0]} = \alpha_{[0]}C_{[0]}A\mu_{[0]} + \alpha_{[0]}\mu_{[0]}^{T}A\mu_{[0]} - \mu_{[0]}$$

(4.9)
$$= \alpha_{[0]}C_{[0]}A\mu_{[0]} + (\alpha_{[0]}\mu_{[0]}^{T}A\mu_{[0]}-1)\mu_{[0]}$$

$$= \alpha_{[0]}C_{[0]}A\mu_{[0]} = \alpha_{[0]}C_{[0]}A(\mu_{[0]}-\mu^{*}) .$$

from the earlier comments made about the properties of the covariance matrix.

As a notational convenience, we will assume that A has been scaled so that $\alpha_{[0]}$ and its successors in the next generations are all practically equal to unity.

4.3. EVOLUTION OF THE MEAN STRATEGY

The relation (4.8 and 4.9) suggests the expression

(4.10a)
$$\frac{d\mu}{dt} = \mu_{[0]}C_{[0]} A(\mu-\mu^{*})$$

for the evolution of the mean strategy in continuous time (effectively when evolution takes place slowly, with the selection effect per generation being very small), and the expression

(4.10b)
$$\mu_{i+1} - \mu^{*} = (Id+\alpha_{[0]}C_{[0]}A)(\mu_{i}-\mu^{*})$$

for the evolution of the mean strategy in discrete times, with the selection effects being appreciable in each generation.

We can interpret these expressions for the changes in mean strategy as follows:

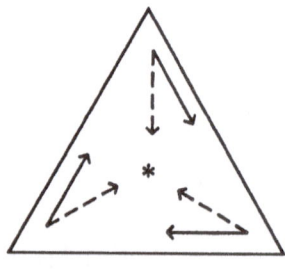

Figure 3

- $\mu - \mu^*$ represents deviations of mean strategies from the ESS.
- For every deviation, $\mu - \mu^*$, there is a vector $A(\mu-\mu^*)$, as in Figure 3, which indicates potential directions in which strategies *could* be changed to benefit from the difference between the population's current mean strategy and the ESS. The assignment of a vector to each strategy in the probability simplex Δ results in a *vector field* over Δ.
- While directions for potential improvement in strategies do exist, they may not be exploitable for a given population. The covariance matrix $C_{[0]}$ measures the variability of the strategies that actually exists, and the population's mean strategy tends to evolve in the directions in which there is both substantial existing variability and substantial potential for improvement of return.
- Since A and C are negative-definite and positive semi-definite, respectively (in the restricted sense of their effects on deviations x which are possible changes in strategy which lie in the support of μ^*), the effect of the matrix $-A$ operating on $\mu - \mu^*$, $-A(\mu-\mu^*)$, is to change the direction of $\mu - \mu^*$ by *less* than 90°, while the effect of C operating on $A(\mu-\mu^*)$, $CA(\mu-\mu^*)$, is to change the direction of $A(\mu-\mu^*)$ by *no more than* 90°.
- Equations (4.8 and 4.9) imply that if $\mu = \mu^*$, then no further change in the evolution of the population's mean strategy will occur.

    Together, these suggest the following pattern of evolution:
- A population not at equilibrium changes in mean strategy, in a direction within 90° of an indicated direction of strategy improvement (given by $A(\mu-\mu^*)$). This direction, in turn, is within 90° of the direction from the current mean strategy, μ, towards the ESS.
- The initial change in mean strategy can be almost exactly *away from* the ESS, but prolonged movement away from the ESS inevitably brings the population's mean strategy close to limits such as the boundary, $\delta\Delta$, thereby substantially reducing the possible strategy diversity in that direction.
- If the population's mean continues to evolve near the boundary of Δ, or near other surfaces which influence the diversity of strategies possible, the covariance matrix may well evolve, and its evolution may well have substantial implications for the evolution of the mean strategy.
- In contrast, if the population's evolution does *not* result in appreciable changes in the covariance matrix, then the mean might be expected to move to the ESS, perhaps in a spiral, or at least to do so to the extent allowed by the diversity of strategies in the population (as measured by the covariance matrix).

    Figure 4 illustrates two possibilities for the case of $n = 3$ tactics, for which Δ is two dimensional. In Figure 4(a), the mean strategy can evolve from the interior of the shaded region to a boundary, with a concurrent disappearance of some strategies from the population. The central arrow in that figure represents a normal to the boundary, with a mean strategy at its base which would be stable against [small] perturbations of the frequencies of the various strategies present in the population. In Figure 4(b), the evolution of the mean eventually leads to a complete disappearance of strategy diversity.

(a)   (b)

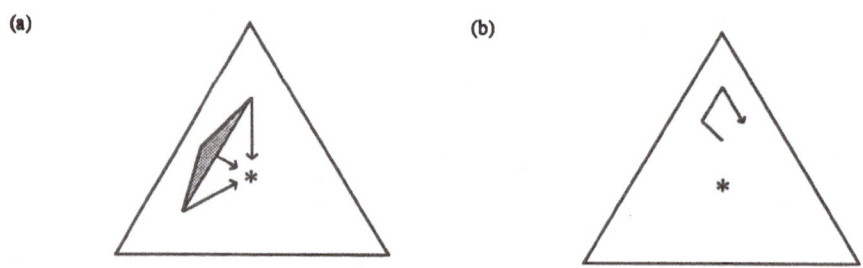

Figure 4

[N.B. A routine calculation shows that not only are the indicated and actual directions within 90° of each other within the space containing Δ, but also that the projections of these directions onto the space spanned by the columns of C (the directions in which strategy variability does exist in the population) are also within 90° of each other – the more relevant result.]

Despite the suggestive nature of these findings, the extent to which counterexamples to the qualitative statements above exist has not been adequately explored, and these findings should be viewed as providing an intuitive but still incomplete picture of the implied evolutions of the populations modelled. In particular, Figure 4 should be viewed with some caution, because it is two dimensional, which precludes the rich dynamics possible in higher dimensional systems. In addition, we have merely established that the actual changes of mean strategies are within 90° of the indicated changes, with the magnitude and direction of the difference between indicated and actual changes depending on the population composition in effect at that time. This does not, of itself, establish the desired conclusions about convergence, although it does provide intuitive insight.   Other approaches such as computer simulations and mathematical analyses provide an important complementary view of the subject.

In fact, for the [asexual] polymorphic populations being considered, a Lyapunov function [a function of the population composition which is bounded below and which persistently decreases when the population is not at equilibrium] can be constructed which shows that the mean strategy will converge to the ESS if a sufficient initial diversity of strategies is present. [Specially, the convex hull of the set of existing strategies must contain the ESS if convergence to that strategy is even to be possible without the introduction of new strategies into the population. Given that this condition is satisfied, convergence does occur.]

The interested reader might wish to explore the effects of discrete time dynamics, including investigating the effect of replacing the payoff matrix A by an affine transformation of it $Id + m^{-1}A$, for m very large. Also of interest is the possibility of divergence rather than convergence with discrete dynamics.

[Added in preparation: A simple argument shows that for at least one case – continuous time evolution with A symmetric and with a strong stability condition – convergence to the ESS *does* occur, unless or until no strategy variability in the direction $\Pi A(\mu-\mu^*)$ exists. A simple result by Hines [1987b] implies that all evolution of strategy frequencies then ceases.

Consider the change in one measure of the size of the deviation of μ_i from μ^*, $(\mu_i-\mu^*)^T (-A) (\mu_i-\mu^*)$. The stated properties of A imply that this measure is positive unless $\mu_i = \mu^*$. Under the assumption of symmetry, equation (4.10b) implies that

$$\frac{d(\mu-\mu^*)^T A(\mu-\mu^*)}{dt} = (\mu-\mu^*)^T(A+A^T)\frac{d\mu}{dt}$$

$$= \alpha_{[0]}(\mu-\mu^*)^T(A+A^T) \, \Pi \, C_{[0]} \, \Pi \, A(\mu-\mu^*)$$

$$= 2\,\alpha_{[0]}[\Pi \, A(\mu-\mu^*)]^T \, C_{[0]} \, [\Pi \, A(\mu-\mu^*) \leq 0$$

Equality occurs only if $[\Pi A(\mu-\mu^*)]^T \, C_{[0]} \, [\Pi A(\mu-\mu^*)] = 0$, or if there is no variability in strategies in the direction $\Pi A(\mu-\mu^*)$.

Note that $\Pi A(\mu-\mu^*) \neq 0$ unless $\mu = \mu^*$, since

$$(\mu-\mu^*)^T \, \Pi \, A \, (\mu-\mu^*) = (\mu-\mu^*)^T A \, (\mu-\mu^*) > 0 \,,$$

by equation (3.4) unless $\mu = \mu^*$.

We also note that if $C_{[0]}$ is of the form $C_{[0]} = \varepsilon\Pi(\mu^*)$ and if μ is in the support of μ^*, so that $\Pi(\mu^*) (\mu-\mu^*) = \mu - \mu^*$, then the standard Euclidean norm of the distance from μ to μ^* decreases, since

$$\frac{d(\mu-\mu^*)^T(\mu-\mu^*)}{dt} = 2(\mu-\mu^*)^T \frac{d\mu}{dt}$$

$$= 2\,\alpha_{[0]}(\mu-\mu^*) \, \Pi \in \Pi(\mu^*) \, \Pi \, A(\mu-\mu^*)$$

$$= 2\in \alpha_{[0]}[\Pi(\mu^*)(\mu-\mu^*)]^T \, A \, [\Pi(\mu^*)(\mu-\mu^*) < 0 \,,$$

using the fact that $\Pi(\mu^*) \, \Pi = \Pi(\mu^*) = \Pi(\mu^*) \, \Pi(\mu^*)$.

Finally, we note that if strategy diversity exists in a single dimension, a line segment, changes within that diversity which are within $90°$ of one another are necessarily in exactly the same direction, producing a drastic simplification.

For a further discussion of the use of the mean-covariance to investigate possible convergence, see a report of work currently in progress by Hines [1988a].]

5.   ESSs and sexual populations

So far, we have considered the applicability of ESS theory to populations with individuals which have a simple reproductive mechanism which guarantees that the offspring are (behaviourally) exact copies of their parents. Such reproduction might occur, for example, if individuals were asexual and reproduced parthenogenetically, if individual strategists bred only with others with the same genetic makeup, or if a single parent contributed all of the (relevant) genetic material, for example in a single-locus haploid sexual population. Naturally, an extension of ESS theory to other biological possibilities is quite desirable – for instance the extension to populations which reproduce sexually.

Early models by Oster and colleagues [Auslander, Guckenheimer and Oster 1978, Mirmirani and Oster 1978] introduce genetic considerations. In a multi-species model, their results show that there can exist equilibrium strategies which are not stable, with the population mean cycling about the equilibrium. The results of these analyses, which were initially interpreted as suggesting that genetics and the ESS approach might be inconsistent, can be at least partially explained by the observation that the models used did not include intraspecific as well as interspecific competition, which proves to be necessary for stability. Following their analyses, a serious question remained as to whether genetics and ESS models are compatible.

A One Locus Multi-Allele Model. Assume a randomly mating population (one at Hardy-Weinberg equilibrium) with diploid individuals differing only in genetic material (alleles) for the one locus controlling their particular strategies. An individual of type (u,v), with, for example, allele u from the mother and v from the father, would have a strategy defined arbitrarily as $s(u,v)$ [$= s(v,u)$ by assumption]. Then, given the assumed random mating, if $p(u)$ is the fraction of alleles of type u in the population, $p(u)p(v)$ is the fraction of individuals of type (u,v) in that same population. Given this, the mean strategy associated with allele u is defined as the weighted sum of the possible strategies to which u contributes:

$$(5.1) \qquad \mu(u) = \sum_v s(u,v)p(v) ,$$

while the covariance matrix of strategies associated with u, $C(u)$, is

$$(5.2) \qquad C(u) = \sum_v (s(u,v) - \mu(u))(s(u,v,-\mu(u))^T p(v) ,$$

and the mean strategy, μ, of the population is

$$(5.3) \qquad \mu = \sum_u \sum_v s(u,v)p(v)p(u) = \sum_u \mu(u) p(u) .$$

The covariance matrix measuring the variability of the mean strategy $\mu(u)$ is given by

$$(5.4) \qquad C^\sim = \sum_u (\mu(u)-\mu)(\mu(u)-\mu)^T p(u) ,$$

while the variability of strategies in the population is

$$C = \sum_u \sum_v (s(u,v)-\mu)(s(u,v)-\mu)^T p(v)p(u)$$

$$= \sum_u \sum_v (s(u,v)-\mu(u)+\mu(u)-\mu)(s(u,v)-\mu(u)+\mu(u)-\mu)^T p(v)p(u)$$

$$= \sum_u \{\sum_v (s(u,v)-\mu(u))(s(u,v)-\mu(u))^T p(v)\} p(u)$$

(5.5) $$+ \sum_u (\mu(u)-\mu) \{\sum_v (s(u,v)-\mu(u))^T p(v)\} p(u)$$

$$+ \sum_u \{\sum_v (s(u,v)-\mu(u)) p(v)\} (\mu(u)-\mu)^T p(u)$$

$$+ \sum_u \{\sum_v p(v)\} (\mu(u)-\mu)(\mu(u)-\mu)^T p(u)$$

$$= \sum_u \{C(u) p(u)\} + C^\sim,$$

since, for example, $\sum_v (s(u,v)-\mu(u)) p(v) = 0$ and $\sum_v p(v) = 1$.

(Note: (5.3) and (5.5) are special cases of standard theorems in statistics.)

The geometry of the set of possible mean strategies is more complex in the diploid case than in the previous, haploid, case. Figure 5 shows the set of mean strategies when only two alleles are present (with the set then forming part of a parabolic curve) and when three alleles are present. [Clearly and importantly, the set of possible mean strategies is not convex.] The boundaries of the set of possible mean strategies are not necessarily related to the mean strategies resulting with extreme allelic frequencies, in which at least one allele is absent.

Dynamics. While the set of possible mean strategies in the diploid case is clearly more intricate than for the haploid case, the equation giving the evolution of the mean strategy is quite similar: with a notation similar to that for the haploid situation, Hines and Bishop [1983, 1984a, 1984b] find

$$\mu_{[1]} - \mu^* = \mu_{[0]} - \mu^* + \lambda C^\sim A (\mu_{[0]}-\mu^*)$$

for a scaling factor λ reflecting the differing force of selection for alleles resulting from diploid rather than haploid reproduction.

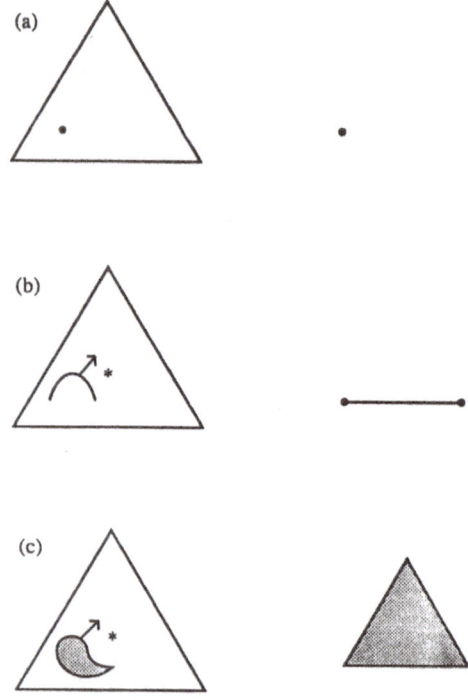

Figure 5

As before, the directions of change in mean strategy $(C^\sim A(\mu-\mu^*))$ are within 90° of $A(\mu-\mu^*)$, which in turn are within 90° of $\mu^* - \mu$. The significant change with the introduction of sexual reproduction into the model is the fact that the set of attainable mean strategies is no longer convex, thereby introducing the possibility of multiple equilibria, for example in Hines and Bishop [1984a]. As well, strategies on the boundary of the set of attainable mean strategies need not correspond to extremes in allelic frequencies. It seems plausible that strategies such as that indicated in Figure 5(c) (for which the surface of the set of attainable mean strategies is strictly convex locally and for which the direction $A(\mu-\mu^*)$ is at right angles to that surface), will correspond to necessarily stable mean strategies, although as noted earlier detailed mathematical analyses are still needed.

Some work on such analyses or on possible counterexamples has been or is being carried out by workers such as Hofbauer, Cressman and Hines. For example, at the 1987 Animal Conflicts workshop at Sheffield, Hofbauer informally reported a finding that divergence from the location of $\mu^\sim$ can occur in one version of "Rock Scissors Paper". We note that the added argument at the end of Section 4.3 applies with equal force (and under the same conditions) for the diploid case, so that convergence to or towards the ESS is again indicated, and it again stops when relevant strategy variability disappears.

6. Finite population model

Recall that in the infinite population case:

(i) a u-user playing a v-user will have a payoff of $\alpha u^T A v$ (measured in terms of expected number of viable offspring).

(ii) averaged over all expected contests, the u-user will have a mean fitness of

$$\alpha E(u^T A v) = \alpha u^T A E(v) = \alpha u^T A \mu ,$$

where $E(v)$ is the expected values of v which is the mean strategy.

The assumption of finite population size raises a number of questions regarding those results which have to be examined.

- Is the individual's choice of tactic influenced by the opponent's strategy?
- Is the mean strategy of the opponent independent of the strategy of the first player (i.e. is $E(v|u) = \mu|_u = \mu$) ?
- Is the average number of contests fixed?
- Is the average payoff fixed?

6.1. CONSIDERATIONS

The effects of finite population size can be classified as deterministic and stochastic. One obvious deterministic effect which is important in finite populations is that individuals do not have themselves as opponents. This implies that the mean strategy faced by a u-user, $E(v|u)$, is not equal to μ. Perhaps less obvious is the need for a mechanism for generating the successive generations. With a finite number of individuals possible in each generation, the fitness resulting from contests with opponents cannot in general refer to an actual fitness, but at best, to an average fitness. In the finite population case, the nature of the mechanism generating successive generations becomes important. An important standard of comparison is provided by the Wright-Fisher model [Ewens, 1979].

Assume that generations are non-overlapping, and are asexually created. Suppose that individuals produce effectively infinite numbers of potential offspring, and that the creation of the next generation must involve a stochastic process in which a finite number of individuals are selected. The Wright-Fisher model predicts that for a population of size N the probability that two individuals will have had the same parent is $1/N$. Taking this as a standard, we identify three cases

- a probability of $1/N$, indicating a process similar to the Wright-Fisher model,
- a probability less than $1/N$, indicating that individuals produce limited numbers of offspring (almost all [one-parent] families have one offspring) relative to the Wright-Fisher model, that is, a comparatively deterministic selection process.
- a probability more than $1/N$ indicates that individuals of equal expected fitness have considerable variability in family size, relative to the Wright-Fisher model, that is, a comparatively variable or stochastic selection process.

Information Asymmetries. Another possibility in a finite population (or in a population in which individuals do not choose opponents randomly from an infinite pool) is that players might make use of previous experience, or information available "on site" in their tactic selection processes. This could, for example, involve knowledge of the current general population (general fitness, typical behaviour, variability of behaviour, ...) or awareness of the opponent's individual strategy (ability, needs,...). Such information might be "perfect" in the sense of being complete, accurate, and universally accessible in which case it can be deterministically modelled, or "imperfect" and it could then be treated as a stochastic effect.

Fitness. In the infinite population model, it was reasonable to compare various strategies by comparing their expected fitnesses. This however is not appropriate in the finite population case. Instead, we must consider both the expected value and the variability of individual fitnesses, as the following discussion of "log-fitness" indicates.

Consider the fortunes of the descendants of an individual. Suppose that the individual and its descendants experience successive fitnesses $\phi_{[1]}, \phi_{[2]}, \ldots, \phi_{[t]}$ in generations $1, 2, \ldots, t$, with the ϕ's being identically distributed independent random variables. The expected number of offspring after t generations would be

$$E(\phi_{[1]} \phi_{[2]} \ldots \phi_{[t]}) = E(\phi_{[1]}) \, E(\phi_{[2]}) \ldots E(\phi_{[t]}) = [E(\phi)]^t .$$

However the actual number is

$$\phi_{[1]} \phi_{[2]} \ldots \phi_{[t]} = \exp\left(\sum_{k=1}^{t} \ln(\phi_k) \right) .$$

For large t, this latter product is likely to be close to $\exp(t[E(\ln \phi)])$ by the law of large numbers, given standard conditions from the theory of statistics. The two per-generation growth rates implied by these arguments are

(i) $E(\phi)$, in the first case;
(ii) $\exp[E(\ln \phi)] \cong E(\phi) - 0.5 \, E(\mathrm{var}(\phi))$,

where the second term is small when the variability in successive effective fitnesses is small. The expected growth rate, $E(\phi)$, therefore overestimates the effective growth rate by an amount reflecting the variability in growth rates from generation to generation.

If descendants of a common ancestor do not all have the same fitness in generation i, then the effective growth for that generation is the *arithmetic* average $\bar{\phi}_i$, which will have lower variability than the variability of individual fitnesses, thereby reducing but not removing the effect of the variability of fitnesses.

6.2. RESULTS OF ANALYSIS

[The material in this selection is based on as yet unpublished work [Hines 1988], previously reported at the 1987 Animal Conflicts workshop at Sheffield.]

Arguments and approaches similar to those presented thus far can be used to establish the relevance of ESS theory to the finite population case. The results derived from these analyses are outlined below, assuming throughout that the ESSs are interior, always employing all n available tactics, and letting N be the population size.

• The basic ESS conditions derived from a monomorphic population argument are almost unchanged. The one exception is that in the first condition, the payoff matrix A is replaced by $A + N^{-1}(A-A^T)$. (This difference, which is zero if $A = A^T$, disappears as $N \to \infty$.)

Thomas and Pohley [1981] obtain a different first condition for the case of finite populations in which only pure strategies are present.

• The absence of self-contesting leads to a difference between the mean strategy of one's opponents and that of the population as a whole. This has an effect proportional to the deviation of the individual's strategy from the mean.

This deterministic effect favours the more extreme (or diverse) strategies.

• The effect of variability in generation-to-generation random sampling (stochastic factor) selects against more extreme strategies.

• Under the Wright-Fisher model, and to the accuracy of approximations used, the overall effect of finiteness of populations is slightly in favour of extreme strategies without being definite.

A tendency towards fairly predictable family sizes for given parental fitness produces a greater advantage for the extreme strategies, while low predictability of family size (say as a result of severe environmental variability) discourages such strategies.

• The benefits of extreme strategies decrease with the increase of variability in the payoffs resulting for any given combination of selection of tactics.

• Both low numbers of possible opponents and low numbers of encounters with opponents deter extreme strategies.

Small population size and small numbers of contests introduce stochastic effects, which were found in this study being described to select against diversity in general (although one minor stochastic effect was found somewhat to favour it).

• The inclusion of possible learning increases the dimension of the problem and so increases the difficulty of analysis. Two important considerations are the amount and quality of information possessed by a given individual, and whether or not optimal use will be made of it. Analysis using the log-fitness approach suggests that imperfect information tends to discourage the use of diverse strategies, while reliable information favours extreme strategies.

6.3. COMPUTER SIMULATIONS

Preliminary results suggest a rate of strategy disappearance apparently little dependent on population size, and perhaps higher than the fixation rate. In the simulations, the ESS was at a disadvantage relative to the pure strategies considered for small stochastic effects, and at an advantage when these effects were large.

Details of the simulations:
• n = 3 possible tactics.
• 4 types of individual strategies: 3 pure types and one mixed interior strategy (1/3, 1/3, 1/3), the ESS for all of the payoff matrices studied.
• an initial population of size N = 4,8,16,...,128, formed by sampling with replacement from the above types of individuals.
• various payoff matrices, all having the ESS just mentioned.
• random pairing of opponents for contests with outcomes determining the exact numbers of offspring contributed to a pool for possible selection [generally without replacement, unless the size of the pool was strictly less than N, in which case the entire pool was replicated as needed].
• the population followed to either 50 or 100 generations, or until fixation or extinction of the ESS occurs.

The theoretical results previously mentioned indicate that deterministic effects favour extreme strategies while less diverse strategies would be encouraged by the presence of considerable stochastic effects. The simulation results seem to show that the ESS advantage relative, at least, to pure strategies increases with variability in payoff (centralizing tendency due to stochasticity) and also increases with population size.

7. ESSs and cooperative behaviour

The previous two sections explored the relevance of ESS theory when assumptions about the structure of the population [polymorphic, sexually reproducing] were introduced. In this section, we return to the basic ESS question of whether specific types of behaviour might persist in stable populations. The particular type of behaviour we consider is cooperative. The examples discussed are considered suggestive, not definitive, and we refer the interested reader to an extensive annotated bibliography on the evolution of cooperation prepared by Axelrod and Dion [1987].

Three major models will be used as examples, respectively:
• Games between Relatives [Grafen 1979, Hines and Maynard Smith 1979]
• Imitator Strategies [Haigh and Hines, 1986]
• The War of Attrition [Bishop and Cannings 1976 & 1978, Bishop et al. 1978, Norman et al. 1977, Hines 1977 & 1978].

Our discussion first introduces modifications to the ESS model which incorporate cooperation, then describes a model with an odd but tractable form of cooperation, and then uses another model to discuss at some length conditions which might encourage cooperation and the long-term fate of such cooperation when mutational effects are allowed for.

7.1. GAMES BETWEEN RELATIVES

As modelled by Grafen and by Hines and Maynard Smith, these games involve infinite populations of asexual individuals who have an increased chance of encountering and contesting with other particular individuals [siblings, neighbours] using the same strategy. Denote the probability of playing against a "sibling" by ρ. (When $\rho = 1$, the effective payoff matrix proves to be symmetric, which tends to encourage cooperative behaviour, while $\rho = 0$ correspond to the standard form of the game.)

Two methods of approach which, confusingly, gave similar results for the standard formulation of the ESS, produce quite distinct conditions for equilibrium if $\rho \neq 0$. One approach is to seek the equilibrium composition for a population consisting solely of *pure* strategists, each using a strategy from one of the vertices of Δ.

The other approach is to seek the mixed strategy, $u^{\#}(\rho)$, which, if used by all members of a population, places any set of invaders using a common strategy at a disadvantage. The conditions to be satisfied by $u^{\#}(\rho)$ turn out to be exactly those that $u^{\#}(\rho)$ be an ESS, but for the payoff matrix $A + \rho A^{T}$ rather than for A. (The effect of the additional term in the payoff matrix is to relocate the ESS; the (strong) stability condition is unchanged.)

Examination of the surface of fitnesses for potential invading strategies that results with a monomorphic population using $u^{\#}(\rho)$ shows that the $u^{\#}(\rho)$-users will have a strict selection advantage over any (sufficiently small, monomorphic) invading population. Contrarily, in a population of pure strategists alone, at equilibrium with respect to the existing strategies, the fitness function shows a selection pressure in favour of mutations towards mixed strategies. In general, the means at equilibrium in the monomorphic-mixed and polymorphic-pure cases are different.

A possibility not examined by Hines and Maynard Smith [but similar to one considered by other authors such as Eshel and Cavalli-Sforza, 1982] is that selection for the value of ρ might occur. One might expect to find such selection pressure (towards cooperation) since, as ρ increases, the fitness of $u^{\#}(\rho)$-users in a monomorphic population also increases. Rather than explore this particular model further, we turn to two other models which do consider the possibility of various degrees of cooperation.

7.2. IMITATOR STRATEGIES

Background. The Prisoner's Dilemma refers to a problem faced by two prisoners being interrogated separately for a joint crime they are (correctly) suspected of having committed. Each prisoner can confess to the crime [Defect] or not confess [and so Cooperate with the other prisoner]. The authorities will reward the prisoner's defection more than his cooperation, regardless of the other prisoner's decision to cooperate or to defect, but the two prisoners do best if both choose to cooperate. This dilemma is modelled by a game with n = 2 pure strategies, D for defect and C for cooperate, and a payoff structure such that

$$E(D,C) > E(C,C) > E(D,D) > E(C,D).$$

Note that, since E(C,C) > E(D,D), if both players do choose the same strategy, they do better choosing C.

Axelrod and Hamilton [1981] modeled a tournament (in two stages) of the Prisoner's Dilemma game, with a number of strategies proposed by participants being used repeatedly against each other. (In the second stage, participants were informed of the outcomes of the first stage.)

We note two findings of the Axelrod and Hamilton study:
• While D is an ESS for the game, so that it was never bested in individual encounters, it was less profitable than some other strategies.
• The simple Tit-for-Tat strategy, which consisted of returning the opponent's tactic used in the previous encounter (and of cooperating initially), routinely did well in the tournament.

The success of the Tit-for-Tat strategy is perhaps not surprising, since it encourages cooperation and discourages defection. On the other hand, Tit-for-Tat, because of its deterministic nature, is a poor choice in some contests. For example, consider a contest with the payoff matrix

$$A = \begin{pmatrix} 0 & -1 & 2 \\ 2 & 0 & -1 \\ -1 & 2 & 0 \end{pmatrix}.$$

The faithful use of the opponent's tactic just previous would condemn the user of Tit-for-Tat to an unending sequence of defeats against an opponent who chooses merely to use the tactic each time which would have defeated its previous tactic.

Viewed as a general deterministic strategy, then, Tit-for-Tat can be quite disadvantageous. One feature of Tit-for-Tat that we do note with interest is its mimicking of the frequency of the opponent's behavioural tactics. An analogue of Tit-for-Tat considered by Haigh and Hines [1986] is that of a strategy which exploits a knowledge of each opponent's strategy (as distinct from knowing the opponent's planned choice of tactic in the next contest). This might occur, for example, if an individual could identify the strategies of opponents, say based on appearance, past encounters, or extensive observation of behaviour in neutral situations or in contests with other individuals. (We ignore the perhaps considerable expense in acquiring such information, regarding it as negligible compared to the subsequent returns from a very long sequence of future encounters.)

The Model. The "imitator strategy" model involves a population composed of individuals with labels which do reliably indicate the inherent strategies of the individuals. The only use allowed of such information, however, is quite restricted: a player of type (u,π) employs its own strategy u with a probability $(1-\pi)$ and employs the opponent's strategy with probability π. Note that this reduces to the usual ESS model if we set $\pi = 0$. If, however, $\pi = 1$, the player's inherent strategy is virtually ignored and only the opponent's strategy used.

Suppose μ^* is an interior strategy $(\mu^* \in \overset{\circ}{\Delta})$ and is an ESS of A, which (necessarily) satisfies the strong stability condition. For a monomorphic population (at equilibrium) using $(\mu^*,0)$, we find that
• Imitators cannot invade the population if and only if μ^* is also an ESS of A^T, and so, it turns out, of $A + A^T$, the symmetric and cooperation-encouraging form of the payoff matrix.

That is, if μ^* is an ESS for the cooperative form of the game as well, the population is then protected from invasion.
• Otherwise, suitable invader strategies $(u,1)$ exist which will lead to the fixation of invader strategies in the population.
• For populations of imitators to be in turn at equilibrium, they must be *monomorphic*, with a common strategy $(\mu^{**},1)$ where μ^{**} is an ESS for A^T, the form; of the game appropriate for contests between imitators. [We say that μ^{**} is the "imitator ESS".]
• For the case in which $\mu^{**} > 0$, if μ^{**} is also an ESS of A, then the equilibrium cannot be displaced, but otherwise, this population can be invaded by suitable non-imitators. (If this condition is met, however, then $\mu^* = \mu^{**}$ and the initial invasion of the original non-imitator monomorphic population by imitators would not have occurred.)
• When the condition $\mu^{**} > 0$ is relaxed, the requirement that μ^{**} be an ESS of $A + A^T$ as well as A^T, for the population to be resistant to invasion, is non-generic, with one exception.

The requirement that the ESSs of A^T and $A + A^T$ remain equal under arbitrary small perturbations of A can only be met if the ESSs for both payoff matrices use the same pure strategy. In that special case, the imitator ESS *is* also a ESS for the cooperative form of the game.
• If in the course of the successful re-invasion of the population of imitators by non-imitators, with $\pi = 0$, the non-imitating population becomes and remains polymorphic, that polymorphism blocks future invasions by imitators, as it places them at a strict disadvantage.

How is this relevant to the Prisoner's Dilemma study? The above results identified one atypical condition under which a population of imitators could be stable, rather than vulnerable to invasion by non-imitators: the ESSs of A^T and of $A + A^T$ must each be a (common) pure strategy. Exactly this condition is satisfied by the Prisoner's Dilemma payoff matrix, raising questions about the degree to which that contest can be regarded as typical of others and so to which the conclusions reached can be generalized.

7.3. THE WAR OF ATTRITION

This two-person contest, first discussed in the context of the ESS by Maynard-Smith [1974], is won by the opponent prepared to wait the longer for a prize of fixed and known value, with each participant in the contest losing previously accumulated resources at a constant rate with time. [More general assumptions can be made and have been studied at length by various authors, especially Bishop and Cannings – see the review article by Hines [1987a] for a more detailed discussion.]

For the two-person contest, the nature of the tactics available is simple: choose a waiting time, T, and quit by that time if you have not already won. [The nature of the tactics possible for three-person of for k-person contests is far more complex, and the investigation of ESSs for such contests is liable to be quite challenging.] Specifically, if T_u and T_v denote the quitting times chosen by a u-user and a v-user, the payoffs to the u-user are taken to be

$$a - bT_v \qquad \text{if } T_u > T_v$$

$$0.5\ a - bT_v \qquad \text{if } T_u = T_v \qquad \text{(each takes one half of the prize on average at least)}$$

$$- bT_u \qquad \text{if } T_u < T_v$$

The simplicity of the set of available tactics implies that a given strategy, say u, can be characterized by a (possibly improper) probability distribution $P_u(t)$ over the continuum $[0,\infty)$ (or all-non-negative finite times), where $P_u(t)$ is the probability that the stopping time chosen will be no more than t.

The ESS is a strategy u^* with probability distribution $P^*(t) = 1 - \exp(-(b/a)t)$, for $t \geq 0$. This strategy gives exponentially distributed times, T, with the scaled times, $(b/a)T$, having a standard exponential distribution. Opponents faced with the ESS strategy find that for (very small) periods of time of length δ, the two immediate decisions "quit within the next δ units of time" and "persist for the next δ units of time" are equally rewarding, regardless of the elapsed amount of time so far – reflecting a well-known memoryless property of the exponential distribution. Since the ESS denies opponents any advantageous possibilities by making all choices exactly equally rewarding, it is not surprising that no strategy does particularly well against the ESS.

In terms of the resulting fitnesses, we find that $E(u^*,u^*) = E(v,u^*) = 0$ for *all* opposing strategies v. [Note: This establishes that u^* satisfies the ESS condition (2.3). Bishop and Cannings have verified that (2.4) is also satisfied.] This finding that the ESS for a contest results in no benefits for those using it, or for their opponents, is somewhat surprising, although it can be interpreted, as Chris Canning pointed out in the Montreal seminar, as indicating that the contest is being used to convert previously acquired resources into forms more immediately relevant to reproduction [such as the acquiring of a mate or a breeding or nesting area]. The model developed by Hines [1977, 1978] was an attempt to understand the original war-of-attrition model as a model for the acquiring, rather than the converting, of resources.

Ecological Formulation. The model uses a naive but plausible approach to the war of attrition. Assumptions and features of the model include the following:
• The rewards are found by a foraging individual at a rate α and are then contested with another individual with probability π, both assumed to be fixed in the earlier paper.

- The animals have fixed total lifespans, and we compare performances on the basis of returns per unit time.
- The number of contests is not fixed, but dependent on the strategy used.
- Losses occur at rates per unit time of v_F while foraging and v_C while contesting.

A standard ESS-style argument, involving a monomorphic population which is presumed to be stable when threatened with a near-negligible monomorphic fraction of mutants, yields a potential equilibrium strategy similar to the original ESS strategy $P^*(t)$, but now with the quitting rate parameter given by $b/a = \alpha(1-\pi) - v_F + v_C$. For notational brevity, we will refer to the ESS-like strategy as an "ESS", in quotes.

As an alternative to the daunting task of exploring the global stability associated with the new equilibrium strategy, its stability relative to other strategies with exponentially distributed times has been considered. These competing strategies can be characterized conveniently as ranging form meek to aggressive, or tending to favour short or long contests respectively.

In the analysis reported by Hines [1977 and 1978], for most environments (characterized by the parameters α, π, v_F and v_C), the "ESS" outperformed other possible exponential strategies in terms of the resulting returns per unit time. The exception to this general finding occurred, unexpectedly, in "harsh" environments, where foraging is hard ($(\alpha-v_F)$ is small) and contests are both likely ($\alpha \pi$ large) and easy (v_C small). In such environments, which result in fairly prolonged contests for individual items with the "ESS", meek strategists proved to have higher fitness when rare or present in modest numbers, with the selective advantage being roughly proportional to the fraction they formed of the population. This suggested that such meek, and effectively cooperative, behaviour would initially tend to increase in the population. This increase was not expected to result in fixation of the meek strategy, however, since in predominantly meek populations, users of the "ESS" would usually win in contests with the meek individuals, while rarely encountering other similarly aggressive individuals and so rarely entering into contests of long duration.

To extend the findings to incorporate the important possibility of changes in strategies over time, perhaps due to mutational or gradual learning effects, numerical studies were carried out (using deterministic calculations, as opposed to Monte Carlo simulations). Findings of these studies are sketched in Figures 6 and 7, which are *not* drawn to scale.

The populations considered had a finite number of possible strategies, each characterized by the mean value, λ, of the stopping time, T, it generated. Meek strategies, for example, had values of λ close to 0, while aggressive strategies had large values of λ. At the beginning of each generation, a small fraction (0.001) of individuals had their values of λ increased or decreased to adjacent values of λ. Fitnesses occurring for each value of λ were then calculated, and used to determine the relative frequencies for the next generation. As a preliminary step, the population was allowed to run for a great many generations, until near-equilibrium was reached. The population, while centered near the "ESS" value, was clearly polymorphic, as might be expected given the persistent mutations of individual strategies. (This polymorphism was found to result in a selection disadvantage for strategies far removed from the "ESS" strategy.)

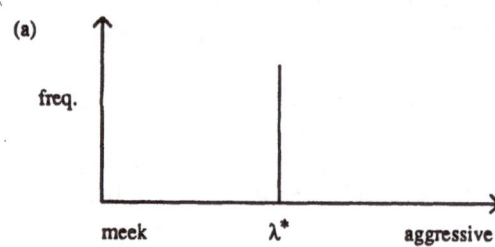

(a)

freq.

meek   λ*   aggressive

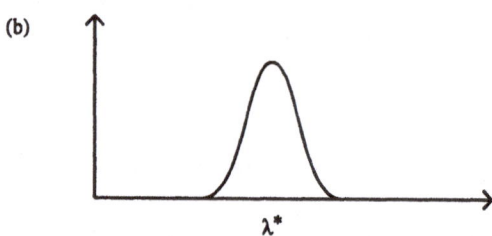

(b)

λ*

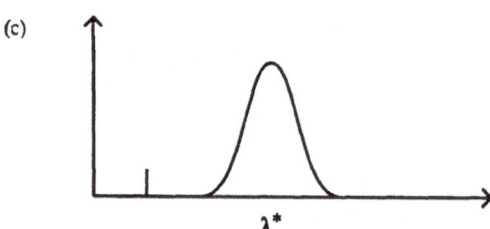

(c)

λ*

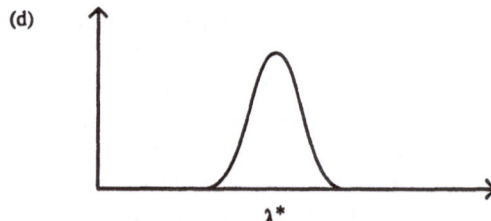

(d)

λ*

Figure 6

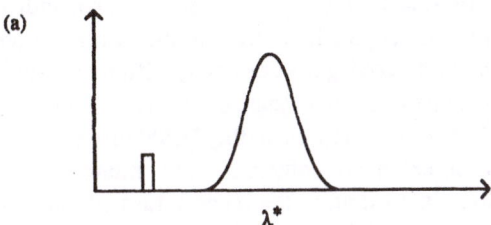

(a)

λ^*

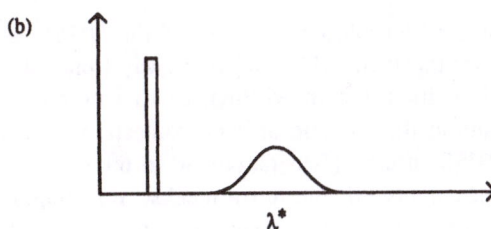

(b)

λ^*

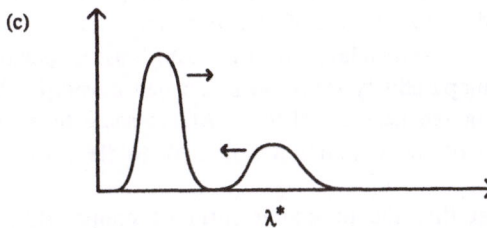

(c)

λ^*

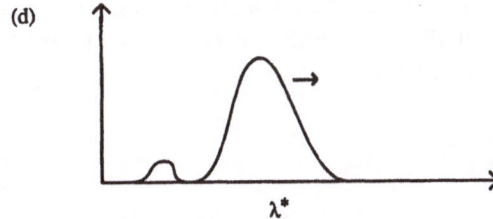

(d)

λ^*

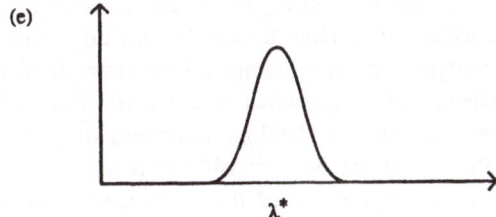

(e)

λ^*

Figure 7

An initial attempt, which introduced a very small fraction (0.1%) of ultra-meeks with $\lambda = 0$, ended with the ultra-meeks rapidly disappearing, and with no other noticeable change in the composition of the population. In contrast, a second attempt with a larger fraction (1%) of ultra-meeks introduced gave drastically different results:

• Initially, the ultra-meeks benefitted from their cooperative nature, and from the fact that they encountered each other more often than the "ESS"-users did, as the latter spent considerable time locked in extended contests. The number of ultra-meeks in the population increased rapidly. Interestingly, the mean return per unit time of all members of the population, including the "ESS"-users, also increased. [Generations 1 - 20, roughly]

• With time, the behaviours of the ultra-meeks and of the "ESS"-users and near-"ESS" users mutated, drawing closer together. This tendency away from the original ultra-meeks is not surprising since slight increases in willingness to continue contests against the numerous ultra-meeks results in the winning of such contests, with only minor additional losses in contests against "ESS" users. [Generations 30 to 60]

• Seriously mutated descendants of the ultra-meeks, no longer protected by the considerable difference between their strategies and those of the descendants of the "ESS"-users, experienced considerable drops in their fitnesses, and that component of the population quickly dropped in size, eventually disappearing. [Generations 70 to 100]

• Further mutations of the descendants of the "ESS"-users occurred, changing the strategies used from the comparatively meek ones that had developed back to the original equilibrium distribution centered near the "ESS". At the same time, the mean return per unit time of the members of the population fell back to its earlier equilibrium value. [Generations 200 to 500]

These findings suggest that the particular form of cooperation considered in this model, at least, is not stable in the long term, even though it does confer considerable advantages to the population in the medium and short terms. The study also emphasizes the well-recognized fact that ESSs and ESS-like behaviour can work very much to the detriment of the population as a whole.

8.   Concluding remarks

The ESS concept is a stimulating and attractive one, perhaps dangerously so. It gains much of its power and elegance by setting aside many characteristics of biological populations, characteristics which might alter the conclusions that would be reached if they were included in a fuller analysis. As the examples here show, however, the ESS model does remain surprisingly relevant when a number of characteristics such as polymorphism, environmental perturbations, sexual reproduction, finiteness of population, and ecology are introduced, at least in the elementary ways considered here.

Perhaps the same caution about the use of the mean-covariance approach should be made. The approach tends to provide quick and intuitively attractive insights. The degree to which these general insights apply in particular cases, however, is not yet well understood, and merits serious investigation, thereby joining a remarkably long list of unanswered questions on the subject.

Attentive participants at the Montreal seminar may have noticed the degree to which this recording of a series of five lectures at the seminar has taken advantage of the second opportunity of presenting the material to add afterthoughts, although the basic structure and level of the series are unaltered. The interested reader is reminded of the presence in this volume of lectures by Cannings, and of the various review articles previously mentioned.

We each would like to express our thank to our hosts and our fellow participants for a rewarding stay in a delightful city.

References

Akin, E. and Hofbauer, J. (1982), Recurrence of the unfit, *Math. Biosci.* **61**, 51-62.

Auslander, D., Guckenheimer, J., and Oster, G. (1978), Random evolutionarily stable strategies, *Theor. Pop. Biol.* **13**, 276-293.

Axelrod, R. and Hamilton, W.D. (1981), The evolution of cooperation, *Science* **211**, 1390-1396.

Axelrod, R. and Dion, D. (1987), *Annotated Bibliography on the Evolution of Cooperation*, Institute of Public Policy Studies, University of Michigan, Ann Arbor.

Barash, D.P. (1982), *Sociology and Behaviour, 2nd. Edition*, Elsevier, New York. Sinauer Associates, Sunderland.

Bishop, D.T. and Cannings, C. (1976), Models of animal conflict, *Adv. in Appl. Probab.* **8**, 616-621.

Bishop, D.T. and Cannings, C. (1978), A generalized war of attrition, *J. Theoret. Biol.* **70**, 85-124.

Bishop, D.T.,Cannings, C. and Maynard Smith, J. (1978), The war of attrition with random rewards, *J. Theoret. Biol.* **74**, 377-388.

Dawkins, R. (1976),*The Selfish Gene*, Oxford University Press, Oxford.

Eshel, I. and Cavalli Sforza, L.L. (1982), Assortment of encounters and the evolution of cooperativeness, *Proc. Nat. Acad. Sci. USA* **79**, 1331-1335.

Ewens, W.J. (1979), *Mathematical Population Genetics*, Springer-Verlag, Berlin.

Grafen, A. (1979), The hawk-dove game played between relatives, *Anim. Behav.* **27**, 905-907.

Haigh, J. and Hines, W.G.S. (1986), Imitator strategies, *Theoret. Population Biol.* **30**, 372-387.

Hines, W.G.S. (1977), Competition with an evolutionary stable strategy, *J. Theoret. Biol.* **67**, 141-153.

Hines, W.G.S. (1978), Mutations and stable strategies, *J. Theoret. Biol.* **72**, 413-428.

Hines, W.G.S. (1982a), Strategy stability in complex randomly mating diploid populations, *J. Appl. Probab.* **19**, 653-659.

Hines, W.G.S. (1982b), Mutations, perturbations and evolutionarily stable strategies, *J. Appl. Probab.* **19**, 204-209.

Hines, W.G.S. (1986), Constancy of mean strategy versus constancy of gene frequencies in a sexual diploid situation, *J. Theoret. Biol.* **121**, 367-369.

Hines, W.G.S. (1987a), Evolutionarily stable strategies: a review of basic theory, *Theoret. Population Biol.* **31**, 195-272.

Hines, W.G.S. (1987b), Can and will a sexual diploid population evolve to an ESS?: The multi-locus case, *J. Theoret. Biol.* **126**, 1-5.

Hines, W.G.S. (1988), Finite population effects in ESS models, Preprint, Department of Mathematics and Statistics, University of Guelph, Guelph, Ontario, Canada.

Hines, W.G.S. and Bishop, D.T. (1983), Evolutionarily stable strategies in diploid populations with general inheritance patterns, *J. Appl. Probab.* **20**, 395-399.

Hines, W.G.S. and Bishop, D.T. (1984a), On the local stability of evolutionarily stable strategies in diploid populations with general inheritance patterns, *J. Appl. Probab.* **20**, 215-224.

Hines, W.G.S. and Bishop. D.T. (1984b), Can and will a sexual diploid population attain an evolutionarily stable strategy, *J. Theoret. Biol.* **111**, 667-686.

Hines, W.G.S. and Maynard Smith, J. (1979), Games between relatives, *J. Theoret. Biol.* **79**, 19-30.

Maynard Smith, J. (1974). The theory of games and the evolution of animal conflict, *J. Theoret. Biol.* **47**, 409-221.

Maynard Smith, J. (1982), *Evolution and the Theory of Games*, Cambridge University Press, Cambridge.

Mirmirani, M. and Oster, G. (1978), Competition, kin selection and evolutionarily stable strategies, *Theoret. Population Biol.* **13**, 304-339.

Norman, R.F., Taylor, P.D. and Robertson, R.J. (1977), Stable equilibrium strategies and penalty functions in a game of attrition, *J. Theoret. Biol.* **65**, 571-578.

Parker, G.A. (1984), Evolutionarily stable strategies, In *Behavioural Ecology, 2nd Edition* (J.R. Krebs and N.B. Davies, eds.), Sinauer Associates, Sunderland.

Riechert, S.E. and Hammerstein, P. (1983), Game theory in the ecological context, *Ann. Rev. Ecol. Syst.* **14**, 377-409.

Taylor, P.D. and Jonker, L.B. (1978), Evolutionarily stable strategies and game dynamics, *Math. Biosci.* **40**, 145-156.

Thomas, B. and Pohley, H.J. (1981), ESS-Theory for finite populations, *Biosystems* **13**, 211-1221.

Zeeman, E.C. (1979), Population dynamics from game theory, in: *Proc. Conf. on Global Theory of Dynamical Systems*, Northwestern Univ., Evanston, 471-497.

ÉVOLUTION DU RAPPORT NUMÉRIQUE DES SEXES ET MODÈLES DYNAMIQUES CONNEXES

Sabin Lessard
Département de mathématiques et de statistique
Université de Montréal
C.P. 6128, Succ. "A"
Montréal, Québec
Canada

ABSTRACT. Several models of sex determination and sex ratio distortion in infinite random mating populations are analysed. Equilibrium structures and evolutionary properties in exact genetic models are studied. The effects of linkage and frequency dependence are also considered.

1. Modèles, exemples et problèmes

1.1. SYSTÈMES "STANDARDS" DE DÉTERMINATION DU SEXE

Il y a encore de nombreuses questions qui se posent sur les conditions qui ont favorisé l'évolution du sexe. On se demande toujours comment les avantages à long terme ont pu finalement compenser les désavantages à court terme. En effet, il est généralement reconnu que le sexe procure un avantage dans des environnements variables dans le temps et l'espace en créant plus de diversité par la recombinaison de gènes. Mais les variations dans l'environnement prennent souvent du temps à se faire sentir alors que le coût associé au sexe (habituellement représenté par une baisse de fertilité par individu) a des conséquences immédiates sur le taux de reproduction.

Quoiqu'il en soit, on observe aujourd'hui une multitude d'espèces sexuées et une grande diversité de systèmes de détermination du sexe. Chez les espèces diploïdes, beaucoup de ces systèmes sont basés sur une partition de génotypes à un locus (qui peut s'étendre sur un chromosome entier dans le cas de chromosomes sexuels). En particulier, la partition en homozygotes et hétérozygotes est très répandue. Ainsi, chez les mammifères et la drosoplile, les femelles sont homozygotes XX alors que les mâles sont hétérozygotes XY, où X et Y représentent des chromosomes de morphologies différentes, le Y étant plus court que le X. Par contre, chez les oiseaux et les papillons, ce sont les mâles qui sont homozygotes ZZ et les femelles hétérozygotes ZW quant à des chromosomes sexuels distincts, Z et W. Chez les espèces haplo-diploïdes comme les hyménoptères dont les abeilles, il n'y pas de chromosomes sexuels proprement dits:

S. Lessard (ed.), Mathematical and Statistical Developments of Evolutionary Theory, 269–325.

les femelles sont diploïdes **XX** alors que les mâles sont haploïdes **XO** (**X** représentant un gamète mâle ou femelle selon le cas et **O** signifiant une absence de gamète d'origine paternelle). Il est à noter que des anomalies de ploïdie et/ou des gènes secondaires situés à d'autres endroits du génome peuvent modifier la détermination du sexe. En l'absence totale de chromosomes sexuels ou de différences de ploïdie particulièrement chez les hermaphrodites, des systèmes moins rigides de détermination du sexe faisant intervenir plusieurs gènes à un ou plusieurs loci autosomiques, sans mentionner les effets de l'environnement, peuvent prendre entièrement la relève. C'est d'ailleurs dans ces formes plus souples de détermination qu'on doit sans doute chercher les origines de la différentiation sexuelle. [1]

1.2. SÉGRÉGATION MENDÉLIENNE ET PARITÉ NUMÉRIQUE DES SEXES

Les lois de Mendel ont marqué la génétique du XXe siècle. Rappelons qu'elles sont obtenues avec la prémisse que les gènes d'origines maternelle et paternelle à un locus se retrouvent avec des chances égales dans les gamètes transmis par un individu. Considérons par exemple le système de détermination du sexe **XX-XY**. Dans ce cas, les lois de Mendel prédisent qu'un mâle **XY** produira en moyenne autant de gamètes porteurs du **X** que de gamètes porteurs du **Y**. Si les gamètes des deux types ont des chances égales de féconder un ovule (nécessairement porteur d'un **X**), alors on devrait observer en moyenne autant de mâles que de femelles dans la progéniture; d'où une parité numérique des sexes dans la population si la population est très grande. Un raisonnement analogue pour le système **ZW-ZZ** conduit à la même conclusion. Une telle conclusion repose cependant sur l'hypothèse que la ségrégation des gènes se fait complètement au hasard lors de la production de gamètes par méiose et qu'aucune sélection ne s'exerce entre les différents types de gamètes avant et pendant la fécondation.

1.3. MODÈLES DE DISTORSION

Il existe des mécanismes génétiques de distorsion qui peuvent affecter le rapport numérique des sexes (*sex ratio*) dans la progéniture de mâles et/ou de femelles. Ces mécanismes sont très variés. Les modèles que nous considérerons et quelques références sont regroupés dans le tableau I. Par exemple, il peut y avoir plusieurs types de chromosome **X** qui ont des taux de ségrégation avec le chromosome **Y** différents chez les mâles **XY** (modèle I). Une telle situation se retrouve chez plusieurs espèces de drosophile. Par contre, chez le moustique *Aedes aegypti*, c'est le **Y** qui semble avoir plusieurs formes avec des taux de ségrégation différents avec le **X** (dans ce cas, les **X** et **Y** sont des allèles particuliers sis au même locus de chromosomes homologues; voir modèle II). Il est à remarquer que la ségrégation entre les différents **X** et **Y** peut se faire non seulement lors de la méiose mais aussi après la production des gamètes sous l'action d'une sélection gamétique précédant la fécondation. Ces deux phénomènes peuvent être difficiles à discerner car les conséquences sur les fréquences des gamètes peuvent être les mêmes. Il peut aussi exister des gènes autosomiques qui modifient les taux de ségrégation entre les **X** et **Y**. Dans le cas de la drosophile, ces gènes ne sont pas liés aux **X** et **Y** (modèle III), alors que dans le cas du moustique *Aedes aegypti*, ces gènes peuvent avoir un taux de recombinaison quelconque $r\,(0 < r < 1/2)$ avec le locus qui est le siège des X et Y (modèle III'). Il peut même exister des combinaisons des modèles précédents: par exemple, des X différents et des gènes autosomiques qui modifient les taux de ségrégation de certains seulement de ces X comme cela a été observé chez quelques espèces de drosophile. Bien que des mécanismes de distorsion de ségrégation analogues pour des systèmes ZW-ZZ soient moins bien documentés, ils existent sans doute tout autant.

Tableau I.

Modèles de distorsion du sex ratio

	génotype	gamète produit	fréquence	exemples
I.	X_iY	X_i	f_i	*Drosophila obscura* (Gershenson, 1928)
		Y	$1 - f_i$	*D. pseudoobscura* (Sturtevant and Dobshansky, 1936)
II.	XY_i	X	$1 - m_i$	*Aedes aegypti* (Hickey and Craig, 1966)
		Y_i	m_i	
III.	$\underline{A_i}\underline{A_j}$ XY	A_i/X	$(1-m_{ij})/2$	*D. melanogaster* (Bell, 1954)
		A_j/X	$(1-m_{ij})/2$	*D. simulans* (Faulhaber, 1967)
		A_i/Y	$m_{ij}/2$	*D. paramelanica* pour une combinaison de I et III
		A_j/Y	$m_{ij}/2$	(Stalker, 1961)
III'.	$\underline{A_i}\underline{A_j}$ XY	A_i/X	$(1-m_{ij})(1-r)$	*Aedes aegypti* (Newton et al., 1978)
		A_j/X	$(1-m_{ij})r$	
		A_i/Y	$m_{ij}r$	
		A_j/Y	$m_{ij}(1-r)$	

	génotype	gamète fécondant	fréquence	exemples
IV.	$\underline{A_i}\underline{A_j}$ XX	A_k/X	f_{ij}	Modèle théorique basé sur des observations suggérant un contrôle maternel
		A_k/Y	$1 - f_{ij}$	
IV'.	$\underline{A_i}\underline{A_j}$ XX	A_k/X	f_{ij}	Modèle précédent pour des espèces haplo-diploïdes
		A_k/O	$1 - f_{ij}$	

génotype	sexe	proportion	exemples
V . A_iA_j	mâle	m_{ij}	m_{ij} = 0 ou 1: Xiphophores (Kallman, 1973), Lemmings
	femelle	$1 - m_{ij}$	(Frega et al., 1976) $0 < m_{ij} < 1$: Hermaphrodisme successif ou simultané
V' . A_iA_j	mâle	m_{ij}	Modèle précédent pour des espèces haplo-diploïdes dont
	femelle	$1 - m_{ij}$	*Habrobracon* (Whiting, 1943)
A_i	mâle	1	

génotype	sexe	valeur sélective	exemples
V'' . A_iA_j	mâle	m_{ij}	Modèle V avec des différences de viabilité et/ou fertilité
	femelle	f_{ij}	(virilité)
VI . A_i	mâle	m_i	Modèle V" pour des espèces haploïdes incluant algues
	femelle	f_i	et champignons

Il existe une autre sorte de distorsion plus difficilement observable mais tout aussi efficace que celle qui modifie la ségrégation des gènes chez les hétérozygotes eux-mêmes. En effet, chez des espèces dont les femelles sont homozygotes, certaines observations suggèrent un contrôle maternel sur la fécondation (ou les résultats de la fécondation). Ainsi une femelle **XX** pourrait accepter (a priori ou a posteriori) que ses ovules soient fécondés par un gamète mâle **X** ou **Y** dans des proportions dépendant de son génotype à un locus autosomique (modèle IV). Un tel contrôle maternel est encore plus plausible chez les espèces haplo-diploïdes qui ont des ovules qui peuvent se développer avec ou sans fécondation (modèle IV'). De tels modèles ont un grand intérêt sur le plan théorique car ils permettent d'étudier l'évolution de "stratégies" reliés à la représentation des deux sexes dans la progéniture.

Un autre modèle qui a un intérêt théorique certain quant aux questions concernant l'évolution dans les espèces sexuées est celui où le génotype à un locus autosomique détermine les proportions de ressources allouées par les individus aux fonctions mâles et femelles respectivement (modèle V). Il peut s'agir ici d'hermaphrodites simultanés (mâles et femelles simultanément comme les escargots et beaucoup de plantes) ou successifs (d'abord mâles puis femelles comme les huîtres et certains poissons ou d'abord femelles puis mâles comme les limaces; chez les plantes, le premier phénomène s'appelle protandrie et le second protogynie). Une autre interprétation est que le locus en question détermine la probabilité pour un individu d'être ou bien mâle, ou bien femelle. Dans le cas où ces probabilités sont toutes 0 ou 1 (système dichotomique), on a une partition de génotypes en deux groupes, celui des femelles et celui des mâles, dont les systèmes **XX-XY** et

ZW-ZZ sont des cas particuliers. Des systèmes dichotomiques à plusieurs allèles ont déjà été observés chez beaucoup d'espèces diploïdes (dont les xiphophores et les lemmings) et même chez les individus diploïdes d'espèces haplo-diploïdes (par exemple, la guêpe *Habrobracon*; voir modèle V').

Le modèle d'allocation sexuelle peut être généralisé en ajoutant des différences de viabilité et/ou de fertilité (ou virilité) selon le génotype et le sexe (modèle V"). On obtient ainsi un modèle analogue à un modèle général de viabilité avec distinction de sexes. Le modèle correspondant pour les espèces haploïdes (incluant algues et champignons chez qui les individus mâles sont généralement caractérisés par une plus grande mobilité) est celui où le gène à un seul exemplaire à un locus donné détermine les valeurs sélectives chez les individus mâles et femelles respectivement (modèle VI).

On ne peut pas passer sous silence de nombreux autres modèles de distorsion du sex ratio qui ne seront pas l'objet de la présente étude. On peut d'abord avoir des modificateurs à plusieurs loci différents dans les modèles déjà mentionnés. Il y a aussi la possibilité d'influences extra-chromosomiques comme des virus transmis via le cytoplasme ou des différences de température et/ou de ressources nutritives dans l'environnement extérieur. D'autre part, il peut y avoir divers conflits entre individus (par exemple, entre le père et la mère ou entre la mère et les descendants de première génération ou les ouvrières chez les haplo-diploïdes) lorsque plusieurs individus sont responsable des distorsions observées dans les progénitures. Enfin il ne faut pas oublier que la polyploïdie est fréquente chez les plantes et que les systèmes de détermination du sexe sont parfois très compliqués pour celles-ci, sans mentionner des systèmes complexes d'incompatibilité.

Pour tous les modèles de distorsion que nous étudierons, nous ferons les hypothèses suivantes:

Hypothèses I. (a) *La population est suffisamment grande pour que les effets de dérive puissent être négligés (population infinie), (b) les générations sont séparées (générations discrètes sans chevauchement), et (c) les croisements se font au hasard (panmixie).*

Nous nous demanderons pour chaque modèle quelles sont les conséquences du mécanisme de distorsion sur la population dans son ensemble et en particulier sur le rapport numérique des mâles et des femelles (sex ratio) et son évolution. Certains modèles (modèles I, II et VI) se prêtent mieux à l'analyse et permettent de poser les problèmes, et même d'entrevoir les conclusions, pour les modèles plus complexes (modèles III, IV, IV', V, V', V"). Les premiers seront donc traités dans cette section (section 1) alors que les seconds seront analysés en détail dans les sections subséquentes (sections 2, 3 et 4). Les effets du linkage entre gènes (modèle III') et d'interactions entre individus seront considérés dans la dernière section (section 5).

1.4. ANALYSE DU MODÈLE II [2]

On considère N types de chromosome **Y**, dénotés $Y_1,...,Y_N$. Ces types sont distincts par leurs taux de ségrégation avec le chromosome **X**. On fait donc l'hypothèse que m_i, la proportion de gamètes Y_i produits par un mâle XY_i, est différent de m_j pour tout $j \neq i$ $(i,j = 1,...N)$.

Si y_i représente la fréquence de Y_i chez les mâles $(i = 1,...N)$, alors on a d'une génération à la suivante

$$(1.1) \qquad y_i' = \frac{m_i y_i}{\sum\limits_{k=1}^{N} m_k y_k}, \qquad i = 1,...,N .$$

On vérifie immédiatement qu'on a équilibre si et seulement si $y_i = 1$ pour un certain i, c'est-à-dire si on a fixation d'un chromosome Y_i dans la population mâle $(i=1,...,N)$. De plus on a la propriété suivante:

Théorème 1.1. [3]

$$(1.2) \qquad \sum_{i=1}^{N} m_i y_i' \geq \sum_{i=1}^{N} m_i y_i$$

avec égalité seulement à l'équilibre.

Donc la proportion de mâles dans la population croît d'une génération à la suivante (en fait jusqu'à la fixation du Y_i correspondant au plus grand m_i représenté dans la population, car la fréquence de ce Y_i croît près de tout autre équilibre vers lequel on ne peut donc pas avoir convergence, alors qu'on doit nécessairement avoir convergence vers un équilibre). A long terme, avec l'introduction de nouveaux Y_i, on aura *évolution vers une population constituée seulement de mâles* (à moins bien sûr que d'autres facteurs n'empêchent cette évolution).

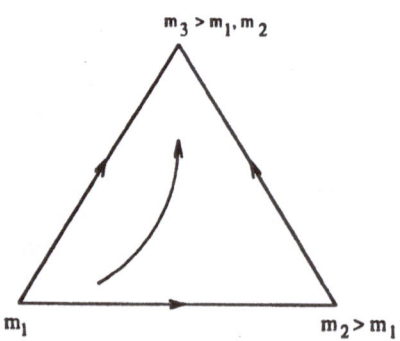

Figure 1.1. Dynamique pour le modèle (1.1) dans le cas $N = 3$ ($\rightarrow$ indique une direction de convergence).

La démonstration du théorème 1.1 repose sur l'inégalité de Schwarz (voir appendice 1.A). Dans le cas de trois types de chromosome Y avec $m_3 > m_2 > m_1$, la dynamique dans le simplexe des vecteurs de fréquence (y_1, y_2, y_3) est illustrée dans la figure 1.1.

1.5. ANALYSE DU MODÈLE I

On suppose que les taux de ségrégation f_i de chromosomes X_i avec le chromosome Y sont tous différents entre eux pour $i = 1,...,N$. En dénotant la fréquence de X_i chez les femelles et les mâles par p_i et q_i, respectivement $(i = 1,...,N)$, on obtient les équations de récurrence

$$q_i' = p_i, \qquad i = 1,...,N,$$

(1.3)
$$p_i' = \frac{p_i}{2} + \frac{f_i q_i}{2 \sum_{k=1}^{N} f_k q_k}, \qquad i = 1,...,N.$$

On vérifie facilement que l'équilibre a lieu seulement lorsque $p_i = q_i = 1$ pour un certain i $(i = 1,...,N)$. On peut prédire la dynamique à long terme avec l'aide du résultat suivant:

Théorème 1.2.

(1.4)
$$\sum_{i=1}^{N} f_i(p_i' + \frac{q_i'}{2}) \geq \sum_{i=1}^{N} f_i(p_i + \frac{q_i}{2})$$

avec égalité seulement à l'équilibre.

Donc, pour un ensemble donné de X_i représentés dans la population, il y aura convergence vers l'état de fixation correspondant au plus grand f_i et, à long terme avec l'introduction de nouveaux X_i, il y aura *évolution vers une population constituée seulement de femelles.*

La démonstration du théorème 1.2 repose également sur l'inégalité de Schwarz (voir appendice 1.B). Il y a une grande similitude entre le théorème 1.2 et le théorème 1.1 puisque $p_i + q_i/2$ dans le théorème 1.2 correspond à la fréquence de X_i dans la population entière, mâles et femelles réunis avec les mâles contribuant deux fois moins que les femelles, comme y_i dans le théorème 1.1 correspond à la fréquence de Y_i dans la population entière car seuls les mâles sont porteurs de tels chromosomes.[4]

1.6. PRÉDICTIONS DE FISHER POUR LES MODÈLES III, IV ET V

Les modèles III, IV est V sont beaucoup plus difficiles à analyser et seront en fait l'objet principal de notre étude.   En fait, pour ces modèles, il n'existe pas de fonctions de Lyapounov connues qui seraient monotones de génération en génération comme c'était le cas pour les modèles I et II (voir théorèmes 1.1 et 1.2).   Mais intuitivement, on peut déjà prédire la direction générale de l'évolution pour ces modèles:

> If we consider the aggregate of an entire generation of [...] offspring it is clear that the total reproductive value of the males in this group is exactly equal to the total value of all the females, because each sex must supply half the ancestry of all future generations of the species.   From this it follows that the sex ratio will so adjust itself, under the influence of Natural Selection, that the total parental expenditure incurred in respect of children of each sex shall be equal. (Fisher, 1930)[5]

Les arguments intuitifs de Fisher font cependant appel à une notion de coût qu'il faut préciser pour donner une interprétation correcte du principe d'évolution sous-jacent.   Chez beaucoup d'espèces les mâles et les femelles à la naissance ne demandent pas les mêmes soins de la part des parents, la même quantité de nourriture, la même protection contre l'environnement ou les prédateurs, etc.   On considère donc en général une période de dépendance de la conception à un âge T, durant laquelle les parents pourvoient aux besoins de leurs petits.   Supposons que les soins parentaux durant cette période entraînent un "coût" s pour un mâle comparé à 1 pour une femelle.   Habituellement on a $s < 1$.   Remarquons que s peut représenter la viabilité relative d'un mâle comparativement à une femelle durant la période de dépendance; il suffit alors que le coût soit proportionnel à la probabilité de survie.   On fait aussi l'hypothèse que le coût total pour les parents est constant, ce qui revient à faire l'hypothèse que les parents consacrent la même quantité totale de ressources à la reproduction.   Dans le cas où les différences de coût entre mâles et femelles reflètent des différences de viabilité durant la période de dépendance, l'hypothèse d'un coût constant est équivalent au remplacement des pertes à cause de la mortalité durant la période de dépendance.   Dans ces conditions et avec cette interprétation, Fisher prédit donc l'évolution vers la parité numérique des sexes dans la population à la fin de la période de dépendance, c'est-à-dire à l'âge T (car c'est seulement alors que les ressources totales allouées aux descendants mâles et femelles par l'ensemble des parents dans la population sont égales).   Le rapport numérique des sexes à la conception dans la population est alors de $1/s$ à 1 pour les mâles comparativement aux femelles.[6]   Il est à remarquer que les mêmes arguments s'appliquent pour les ressources allouées aux fonctions mâles et femelles (production d'ovules versus production de pollen ou sperme) par les hermaphrodites.

1.7. ANALYSE DU MODÈLE HAPLOIDE (MODÈLE VI)

Les arguments de Fisher sont valides lorsque les ressources allouées aux descendants mâles et femelles par les parents (ou aux fonctions mâles et femelles par les hermaphrodites) sont constantes au total; seules les proportions de ressources allouées aux mâles et aux femelles peuvent varier et donc on a un modèle d'allocation sexuelle pure. Dans le cas où les ressources totales allouées à la reproduction sont variables, introduisant ainsi des différences de viabilité et/ou fertilité, il existe un principe d'évolution plus général que celui de la parité entre les sexes. Ce principe sera illustré en utilisant le modèle haploïde qui va nous permettre en même temps de vérifier la justesse des prédictions de Fisher dans un cas relativement simple.

Si x_i représente la fréquence d'un allèle A_i à la conception chez les mâles aussi bien que chez les femelles mais avec une valeur sélective m_i chez les mâles et f_i chez les femelles $(i = 1,...,N)$, alors on aura à la génération suivante

$$(1.5) \qquad x_i' = x_i \left[\frac{f_i}{2f} + \frac{m_i}{2m} \right], \qquad i = 1,...,N ,$$

où

$$(1.6) \qquad f = \sum_{i=1}^{N} f_i x_i , \qquad m = \sum_{i=1}^{N} m_i x_i .$$

La quantité

$$(1.7) \qquad \frac{f_i}{2f} + \frac{m_i}{2m}$$

représente la valeur reproductive de A_i $(i = 1,...,N)$ comparée à une valeur moyenne égale à 1 dans la population. La formule (1.7) a été baptisée *formule de Shaw-Mohler* par Charnov (1982) en l'honneur des auteurs qui l'on introduite.

Le système de récurrence (1.5) admet comme fonction de Lyapounov le produit $m \cdot f$. En effet, ce produit a la propriété suivante:

Théorème 1.3.[7]

$$(1.8) \qquad (m \cdot f)' \geq m \cdot f$$

avec égalité seulement à l'équilibre.

Le théorème 1.3 se démontre en observant que le système (1.5) est aussi celui pour un modèle de viabilité classique pour une population diploïde sans distinction de sexes avec $(f_i m_j + m_i f_j)/2$ comme viabilité pour $A_i A_j$ $(i,j = 1,...,N)$ et $m \cdot f$ comme viabilité moyenne. Ce produit croît donc d'une génération à la suivante par le "théorème fondamental de la sélection naturelle" (voir appendice 1.C). De plus les équilibres stables correspondent aux maxima locaux de m • f (en fait aux maxima globaux de m • f car on a le produit de deux fonctions linéaires) par rapport au simplexe des vecteurs de fréquence à N composantes, c'est-à-dire,

$$(1.9) \qquad \Delta_N = \left\{ (x_1, \ldots, x_N) : x_i \geq 0 \text{ pour tout } i, \sum_{i=1}^{N} x_i = 1 \right\}.$$

En fait, génériquement, il y a seulement deux structures d'équilibre possibles telles qu'illustrées dans la figure 1.2: ou bien le maximum de m • f est atteint à l'intérieur de Δ_N et dans ce cas il y a convergence globale à partir de n'importe quel point à l'intérieur de Δ_N vers la section d'hyperplan correspondant à ce maximum, ou bien le maximum de m • f est atteint à un sommet et dans ce cas la convergence a lieu vers ce sommet. Il est à remarquer que dans le premier cas il existe une surface d'équilibres qui correspond au maximum de m • f et c'est principalement ce qui distingue le modèle VI des modèles I et II quant à la dynamique.

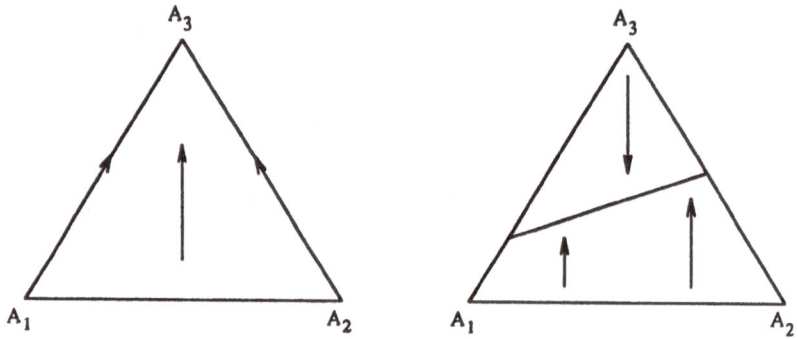

Figure 1.2. Structures d'équilibre pour le modèle (1.5) dans le cas N = 3 : (a) le maximum de m • f est atteint à un état de fixation, ici A_3 ; (b) le maximum de m • f n'est atteint à aucun état de fixation.

Dans le cas particulier où $f_i = 1 - m_i$ pour tout i, on a m • f = m(1-m) dont le maximum absolu est 1/4 qui est atteint lorsque m = 1/2. Tant que ce maximum n'est pas atteint, il existe toujours la possibilité d'un nouvel allèle dont l'introduction conduira après plusieurs générations à un produit m • f plus près de 1/4. Une fois qu'un équilibre avec m • f = 1/4 est atteint, l'introduction de tout nouvel allèle en petite quantité conduira à nouveau à un équilibre avec m • f = 1/4 appartenant à la même surface d'équilibres que le précédent. *Donc dans le cas d'allocation sexuelle pure chez les populations haploïdes, on a évolution vers la parité numérique entre les sexes.*

L'objectif principal des prochains chapitres sera de démontrer des propriétés dynamiques analogues pour des systèmes plus complexes, à savoir des populations sexuées diploïdes.

Appendices

1.A. DÉMONSTRATION DU THÉORÈME 1.1

On a

$$
(1.10) \qquad \sum_{i=1}^{N} m_i y_i' = \frac{\displaystyle\sum_{i=1}^{N} m_i^2 y_i}{\displaystyle\sum_{k=1}^{N} m_k y_k} \geq \sum_{i=1}^{N} m_i y_i
$$

par l'inégalité de Schwarz avec égalité si et seulement si $y_i = 1$ pour un certain i $(i = 1,...,N)$.

1.B. DÉMONSTRATION DU THÉORÈME 1.2

Dans ce cas on a

$$
(1.11) \qquad \sum_{i=1}^{N} f_i p_i' = \sum_{i=1}^{N} f_i \left(\frac{p_i}{2} + \frac{f_i q_i}{2\displaystyle\sum_{k=1}^{N} f_k q_k} \right) \geq \frac{\displaystyle\sum_{i=1}^{N} f_i p_i}{2} + \frac{\displaystyle\sum_{i=1}^{N} f_i q_i}{2}
$$

avec égalité si et seulement si $q_i = 1$ pour un certain i $(i = 1,...,N)$. La conclusion du théorème 1.2 découle alors de la relation $q_i' = p_i$ pour $i = 1,...,N$.

1.C. DÉMONSTRATION DU THÉORÈME 1.3

On vérifie facilement que le système de récurrence (1.5) peut se mettre sous la forme

$$
(1.12) \qquad x_i' = \frac{x_i \displaystyle\sum_{j=1}^{N} v_{ij} x_j}{\displaystyle\sum_{k,l=1}^{N} v_{kl} x_k x_l}, \qquad i = 1,...,N ,
$$

avec

$$(1.13) \qquad v_{ij} = \frac{f_i m_j + m_i f_j}{2}, \qquad i,j = 1,\dots,N,$$

et

$$(1.14) \qquad \sum_{i,j=1}^{N} v_{ij} x_i x_j = m \cdot f.$$

Pour le modèle (1.12), il est bien connu que

$$(1.15) \qquad \sum_{k,l=1}^{N} v_{kl} x'_k x'_l \geq \sum_{k,l=1}^{N} v_{kl} x_k x_l$$

avec égalité seulement à l'équilibre. (C'est le théorème fondamental de la sélection naturelle pour le modèle classique de viabilité à un locus pour une population diploïde sans distinction de sexes; voir, par ex., Crow et Kimura, 1970; Ewens 1979.) En effet, en définissant

$$(1.16) \qquad w_i = \sum_{j=1}^{N} v_{ij} x_j, \qquad i = 1,\dots,N,$$

on a (Kingman, 1961a)

$$\sum_{i,j=1}^{N} v_{ij} x_i w_i x_j w_j = \sum_{i,j,k=1}^{N} v_{ij} v_{ik} x_i x_j x_k w_j = \sum_{i,j,k=1}^{N} v_{ij} v_{ik} x_i x_j x_k \left(\frac{w_j + w_k}{2} \right)$$

$$(1.17) \qquad \geq \sum_{i,j,k=1}^{N} v_{ij} v_{ik} x_i x_j x_k \sqrt{w_j w_k} = \sum_{i=1}^{N} x_i \left(\sum_{j=1}^{N} v_{ij} x_j \sqrt{w_j} \right)^2$$

$$\geq \left(\sum_{i,j=1}^{N} v_{ij} x_i x_j \sqrt{w_j} \right)^2 = \left(\sum_{j=1}^{N} x_j w_j^{3/2} \right)^2 \geq \left(\sum_{j=1}^{N} x_j w_j \right)^3,$$

ce qu'il fallait démontrer.[8]

2.   Structures d'équilibre

Dans cette section nous étudierons les structures d'équilibre pour les modèles III, IV, IV', V et V'. Comme les structures d'équilibre s'avèrent être analogues pour ces modèles, nous nous concentrons sur le modèle V.[9]

Supposons qu'un individu de génotype $A_i A_j$ à un locus donné est mâle avec probabilité m_{ij} et femelle avec probabilité $1 - m_{ij}$ $(0 < m_{ij} = m_{ji} < 1$ pour tout $i,j = 1,...N)$. Si q_i désigne la fréquence de l'allèle A_i chez les mâles et p_i la fréquence de A_i chez les femelles $(i = 1,...N)$, alors on aura d'une génération à la suivante

$$q_i' = \frac{p_i \sum_{j=1}^{N} m_{ij}q_j + q_i \sum_{j=1}^{N} m_{ij}p_j}{2m}, \qquad i = 1,...,N ,$$

(2.1)

$$p_i' = \frac{p_i \sum_{j=1}^{N} (1-m_{ij})q_j + q_i \sum_{j=1}^{N} (1-m_{ij})p_j}{2(1-m)}, \qquad i = 1,...,N ,$$

où

(2.2)
$$m = \sum_{i,j=1}^{N} m_{ij}p_i q_j .$$

La quantité m représente *la proportion de mâles dans la population.*

2.1. CLASSES D'ÉQUILIBRE

Remarquons que, dû à la panmixie, la fréquence de A_i à la conception doit être égale à la fréquence de A_i chez les parents, et ce pour tout i . Donc on doit avoir

(2.3)
$$m q_i' + (1-m)p_i' = \frac{q_i + p_i}{2}, \qquad i = 1,...,N .$$

On en déduit immédiatement qu'à l'équilibre

ou bien (1) $m \neq 1/2$ *et alors* $p_i = q_i$ *pour* $i = 1,...,N$,
ou bien (2) $m = 1/2$.

Dans le cas (1), les équilibres sont dits *symétriques*, et dans le cas (2), *équilibres de parité* (sous-entendu, *numérique des sexes*). Il est à remarquer que les conditions (1) ou (2) sont nécessaires pour avoir équilibre mais non suffisantes.

2.2. ÉQUILIBRES SYMÉTRIQUES

Un équilibre symétrique est décrit par un seul vecteur de fréquence $\mathbf{p^*} = (p_1^*, \ldots, p_N^*)$ où p_i^* représente la fréquence de A_i chez les mâles et les femelles à l'équilibre $(i = 1, \ldots, N)$. La proportion de mâles à l'équilibre est alors

$$(2.4) \qquad m^* = \sum_{i,j=1}^{N} m_{ij} p_i^* p_j^* .$$

Les équilibres symétriques sont caractérisés par le théorème suivant:

Théorème 2.1.[10] (a) *Un équilibre symétrique* $\mathbf{p^*} = (p_1^*, \ldots, p_N^*)$ *correspond à un équilibre pour le modèle classique de viabilité* $M = \|m_{ij}\|_{i,j=1,\ldots,N}$ *défini par*

$$(2.5) \qquad p_i' = \frac{p_i \displaystyle\sum_{j=1}^{N} m_{ij} p_j}{w(p)}, \qquad i = 1, \ldots, N ,$$

où

$$(2.6) \qquad w(p) = \sum_{i,j=1}^{N} m_{ij} p_i p_j ,$$

et aussi pour le modèle correspondant avec la matrice de viabilité $U - M = \|1 - m_{ij}\|_{i,j=1,\ldots,,N}$.

(b) Un équilibre symétrique p^* *avec* $m^* < 1/2$ *(ou* $m^* > 1/2$*) est stable si et seulement s'il est stable pour le modèle* M *(ou* $U-M$, *respectivement).*

Démonstration du Théorème 2.1. (a) Condition d'équilibre.   La condition d'équilibre pour le modèle (2.5) (que ce soit avec la matrice de viabilité M ou $U-M$) est

$$(2.7) \qquad (Mp^*)_i = w(p^*) \text{ dès que } p_i^* > 0$$

où $(Mp^*)_i$ désigne la i^e composante du vecteur Mp^*. La condition (2.7) est aussi la condition d'équilibre d'un équilibre symétrique pour le modèle (2.1) avec $m^* = w(p^*)$.

(b)   Condition de stabilité interne. Supposons d'abord que p^* est un équilibre symétrique polymorphique, c'est-à-dire, avec $p_i^* > 0$ pour $i = 1, \ldots, N$. Considérons de petites perturbations ξ et η sur p^* telles que

$$(2.8) \qquad q = q^* + \eta \quad \text{et} \quad p = p^* + \xi$$

sont des vecteurs de fréquence suffisamment près de p^*. La transformation de (η, ξ) d'une génération à la suivante selon le modèle (2.1) est alors approximée par la matrice

$$(2.9) \quad \begin{bmatrix} R & R \\ S & S \end{bmatrix} \quad \text{avec} \quad R = \frac{1}{2}\left[I + \frac{\text{diag}(p^*)M}{m^*}\right], S = \frac{1}{2}\left[I - \frac{\text{diag}(p^*)M}{1-m^*}\right],$$

où $\text{diag}(p^*)$ dénote la matrice diagonale avec le vecteur p^* sur la diagonale principale et I désigne la matrice identité. A part 0, les valeurs propres de cette matrice sont celles de

$$(2.10) \quad R + S = I + \left[\frac{1-2m^*}{2(1-m^*)}\right]\frac{\text{diag}(p^*)M}{m^*}.$$

D'autre part, la transformation de ξ selon le modèle (2.5) est approximée par

$$(2.11) \quad L = I + \frac{\text{diag}(p^*)M}{m^*}.$$

Dans les deux cas, en supposant $m^* < 1/2$, les valeurs propres affectant les vecteurs ξ admissibles (c'est-à-dire, perpendiculaires à $1 = (1,...,1)$ pour que les vecteurs $p^* + \xi$ soient des vecteurs dont la somme des composantes égale 1; voir (2.8)) sont < 1 en module si et seulement si les valeurs propres de la matrice

$$(2.12) \quad M^* = \frac{\text{diag}(p^*)M}{m^*}$$

qui sont différentes de 1 (avec vecteur propre à droite p^* par la condition d'équilibre (2.7) et le fait que $m^* = w(p^*)$) sont toutes négatives. Il est à remarquer que 1 est une valeur propre simple de M^* qui domine toutes les autres valeurs propres de M^* en module (car il lui correspond un vecteur propre positif; l'affirmation découle alors de la théorie de Perron-Frobenius pour les matrices positives, c'est-à-dire, les matrices dont toutes les entrées sont positives; voir, p. ex., Gantmacher, 1959). De plus, la matrice M^* est symétrique par rapport au produit scalaire

$$(2.13) \quad \ll x,y \gg = \langle \text{diag}(p^*)^{-1}x,y \rangle\, m^*$$

où $\langle \cdot, \cdot \rangle$ dénote le produit scalaire standard, c'est-à-dire,

$$(2.14) \quad \langle x,y \rangle = \sum_{i=1}^{N} x_i y_i$$

pour les vecteurs $x = (x_1,...,x_N)$ et $y = (x_1,...,x_N)$. En effet, on vérifie aisément que

$$(2.15) \quad \ll M^*x,y \gg = \langle Mx,y \rangle = \langle x,My \rangle = \ll x,M^*y \gg.$$

La condition (2.12) devient donc

(2.16) $\langle\langle \xi, M^*\xi \rangle\rangle < 0$ dès que $\langle\langle \xi, p^* \rangle\rangle = 0$, $\xi \neq 0$,

ce qui est équivalent à

(2.17) $\langle \xi, M\xi \rangle < 0$ dès que $\langle \xi, 1 \rangle = 0$, $\xi \neq 0$.

La condition (2.17) est la condition de stabilité pour un équilibre polymorphique avec $m^* < 1/2$. Lorsque l'équilibre n'est pas polymorphique, la condition (2.17) pour la sous-matrice principale de M correspondant aux composantes positives de l'équilibre est la condition de stabilité interne. Lorsque $m^* > 1/2$, on remplace M par $U - M$ en utilisant la symétrie du modèle pour les deux sexes.

Condition de stabilité externe. Supposons que $p_i^* = 0$ pour un certain i. Alors, près de l'équilibre symétrique p^*, on a selon le modèle (2.1)

$$(2.18) \quad \begin{bmatrix} q_i' \\ p_i' \end{bmatrix} = \begin{bmatrix} \dfrac{(Mp^*)i}{2m^*} & \dfrac{(Mp^*)_i}{2m^*} \\ \dfrac{1-(Mp^*)_i}{2(1-m^*)} & \dfrac{1-(Mp^*)_i}{2(1-m^*)} \end{bmatrix} \begin{bmatrix} q_i \\ p_i \end{bmatrix} + \text{termes d'ordre supérieur.}$$

Les valeurs propres de la matrice dans (2.18) sont 0 et

$$(2.19) \quad \frac{(Mp^*)_i}{2m^*} + \frac{1-(Mp^*)_i}{2(1-m^*)} = \frac{(1-2m^*)(Mp^*)_i + m^*}{(1-2m^*) + m^*}.$$

Dans le cas où $m^* < 1/2$, la quantité (2.19) est < 1 si et seulement si

$$(2.20) \quad \frac{(Mp^*)_i}{w(p^*)} < 1,$$

en se rappelant que $m^* = w(p^*)$. Mais selon le modèle (2.5), on a près de p^*

$$(2.21) \quad p_i' = \left[\frac{(Mp^*)_i}{w(p^*)} \right] p_i + \text{termes d'ordre supérieur.}$$

Pour les deux modèles, on a donc stabilité externe de p^* avec $m^* < 1/2$ si et seulement si on a l'inégalité (2.20) pour tout i pour lequel $p_i^* = 0$. Dans le cas où $m^* > 1/2$, on remplace M par $U - M$ par symétrie entre les sexes.

Pour éviter des dégénérescences dans les conditions d'existence et de stabilité des équilibres symétriques de (2.1), nous ferons toujours les hypothèses suivantes:

Hypothèses II.
(1) $M = \|m_{ij}\|_{i,j=1,...,N}$ *où* $0 < m_{ij} = m_{ji} < 1$ *pour* $i,j = 1,...,N$.
(2) M *et toutes les sous-matrices principales de* M *sont non-singulières.*
(3) *Toutes les valeurs propres à un équilibre symétrique sont différentes entre elles et différentes de* 1.
(4) $m \neq 1/2$ *à tout équilibre symétrique.*

Selon l'hypothèse (1), M et U - M sont des matrices symétriques positives. L'hypothèse (2) garantit qu'il existe au plus un équilibre symétrique par sous-ensemble d'allèles, car la condition d'équilibre (2.7) admet alors au plus une solution par sous-ensemble de composantes positives. L'hypothèse (3) assure que des approximations linéaires près des équilibres symétriques sont suffisantes pour étudier leur stabilité. Enfin l'hypothèse (4) écarte la possibilité d'un équilibre symétrique qui serait aussi de parité. Ces hypothèses interdisent certaines relations fonctionnelles particulières entre les entrées de M. Ces hypothèses sont génériques dans le sens que si elles ne sont pas satisfaites alors de petites perturbations sur les entrées de M feront en sorte qu'elles le seront, et si elles sont satisfaites alors toutes perturbations suffisamment petites sur les entrées de M feront en sorte qu'elles le seront encore.

Remarque. Dans le modèle de viabilité M, *un équilibre* p* *correspond à un point critique de* w(p) *par rapport aux perturbations sur les composantes positives de* p* *et un équilibre stable* p* *à un maximum local de* w(p) *par rapport aux perturbations sur toutes les composantes de* p*. En effet, la première affirmation découle de la relation

$$(2.22) \qquad \frac{\partial}{\partial p_i} w(p) = 2(Mp)_i, \qquad i = 1,...,N$$

et de la condition d'équilibre (2.7). Quant à la seconde affirmation sur la stabilité d'un équilibre p*, elle se justifie par les égalités

$$w(p^*+\xi) = \langle p^*+\xi, M(p^*+\xi)\rangle$$

$$= \langle p^*, Mp^*\rangle + \langle \xi, Mp^*\rangle + \langle p^*, M\xi\rangle + \langle \xi, M\xi\rangle$$

$$(2.23) \qquad = \langle p^*, Mp^*\rangle + 2\langle \xi, Mp^*\rangle + \langle \xi, M\xi\rangle$$

$$= w(p^*) + 2 \sum_{i:p_i^*=0} \xi_i[(Mp^*)_i - w(p^*)] + \langle \xi, M\xi\rangle$$

et les conditions de stabilité interne et externe (2.17) et (2.20). Il est à remarquer que *le maximum de* w(p) *est global par rapport aux perturbations sur les composantes positives d'un équilibre stable* p*.

2.3. ÉQUILIBRES DE PARITÉ

Pour décrire les équilibres de parité, on introduit la matrice

$$(2.24) \qquad B_M(p) = \text{diag}(p)M + \text{diag}(Mp)$$

et la fonction spectrale

$$(2.25) \qquad \rho_M(p) = \text{la plus grande valeur propre en module de } B_M(p)$$

où $M = \|m_{ij}\|_{i,j=1,\ldots,N}$ est une matrice positive qui satisfait aux hypothèses de la section 2.2 et $p = (p_1,\ldots,p_N)$ est un vecteur de fréquence. Un équilibre de parité de (2.1) est alors représenté par deux vecteurs de fréquence $p = (p_1,\ldots,p_N)$ et $q = (q_1,\ldots,q_N)$ satisfaisant

$$(2.26a) \qquad B_M(p)q = q \, ,$$

ou encore, ce qui est équivalent,

$$(2.26b) \qquad B_{U-M}(q)p = p \, ,$$

c'est-à-dire,

$$(2.27a) \qquad \rho_M(p) = 1 \text{ avec } q \text{ comme vecteur propre à droite,}$$

ou encore, ce qui est équivalent,

$$(2.27b) \qquad \rho_{U-M}(q) = 1 \text{ avec } p \text{ comme vecteur propre à droite.}$$

Il est à remarquer que dans chaque cas il peut exister un seul vecteur propre à droite qui soit un vecteur de fréquence pour la valeur propre 1 et que cette valeur propre est alors nécessairement la plus grande valeur propre en module. Cela découle de la théorie de Perron-Frobenius (voir, p.ex., Gantmacher, 1959) en observant que les vecteurs de fréquence p et q d'un équilibre de parité doivent avoir les mêmes composantes nulles, que les sous-matrices de $B_M(p)$ et $B_{U-M}(q)$ correspondant aux composantes positives sont positives et que celles correspondant aux composantes nulles sont des matrices diagonales avec des quantités positives strictement < 1 sur la diagonale (en fait, les composantes de Mp et $(U-M)q$, respectivement, correspondant aux composantes nulles de p et q).

A partir de maintenant, un équilibre de parité sera représenté par un vecteur de fréquence p satisfaisant (2.27a) ou un vecteur de fréquence q satisfaisant (2.27b).

Critère pour l'existence d'équilibres de parité.[11] L'existence de vecteurs de fréquence p satisfaisant (2.27a) est difficile à vérifier directement. Heureusement, il existe un moyen indirect simple pour savoir si oui ou non des équilibres de parité existent pour le système

(2.1) sous les hypothèses de la section 2.2. *Ce critère d'existence d'équilibres de parité est l'existence de deux équilibres symétriques,* p* *et* p** *, tels que*

$$(2.28) \qquad\qquad w(\mathbf{p}^*) < \frac{1}{2} < w(\mathbf{p}^{**}) \; .$$

Dans ce cas, p* *et* p** *sont complètement séparés par une surface d'équilibres de parité, c'est-à-dire que toute courbe continue de vecteurs de fréquence allant de* p* *à* p** *rencontre un équilibre de parité* p (voir figure 2.1).

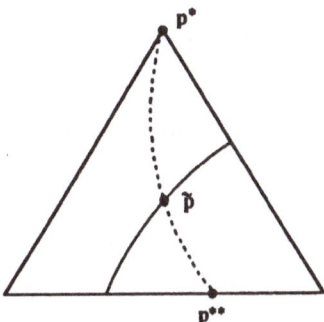

Figure 2.1. Représentation d'une surface d'équilibre de parité séparant deux équilibres symétriques pour le modèle 2.1.

La démonstration du critère ci-dessus repose essentiellement sur le résultat suivant qui permet de caractériser les équilibres des modèles de viabilité M et U - M , et donc aussi les équilibres symétriques de (2.1) par le théorème 2.1, à l'aide des fonctions $\rho_M(\mathbf{p})$ et $\rho_{U-M}(\mathbf{p})$:

Théorème 2.2. (Karlin et Lessard, 1984) *Pour le modèle de viabilité* M *avec les hypothèses* II, (a) *un équilibre polymorphique* p* *correspond à un point critique de* $\rho_M(\mathbf{p})$ *par rapport aux perturbations sur les composantes de* p* *, et* (b) *un équilibre stable* p* *(polymorphique ou non) correspond à un maximum local strict de* $\rho_M(\mathbf{p})$ *par rapport aux perturbations sur les composantes de* p* *. De plus, à un équilibre stable* p* *, on a*

$$(2.29) \qquad\qquad \rho_M(\mathbf{p}^*) = 2w(\mathbf{p}^*) \; .$$

(Comparer avec la caractérisation basée sur la fonction $w(\mathbf{p})$ selon la remarque de la section 2.2; voir l'appendice 2.A pour la démonstration du théorème 2.2.)

   La démonstration du critère est alors immédiate. En effet, la condition (2.28) entraîne

$$(2.30) \qquad\qquad \rho_M(\mathbf{p}^*) = 2w(\mathbf{p}^*) < 1 < 2w(\mathbf{p}^{**}) = \rho_M(\mathbf{p}^{**})$$

et la fonction $\rho_M(p)$ étant continue par rapport à p prend nécessairement la valeur 1 entre p^* et p^{**}. Inversement, s'il existe un équilibre de parité, alors nécessairement il existe des équilibres symétriques p^* et p^{**} satisfaisant

(2.31a) $$\max \rho_M(p) = \rho_M(p^{**}) = 2w(p^{**}) \geq 1 ,$$

(2.31b) $$\max \rho_{U\text{-}M}(p) = \rho_{U\text{-}M}(p^*) = 2[1-w(p^*)] \geq 1 ,$$

le maximum étant pris dans les deux cas sur tous les vecteurs de fréquence p. Les équations (2.31a) et (2.31b) impliquent

$$w(p^*) \leq \frac{1}{2} \leq w(p^{**}) .$$

Mais en fait les inégalités ci-dessus sont strictes étant donné les hypothèses II sur les équilibres symétriques.

Appendices

2.A. DÉMONSTRATION DU THÉORÈME 2.2

Pour simplifier la notation, on fait

(2.32) $$B(p) = B_M(p) \text{ et } \rho(p) = \rho_M(p)$$

dans (2.26a) et (2.27a) où M est une matrice de viabilité satisfaisant aux hypothèses II. De plus, remarquons que

(2.33) $$\rho(p) = \max\{\rho^\circ(p),(Mp)_i \text{ pour tout } i \text{ tel que } p_i = 0\}$$

où $\rho^\circ(p)$ désigne la plus grande valeur propre en module de la sous-matrice principale de $B(p)$ correspondant aux composantes positives de p. La démonstration du théorème 2.2 va utiliser deux lemmes qui seront démontrés ultérieurement.

Lemme 1. p^* est un équilibre pour le modèle de viabilité M si et seulement si

(2.34) $$B(p^*)p^* = 2w(p^*)p^* .$$

Dans ce cas

(2.35) $$\rho^\circ(p^*) = 2w(p^*) .$$

Lemme 2.

$$(2.36) \qquad \rho(p^*) = \rho°(p^*)$$

si l'équilibre p^* *est un point de maximum local strict de* $\rho(p)$ *par rapport aux vecteurs de fréquence* p.

Démonstration de la partie (a) du Théorème 2.2.   Considérons un équilibre $p^* = (p_1^*, \ldots, p_N^*)$ avec $p_i^* > 0$ pour tout i et définissons

$$(2.37) \qquad p(s) = p^* + s\xi$$

où $\langle \xi, 1 \rangle = 0$ et s est suffisamment petit pour que $p(s)$ soit un vecteur de fréquence positif. La matrice $B(p(s))$ étant alors positive, on peut trouver un et un seul vecteur de fréquence positif $q(s)$ satisfaisant

$$(2.38) \qquad B(p(s))q(s) = \rho(s)q(s)$$

où $\rho(s) = \rho(p(s))$ dans la notation (2.33). En dérivant par rapport à s, on obtient

$$(2.39) \qquad B(q(s))\dot{p}(s) + B(p(s))\dot{q}(s) = \dot{\rho}(s)q(s) + \rho(s)\dot{q}(s) .$$

Le produit scalaire avec le vecteur

$$(2.40) \qquad \frac{q(s)}{p(s)} = \left(\frac{q_1(s)}{p_1(s)}, \ldots, \frac{q_N(s)}{p_N(s)} \right)$$

donne alors

$$(2.41) \qquad \langle \frac{q(s)}{p(s)} B(q(s)), \dot{p}(s) \rangle = \dot{\rho}(s) \langle \frac{q(s)}{p(s)}, q(s) \rangle .$$

D'autre part, on a

$$(2.42) \qquad \langle \frac{p(s)}{q(s)} B(q(s)), \dot{p}(s) \rangle = 0 .$$

Par le lemme 1, on doit montrer que $\rho(s)$ a un point critique à $s = 0$ quelque soit la direction ξ si et seulement si $q(0) = p(0)$. Mais si

$$(2.43) \qquad \dot{\rho}(0) = 0 \text{ pour tout } \xi = \dot{p}(0) ,$$

alors

(2.44) $$\frac{p(0)}{q(0)} B(q(0)) = \text{un multiple de } \frac{q(0)}{p(0)} B(q(0)) ,$$

ce qui implique

(2.45) $$\frac{p(0)}{q(0)} = \text{un multiple de } \frac{q(0)}{p(0)} ,$$

car la matrice

(2.46) $$B(q(0)) = \text{diag}(Mq(0)) \ [I + \text{diag}(Mq(0))^{-1}\text{diag}(q(0))M]$$

est inversible, la seconde matrice entre les crochets dans (2.46) étant positive avec vecteur propre à gauche positif **1** pour la valeur propre 1 qui domine donc toutes les autres valeurs propres en module par la théorie de Perron-Frobenius. Enfin on déduit facilement de (2.45) que

(2.47) $$p(0) = \text{un multiple de } q(0) ,$$

ce qui n'est possible que si $p(0) = q(0)$. Les implications inverses sont immédiates.

Démonstration de la partie (b) du théorème 2.2.

Stabilité interne. On suppose d'abord que l'équilibre **p*** est polymorphique et on prend la dérivée seconde par rapport à s dans (2.38) évaluée à $s = 0$ pour obtenir

(2.48) $$B(p^*)\ddot{p} + B(p^*)\ddot{q} + 2\text{diag}(\dot{p})M\dot{q} + 2\text{diag}(\dot{q})M\dot{p} = \rho\ddot{q} + \ddot{\rho}p^* .$$

(A partir de maintenant, les évaluations à 0 des fonctions de s ne sont pas indiquées.) Le produit scalaire avec **1** donne dans ce cas

(2.49) $$\ddot{\rho} = 4\langle \dot{p}, M\dot{q} \rangle = 4\langle\!\langle \dot{p}, M^*\dot{q} \rangle\!\rangle$$

où M* est la matrice de (2.12) et $\langle\!\langle \cdot, \cdot \rangle\!\rangle$ le produit scalaire de (2.13) pour lequel M* est symétrique. D'autre part, la dérivée première par rapport à s dans (2.38) évaluée à $s = 0$ donne

(2.50) $$(I - M^*)\dot{q} = (I + M^*)\dot{p} .$$

Soit maintenant $\{x_1, \ldots, x_{N-1}, p^*\}$ une base de vecteurs propres à droite de M* orthonormée par rapport à $\langle\!\langle \cdot, \cdot \rangle\!\rangle$ avec valeurs propres (réelles) correspondantes $\lambda_1, \ldots, \lambda_{N-1}, 1$. On a $|\lambda_i| < 1$ pour $i = 1, \ldots, N-1$ et

(2.51) $$\dot{p} = \sum_{i=1}^{N-1} \lambda_i \langle\!\langle \dot{p}, \xi_i \rangle\!\rangle , \quad \dot{q} = \sum_{i=1}^{N-1} \lambda_i \langle\!\langle \dot{q}, \xi_i \rangle\!\rangle ,$$

car ces vecteurs sont perpendiculaires à $\mathbf{p}^*$ par rapport au produit scalaire $\langle\!\langle \cdot,\cdot \rangle\!\rangle$ étant perpendiculaires à $\mathbf{1}$ par rapport au produit scalaire $\langle \cdot,\cdot \rangle$. Par conséquent, on trouve

$$
\begin{aligned}
\ddot{\rho} &= 4 \sum_{i=1}^{N-1} \lambda_i \langle\!\langle \dot{\mathbf{p}},\xi_i \rangle\!\rangle \, \langle\!\langle \dot{\mathbf{q}},\xi_i \rangle\!\rangle \\
(2.52) \qquad &= 4 \sum_{i=1}^{N-1} \lambda_i \langle\!\langle \dot{\mathbf{p}},\xi_i \rangle\!\rangle^2 \left[\frac{1+\lambda_i}{1-\lambda_i} \right] \\
&< \quad 0
\end{aligned}
$$

dans tous les cas si et seulement si $\lambda_i < 1$ pour $i = 1,\dots,N-1$, ce qui est équivalent à la condition de stabilité interne (2.16). Lorsque $\mathbf{p}^*$ a des composantes nulles, alors on a la stabilité interne si et seulement si on est à un point de maximum local strict de $\rho^\circ(\mathbf{p})$ par rapport aux perturbations sur les composantes positives de $\mathbf{p}^*$. Mais si $\mathbf{p}^*$ possède aussi la propriété de stabilité externe (2.20), alors on peut remplacer $\rho^\circ(\mathbf{p})$ par $\rho(\mathbf{p})$ dans l'énoncé précédent par le lemme 2.

Stabilité externe. On considère l'équation

$$(2.53) \qquad B(\mathbf{p}(s))\mathbf{q}(s) = \rho(s)\mathbf{q}(s)$$

pour le vecteur de fréquence $\mathbf{p}(s) = \mathbf{p}^* + s\xi$ où $\mathbf{p}^*$ est un équilibre quelconque et $\xi_i > 0$ si $p_i^* = 0$. En supposant (2.36), on a par le lemme 1

$$(2.54) \qquad \mathbf{q}(0) = \mathbf{p}^* \quad \text{avec} \quad \rho(0) = 2w(\mathbf{p}^*).$$

La dérivée par rapport à s dans (2.53) évaluée à $s = 0$ donne alors

$$(2.55) \qquad B(\mathbf{p}^*)[\dot{\mathbf{p}}+\dot{\mathbf{q}}] - 2w(\mathbf{p}^*)\dot{\mathbf{q}} = \dot{\rho}\mathbf{p}^*.$$

En faisant le produit scalaire avec $\mathbf{1}$, on obtient

$$
\begin{aligned}
\dot{\rho} &= 2\langle M\mathbf{p}^*,\dot{\mathbf{p}}+\dot{\mathbf{q}}\rangle \\
&= 2 \sum_{i:p_i^*=0} [(M\mathbf{p}^*)_i - w(\mathbf{p}^*)][\dot{p}_i+\dot{q}_i] \\
(2.56) \qquad &= -4w(\mathbf{p}^*) \sum_{i:p_i^*=0} \left[\frac{w(\mathbf{p}^*)-(M\mathbf{p}^*)_i}{2w(\mathbf{p}^*)-(M\mathbf{p}^*)_i} \right] \dot{p}_i \\
&< \quad 0
\end{aligned}
$$

dans tous les cas si et seulement si $w(p^*) > (Mp^*)_i$ pour tout i pour lequel $p_i^* = 0$, ce qui est équivalent à la condition de stabilité externe (2.20). Il reste maintenant à démontrer les lemmes 1 et 2.

Démonstration du lemme 1. On a

$$B(p^*)p^* = [\text{diag}(p^*)M + \text{diag}(Mp^*)]p^*$$

(2.57)

$$= 2\text{diag}(p^*)Mp^* .$$

La conclusion du lemme découle alors de la condition d'équilibre (2.7) pour la première partie et de la théorie de Perron-Frobenius pour les matrices positives pour la seconde.

Démonstration du lemme 2. Pour un équilibre stable, (2.36) est vrai par le lemme 1 et la condition de stabilité externe (2.20). Pour un point de maximum local strict p^* de $\rho(p)$, supposons que (2.36) n'est pas vérifié. Alors

(2.58) $\rho(p^*) = (Mp^*)_i$ pour un certain i tel que $p_i^* = 0$, disons i = 2 ,

et p^* est un point de maximum local strict de la fonction linéaire $(Mp)_2$. Ceci est possible seulement si p^* a une seule composante positive, disons $p_1^* = 1$, et alors

(2.59) $$m_{12} = \rho(p^*) > \rho°(p^*) = 2m_{11}$$

et

(2.60) $$m_{12} = (Mp^*)_2 > (Mp^*)_k = m_{1k} \text{ pour } k \geq 3$$

par le lemme 1 et les hypothèses II. Par continuité et l'expression (2.33) pour $\rho(p)$, ceci entraîne que $\rho(p) = \rho°(p)$ pour $p = (p_1, p_2, 0, ..., 0)$ avec $p_2 > 0$ suffisamment petit. On suppose donc $M = \|m_{ij}\|_{i,j=1,2}$ et $p = (p_1, p_2)$ pour lesquels

(2.61) $$B(p) = \text{diag}(p)M + \text{diag}(Mp) = \begin{bmatrix} 2p_1 m_{11} + p_2 m_{12} & p_1 m_{12} \\ p_2 m_{12} & 2p_2 m_{22} + p_1 m_{12} \end{bmatrix} .$$

Puis on choisit $p(s) = (1,0) + s(-1,1)$ dans (2.38). Alors on a

(2.62) $$B(p(0)) = \begin{bmatrix} 2m_{11} & m_{12} \\ 0 & m_{12} \end{bmatrix}$$

dont la plus grande valeur propre est $\rho(0) = m_{12}$ avec vecteur propre a droite

$$(2.63) \qquad q(0) = (q_1(0), q_2(0)) = \left(\frac{m_{12}}{2m_{12} - 2m_{11}} , \frac{m_{12} - 2m_{11}}{2m_{12} - 2m_{11}} \right).$$

En substituant dans (2.39) évalué à $s = 0$, on obtient

$$(2.64) \qquad B(q(0))\dot{p}(0) + B(p(0))\dot{q}(0) = \dot{p}(0)q(0) + m_{12}\dot{q}(0)$$

et l'égalité pour la seconde composante donne

$$(2.65) \qquad -q_2(0)m_{12} + 2q_2(0)m_{22} + q_1(0)m_{12} = \dot{p}(0)q_2(0) ,$$

c'est-à-dire,

$$\dot{p}(0) = 2m_{22} - m_{12} + [q_1(0)/q_2(0)]m_{12}$$

$$= \frac{[2m_{22} - m_{12}][m_{12} - 2m_{11}] + m_{12}^2}{m_{12} - 2m_{11}}$$

$$(2.66) \qquad\qquad = \frac{2m_{12}[m_{11} + m_{22}] - 4m_{11}m_{22}}{m_{12} - 2m_{11}}$$

$$> \frac{4m_{11}^2}{m_{12} - 2m_{11}} > 0 .$$

Ceci contredit l'hypothèse que $\rho(p)$ atteint un maximum local à $(1,0)$.

3. Problème de la convergence

Il ne suffit pas de décrire la structure d'équilibre pour connaître la dynamique dans un système d'équations. En particulier pour le modèle V, on n'a pas encore démontré la stabilité des surfaces d'équilibres de parité et encore moins la convergence vers des états d'équilibre stables à partir d'états quelconques qui ne sont pas des équilibres. Une analyse exhaustive est cependant possible dans le cas de 2 allèles en utilisant des propriétés de monotonicité. En fait le modèle plus général V" est accessible à une analyse globale de cette façon. Les mêmes techniques peuvent être appliquées aux autres modèles dans le cas de 2 allèles.

3.1. MODÈLE DE DÉTERMINATION DU SEXE V AVEC 2 ALLÈLES, A_1 ET A_2 [12]

On considère le modèle (2.1) avec les matrices de paramètres

$$(3.1) \qquad M = \begin{bmatrix} m_{11} & m_{12} \\ m_{21} & m_{22} \end{bmatrix}, \quad U-M = \begin{bmatrix} 1-m_{11} & 1-m_{12} \\ 1-m_{21} & 1-m_{22} \end{bmatrix}.$$

On vérifie facilement qu'un équilibre symétrique polymorphique, $p^* = (p^*, 1-p^*)$ avec $0 < p^* < 1$, existe si et seulement si

$$(3.2) \qquad m_{12} > m_{11}, m_{22} \quad \text{ou} \quad m_{12} < m_{11}, m_{22}.$$

Dans ce cas

$$(3.3) \qquad p^* = \frac{m_{22}-m_{12}}{m_{11}+m_{22}-2m_{12}}$$

et la proportion de mâles à l'équilibre est

$$(3.4) \qquad w(p^*) = \frac{m_{11}m_{22}-m_{12}^2}{m_{11}+m_{22}-2m_{12}}.$$

Sur la base des théorèmes 2.1 et 2.2, l'existence d'équilibres de parité et la stabilité des équilibres symétriques peuvent être étudiées avec l'aide de la fonction $\rho_M(p)$ qui donne la plus grande valeur propre en module de la matrice

$$(3.5) \qquad \begin{bmatrix} 2pm_{11}+(1-p)m_{12} & pm_{12} \\ (1-p)m_{12} & 2(1-p)m_{22}+pm_{12} \end{bmatrix}$$

pour $0 \le p \le 1$, et de ses valeurs aux équilibres symétriques qui sont

$$\rho_M(0) = \max\{m_{12}, 2m_{22}\}, \quad \rho_M(1) = \max\{2m_{11}, m_{12}\}$$

$$(3.6) \qquad \text{et}$$

$$\rho_M(p^*) = 2w(p^*) \quad \text{quand } p^* \text{ existe.}$$

Remarquons qu'on n'a pas besoin de considérer la fonction $\rho_{U-M}(p)$ pour étudier la stabilité des équilibres symétriques car, dans le cas de 2 allèles, un équilibre est stable pour le modèle de viabilité M si et seulement s'il est instable pour le modèle de viabilité U-M, est vice versa. D'autre part, sans perte de généralité, on peut supposer

$$(3.7) \qquad m_{12} > 1/2 \quad \text{et} \quad m_{11} < m_{22}.$$

La figure 3.1 donne une liste complète des structures d'équilibre possibles sous ces conditions. On constate qu'il peut exister *au plus 2 équilibres de parité. La stabilité des équilibres de parité* quand ils existent et *la convergence vers les équilibres stables* découlent de propriétés de monotonicité qui seront étudiées dans la section suivante.

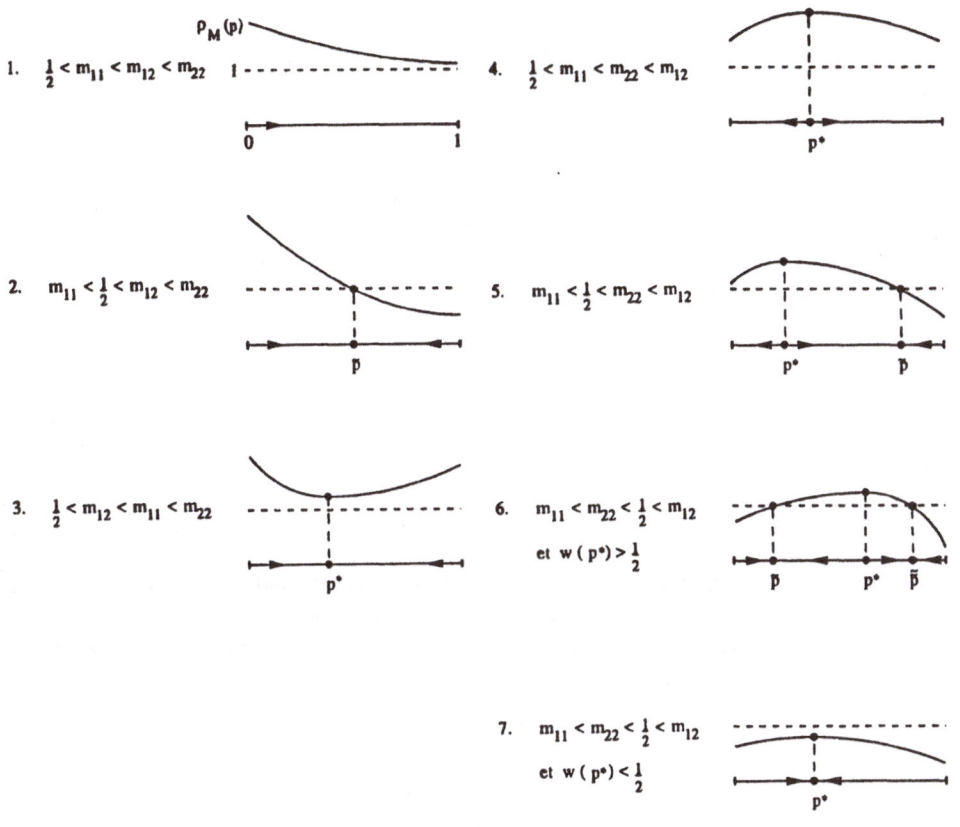

Figure 3.1. Structures d'équilibre pour le modèle (3.1) selon le graphe de $\rho_M(p)$, la plus grande valeur propre en module de la matrice (3.5). (p^* dénote un équilibre symétrique, $\tilde{p}$ et q des équilibres de parité.)

3.2. CONVERGENCE DANS LE MODÈLE GÉNÉRAL DE VIABILITÉ AVEC DISTINCTION DE SEXES V'' DANS LE CAS DE 2 ALLÈLES

On suppose que les viabilités relatives de A_1A_1, A_1A_2 et A_2A_2 sont $m_1/2$, 1 et $m_2/2$ chez les mâles et $f_1/2$, 1 et $f_2/2$ chez les femelles $(m_1, m_2, f_1, f_2 > 0)$. Si p désigne la fréquence de A_1 chez les femelles et q cette fréquence chez les mâles, alors on aura d'une génération à la suivante, en supposant les hypothèses I vérifiées,

$$p' = \frac{f_1 pq+(1-p)q+p(1-q)}{f_1 pq+2(1-p)q+2p(1-q)+f_2(1-p)(1-q)} ,$$

(3.8)

$$q' = \frac{m_1 pq+(1-p)q+p(1-q)}{m_1 pq+2(1-p)q+2p(1-q)+m_2(1-p)(1-q)} .$$

En définissant

(3.9)
$$x_1 = \frac{p}{1-p} , x_2 = \frac{q}{1-q} ,$$

on obtient d'une génération à la suivante

$$x'_1 = \frac{f_1 x_1 x_2 + x_1 + x_2}{f_2 + x_1 + x_2} ,$$

(3.10)

$$x'_2 = \frac{m_1 x_1 x_2 + x_1 + x_2}{m_2 + x_1 + x_2} .$$

Équilibres. A part $0 = (0,0)$ qui correspond à l'état de fixation de l'allèle A_2 dans la population et $\infty = (\infty,\infty)$ qui correspond à la fixation de A_1 , il existe *au plus trois équilibres* $x = (x_1,x_2)$ pour (3.10). Ces équilibres sont *polymorphiques* $(0 < x_1,x_2 < \infty)$ et à ces équilibres $t = x_2/x_1$ satisfait

(3.11)
$$t^3 -3\alpha t^2 + 3\beta t - 1 = 0$$

où

$$\alpha = (\alpha_1\beta_2+\alpha_2+\beta_1)/3 , \qquad \beta = (\alpha_2\beta_1+\alpha_1+\beta_2)/3 ,$$

(3.12)
$$\alpha_1 = f_1 - 1 , \qquad \beta_1 = m_1 - 1 ,$$

$$\alpha_2 = f_2 - 1 , \qquad \beta_2 = m_2 - 1 .$$

Dans ce cas

(3.13)
$$x_1 = \frac{t-\alpha_2}{1-\alpha_1 t} , x_2 = \frac{\beta_2 t-1}{\beta_1 -t} .$$

De plus, une étude détaillée de cas révèle qu'il existe *au plus deux équilibres polymorphiques si 0 est stable.* (Voir Karlin et Lessard, 1986, pour plus de détails.)

Propriété de monotonicité. Avec les définitions d'inégalité suivantes entre deux vecteurs:

$\leq$ pour inégalité composante par composante,

(3.14) $<$ pour inégalité composante par composante mais avec inégalité stricte pour au moins une composante,

$\ll$ pour inégalité stricte composante par composante,

la transformation $T = (T_1, T_2)$ définie par

$$(3.15) \qquad\qquad T(x_1, x_2) = (x'_1, x'_2)$$

est strictement monotone dans le sens que

$$(3.16) \qquad\qquad Tx \ll Ty \quad \text{si} \quad x < y .$$

Ceci est lié au fait que la matrice des dérivées partielles

$$(3.17) \qquad\qquad L = \begin{bmatrix} \dfrac{\partial T_1}{\partial x_1} & \dfrac{\partial T_1}{\partial x_2} \\[2ex] \dfrac{\partial T_2}{\partial x_1} & \dfrac{\partial T_2}{\partial x_2} \end{bmatrix}$$

est positive en tout point du domaine de définition. En effet

$$(3.18) \qquad \frac{\partial T_1}{\partial x_1} = \frac{f_1 f_2 x_2 + f_1 x_2^2 + f_2}{(f_2 + x_1 + x_2)^2} > 0 \quad \text{pour tout} \quad x_1, x_2 \geq 0$$

et de même par analogie pour les autres dérivées. Donc les deux composantes de T, T_1 et T_2, sont des fonctions strictement croissantes en x_1 et x_2 du domaine de définition, ce qui assure que T est strictement monotone. Dans ce qui suit, on fera l'hypothèse suivante:

Hypothèse III. A tout équilibre de (3.10), les valeurs propres de L sont $\neq 1$.

Théorème 3.1. (Karlin, 1972; Karlin et Lessard, 1986)[13] Pour le système (3.10) avec l'hypothèse III, (a) on a convergence vers un équilibre stable à partir de n'importe quel point initial sauf sur les courbes séparant les domaines de convergence et contenant les équilibres instables, (b) les équilibres sont ordonnés par la relation $\ll$, et (c) les équilibres stables et instables alternent. (Voir la figure 3.2.)

a) 0 instable

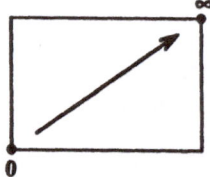
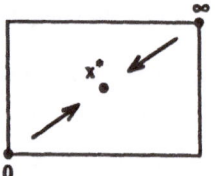

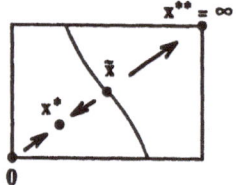
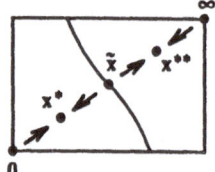

situations impossibles

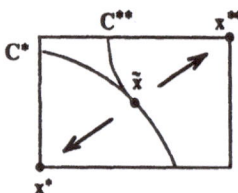
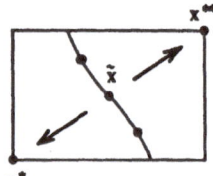

b) **0** stable: faire **x* = 0** dans (a).

Figure 3.2. Dynamique globale pour le modèle (3.10).

Démonstration du Théorème 3.1. Si **0** est instable, alors il existe un vecteur **z ≫ 0** tel que

(3.19) $Lz = \lambda z \quad (\lambda > 1)$

où L est la matrice (positive) des dérivées premières de T évaluées à **0**. Si **z** est choisi suffisamment petit, alors

(3.20) $Tz \gg z$.

Par la monotonicité de T, on a aussi

(3.21) $T^2z = T(Tz) \gg Tz$

et par des itérations successives

(3.22) $T^nz \gg T^{n-1}z$ pour tout $n \geq 3$.

On en déduit l'existence d'un point x^* tel que

(3.23) $T^nz \uparrow x^* \gg 0$ et $Tx^* = x^*$.

($\uparrow$ pour croissance composante par composante.) De plus, si $z < y < x^*$, alors

(3.24) $T^nz \ll T^ny \ll T^nx^* = x^*$ pour tout $n \geq 1$,

et conséquemment $T^ny \to x^*$ lorsque $n \to \infty$. Donc x^* est stable, car sinon le même argument à partir de x^* au lieu de 0 mais dans la direction opposée (en prenant $z \ll 0$) conduirait à une contradiction. Remarquons aussi que pour tout $y \gg 0$ (en fait $y > 0$ car alors $Ty \gg 0$), T^ny converge vers l'ensemble $\geq x^*$ car on peut prendre $z < y$ dans (3.23) et alors $T^nz \ll T^ny$ pour tout $n \geq 1$ avec $T^nz \to x^*$ lorsque $n \to \infty$.

Si $x^* = \infty$, on a terminé. Sinon, on considère le domaine d'attraction de x^*. Remarquons que si x appartient à ce domaine et $x^* < y < x$, alors y appartient aussi à ce domaine. De plus la frontière de ce domaine restreinte à l'ensemble $> x^*$, dénotée C^*, est invariante sous T. On en conclut que C^* contient un équilibre $\tilde{x}$. Cet équilibre est instable et satisfait $\tilde{x} > x^*$ (en fait $\tilde{x} \gg x^*$ car $\tilde{x} = T\tilde{x} \gg Tx^* = x^*$).

Si $\tilde{x} = \infty$, on a terminé. Sinon, en adaptant les arguments précédents, on peut trouver un équilibre stable $x^{**} \gg \tilde{x}$ tel quel $\tilde{x}$ appartient à la frontière du domaine d'attraction de x^{**} restreinte à l'ensemble $< x^{**}$, dénotée C^{**}. Dans le cas où $x^{**} \neq \infty$ il n'existe pas d'autres équilibres (car il peut exister au plus 3 équilibres polymorphiques), l'état de fixation ∞ est instable et on a $\infty \gg x^{**}$. Dans tous les cas, il y a convergence vers l'ensemble Q de tous les points $\geq x^*$ et $\leq x^{**}$ à partir de n'importe quel point $0 < x < \infty$.

Il reste à montrer que $C_1 = C^* \cap Q$ et $C_2 = C^{**} \cap Q$ coïncident. Soit $C = C_1 \cap C_2$. Si $C_1 - C$ et $C_2 - C$ ne sont pas vides, ils doivent être tous les deux invariants sous T et contenir chacun au moins un équilibre. (A priori cet équilibre pourrait appartenir à la partie commune C en étant à un point de jonction de $C_1 - C$ et $C_2 - C$, mais une telle configuration est impossible à un équilibre avec l'hypothèse III.) Mais alors, avec $\tilde{x}$, cela ferait au moins trois équilibres polymorphiques instables en plus de l'équilibre polymorphique stable x^*, ce qui entre en contradiction avec au plus trois équilibres polymorphiques. Donc $C_1 = C_2 = C$. Notons enfin que C ne peut pas contenir plus qu'un équilibre car alors C devrait contenir au moins trois équilibres (la transformation T sur la courbe C se ramenant à une transformation d'un intervalle sur lui-même), ce qui est impossible pour la même raison que précédemment.

Si 0 est stable, il suffit de remplacer x^* par 0 dans le traitement ci-dessus et utiliser le fait que dans ce cas il peut exister au plus deux équilibres polymorphiques.

4. Dynamique à long terme

Par dynamique à long terme, nous entendons évolution de la population lorsque des allèles mutants sont introduits "de temps en temps". Par "de temps en temps", nous sous-entendons qu'un nouvel allèle est introduit seulement après que la population ait atteint un équilibre. Cette approche nous permet d'une part de mettre à l'épreuve les équilibres existants en les confrontant à de nouvelles combinaisons génétiques possibles et d'autre part de dégager des principes d'évolution à partir des caractéristiques des équilibres successifs qui seront ainsi atteints.

Nous considérerons le modèle V" pour lequel $m_{ij} > 0$ et $f_{ij} > 0$ représentent les viabilités relatives de A_iA_j chez les mâles et les femelles respectivement $(i,j = 1,...,N)$. Pour tenir compte de contraintes biologiques et/ou écologiques, nous ferons l'hypothèse que tous les (m_{ij},f_{ij}) appartiennent à un certain ensemble S. Par exemple, Charlesworth et Charlesworth (1981) et Charnov (1982) ont considéré des ensembles S de la forme

$$(4.1) \qquad S = \{(s^\alpha,(1-s^\beta):0 < s < 1\} \quad \text{où} \quad \alpha > 0, \ \beta > 0.$$

Mais en général, l'ensemble S peut être n'importe quel sous-ensemble borné du quadrant positif.

En dénotant la fréquence de A_i par q_i chez les mâles et p_i chez les femelles $(i = 1,...,N)$, on obtient d'une génération à la suivante dans une population infinie panmictique

$$(4.2) \qquad q_i' = \frac{q_i \sum_{j=1}^{N} m_{ij}p_j + p_i \sum_{j=1}^{N} m_{ij}q_j}{2m}, \quad i = 1,...,N,$$

$$p_i' = \frac{q_i \sum_{j=1}^{N} f_{ij}p_j + p_i \sum_{j=1}^{N} f_{ij}q_j}{2f}, \quad i = 1,...,N,$$

$$(4.3) \qquad m = \sum_{i,j=1}^{N} m_{ij}p_iq_j, \quad f = \sum_{i,j=1}^{N} f_{ij}p_iq_j.$$

Considérons la population à un équilibre $p_1^*, \ldots, p_N^*$, $q_1^*, \ldots, q_N^*$ qui est polymorphique (c'est-à-dire, avec $p_i^* q_i^* > 0$ pour $i = 1, \ldots N$ pour que les allèles $A_1, \ldots, A_N$ soient tous représentés à l'équilibre) et stable (car seul un équilibre stable est observable dans la réalité) avec m^* et f^* comme viabilités moyennes chez les mâles et les femelles calculées selon (4.3) à l'équilibre. Supposons maintenant qu'un allèle mutant A_{N+1} est introduit en petite quantité dans la population. Ses fréquences q_{N+1} et p_{N+1} chez les mâles et les femelles, respectivement, seront reliées sur deux générations successives par les équations

$$(4.4) \quad \begin{bmatrix} q_{N+1} \\ p_{N+1} \end{bmatrix}' = \begin{bmatrix} \dfrac{(Mp^*)_{N+1}}{2m^*} & \dfrac{(Mq^*)_{N+1}}{2m^*} \\ \dfrac{(Fp^*)_{N+1}}{2f^*} & \dfrac{(Fq^*)_{N+1}}{2f^*} \end{bmatrix} \begin{bmatrix} q_{N+1} \\ p_{N+1} \end{bmatrix} + \text{termes d'ordre supérieur}$$

où

$$(4.5) \quad M = \|m_{ij}\|_{i,j=1,\ldots,N+1}, \qquad F = \|f_{ij}\|_{i,j=1,\ldots,N+1},$$

$$p^* = (p_1^*, \ldots, p_N^*, 0), \qquad q^* = (q_1^*, \ldots, q_N^*, 0).$$

En dénotant la matrice (positive) dans (4.4) par V, on aura *invasion* de A_{N+1} (à un taux géométrique) si et seulement si $\rho(V) > 1$ où $\rho(V)$ désigne la plus grande valeur propre en module de V. Dans le cas $\rho(V) < 1$, on aura *extinction* de A_{N+1} (à un taux géométrique). Dans le cas dégénéré $\rho(V) = 1$, la condition exacte pour l'invasion ou l'extinction de A_{N+1} (mais à un taux algébrique) peut être obtenue en considérant les termes quadratiques dans (4.4).

Interprétation de $\rho(V)$. On a

$$(4.6) \quad \rho(V) = v_1 \left[\frac{(Mp^*)_{N+1}}{2m^*} + \frac{(Fp^*)_{N+1}}{2f^*} \right] + v_2 \left[\frac{(Mq^*)_{N+1}}{2m^*} + \frac{(Fq^*)_{N+1}}{2f^*} \right]$$

où

$$(4.7) \quad V \begin{bmatrix} v_1 \\ v_2 \end{bmatrix} = \rho(V) \begin{bmatrix} v_1 \\ v_2 \end{bmatrix} \quad \text{avec } v_1 + v_2 = 1 \text{ et } v_1, v_2 > 0.$$

L'équation (4.6) peut s'écrire sous la forme

(4.8)
$$\rho(V) = \frac{m^*_{N+1}}{2m^*} + \frac{f^*_{N+1}}{2f^*}$$

où

(4.9)
$$m^*_{N+1} = v_1(Mp^*)_{N+1} + v_2(Mq^*)_{N+1} \, ,$$

$$f^*_{N+1} = v_1(Fp^*)_{N+1} + v_2(Fq^*)_{N+1} \, .$$

Mais en itérant (4.4), on obtient pour $k \geq 2$

(4.10)
$$\begin{bmatrix} q_{N+1} \\ p_{N+1} \end{bmatrix}^{(k)} = V^k \begin{bmatrix} q_{N+1} \\ p_{N+1} \end{bmatrix} + \text{termes d'ordre supérieur,}$$

ce qui donne approximativement un multiple de (v_1, v_2) pour k assez grand, car alors

(4.11)
$$V^k \cong \rho(V)^k \begin{bmatrix} v_1 \\ v_2 \end{bmatrix} [w_1 \ \ w_2]$$

où

(4.12)
$$[w_1 \ \ w_2]V = \rho(V)[w_1 \ \ w_2] \text{ et } v_1w_1 + v_2w_2 = 1 \, .$$

Donc v_1 et v_2 représentent les proportions relatives de A_{N+1} chez les mâles et les femelles à long terme près de l'équilibre. Par conséquent, m^*_{N+1} et f^*_{N+1} représentent les *viabilités marginales* de A_{N+1} chez les mâles et les femelles, respectivement, et $\rho(V)$ la *valeur reproductive marginale* de A_{N+1} près de l'équilibre.

Remarque. La condition pour invasion à un taux géométrique

(4.13)
$$\frac{m^*_{N+1}}{2m^*} + \frac{f^*_{N+1}}{2f^*} > 1$$

peut être généralisée pour le cas de plusieurs loci au lieu d'un seul et pour le cas de paramètres de viabilité qui dépendent des fréquences des allèles. Il suffit de définir les viabilités marginales selon les proportions relatives des différents gamètes mutants chez les mâles et les femelles à long terme près de l'équilibre. (Voir Eshel et Feldman, 1984; Lessard, 1988.)

Le lemme suivant va nous permettre de définir des propriétés d'optimalité basées sur la condition pour invasion (4.13):

Lemme 4.1.[14] *Soit* $\mathcal{E}(S)$ *l'enveloppe convexe de* S , *c'est-à-dire l'ensemble de toutes les combinaisons linéaires convexes de points de* S . *Les propriétés suivantes sont équivalentes:*

(4.14) (i) $\hat{m} \cdot \hat{f} = \max\{m \cdot f : (m,f) \text{ dans } \mathcal{E}(S)\}$;

(4.15) (ii) $\dfrac{m}{2\hat{m}} + \dfrac{f}{2\hat{f}} \le 1$ *pour tout* (m,f) *dans* S;

(4.16) (iii) $\dfrac{\hat{m}}{2\tilde{m}} + \dfrac{\hat{f}}{2\tilde{f}} \ge 1$ *pour tout* $(\tilde{m},\tilde{f})$ *dans* $\mathcal{E}(S)$

avec égalité seulement si $(\tilde{m},\tilde{f}) = (\hat{m},\hat{f})$.

De plus $(\hat{m},\hat{f})$ *est unique.* (Voir la figure 4.1.)

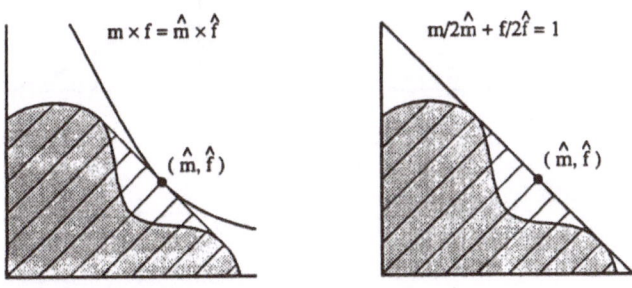

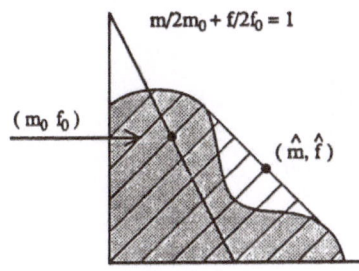

Figure 4.1. Illustration des propriétés de $(\hat{m},\hat{f})$ du lemme 4.1.

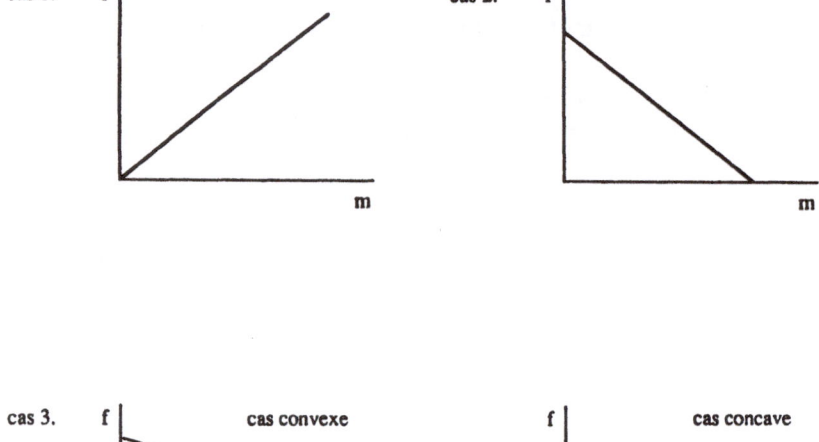

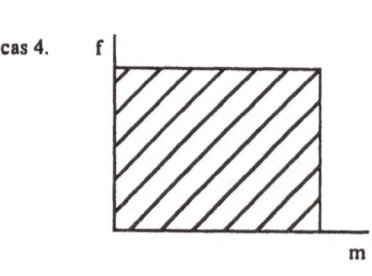

Figure 4.2. Ensembles S pour les paramètres (m$_{ij}$,f$_{ij}$) du modèle V".

Le lemme 4.1 a deux conséquences immédiates:

Corollaire 4.1. (a) *Un équilibre de* (4.2) *avec* $(m^*,f^*) = (\hat{m},\hat{f})$ *ne peut être envahi à un taux géométrique par aucun allèle mutant, et* (b) *un allèle mutant* A_{N+1} *avec* (m^*_{N+1},f^*_{N+1}) = $(\hat{m},\hat{f})$ *envahit tout autre équilibre à un taux géométrique.*

Remarque. La propriété (b) du corollaire 4.1 est la propriété d'optimalité pour un individu telle que proposée par MacArthur (1965) alors que la propriété (a) est la propriété d'optimalité pour une population considérée par Charnov et al. (1976) et Charnov (1982).

La démonstration du lemme 4.1 qui est à la base du corollaire 4.1 est présentée en appendice (appendice 4.A.). La question qu'on peut maintenant se poser est la suivante: étant donné que l'état représenté par $(\hat{m},\hat{f})$ est optimal pour le modèle (4.2), va-t-on nécessairement atteindre cet état en introduisant successivement au cours du temps des allèles mutants? Et si oui, de quelle manière? Pour répondre à cette question, nous allons considérer quatre cas (voir figure 4.2).

Cas 1. $f_{ij} = cm_{ij}$ *pour tout* i,j $(c>0)$.

On est alors en présence d'un modèle de viabilité sans distinction réelle de sexes (les paramètres de viabilité pour les mâles et les femelles étant les mêmes à une constante positive c près). Dans ce cas, *le produit des viabilités moyennes* $m \cdot f$ égale cm^2 qui *croît d'une génération à la suivante* par le théorème fondamental de la sélection naturelle appliqué au modèle de viabilité $M = \|m_{ij}\|$ (voir appendice 1.C).

Cas 2. $f_{ij} = a - bm_{ij}$ *pour tout* i,j $(a,b>0)$.

En définissant

$$(4.17) \qquad \tilde{f}_{ij} = \frac{f_{ij}}{a}, \ \tilde{m}_{ij} = \frac{bm_{ij}}{a} \quad \text{pour tout } i,j,$$

on a

$$(4.18) \qquad \tilde{f}_{ij} = 1 - \tilde{m}_{ij} \quad \text{pour tout } i,j.$$

Donc on peut supposer $a = b = 1$. Mais on est alors en présence d'un modèle de détermination de sexe (modèle V) pour lequel $m \cdot f = m(1-m)$ où m représente la proportion de mâles dans la population. De plus on a le résultat suivant:

Théorème 4.1. (Karlin et Lessard, 1984)[15] *Avec les hypothèses* II *pour le modèle de détermination de sexe* (2.1), *la proportion de mâles dans la population,* m, *croît d'un équilibre au suivant "attaignable" suivant l'invasion d'un allèle mutant tant que* $m = 1/2$ *n'est pas atteint.* (Le sens exact à donner à "attaignable" sera précisé plus loin.)

Les hypothèses II garantissent en particulier que l'invasion d'un allèle mutant se fait nécessairement à un taux géométrique. Que ce soit pour le modèle de détermination de sexe ou le modèle plus général (4.2) dans le cas 2 ci-dessus, on conclut immédiatement du théorème 4.1 que, sous des hypothèses de non-dégénérescence, le produit $m \cdot f$ se rapproche toujours plus de sa valeur optimale lorsqu'on considère la suite d'équilibres atteints en introduisant successivement des allèles mutants.

Corollaire 4.2. En général dans le cas 2, le produit $m \cdot f$ croît d'un équilibre au suivant "attaignable" suivant l'invasion d'un allèle mutant tant que le maximum absolu $\hat{m} \cdot \hat{f}$ n'est pas atteint à l'équilibre.

La démonstration du théorème 4.1 repose sur la propriété suivante pour le modèle de viabilité M (voir appendice 4.B pour la démonstration):

Théorème 4.2. (Karlin, 1978)[16] Dans le modèle de viabilité (2.5) avec matrice de viabilité $M = \|m_{ij}\|_{i,j=1,\ldots,N+1}$ et les hypothèses II, si $p^* = (p_1^*,\ldots,p_N^*,0)$ avec $p_1^*,\ldots,p_N^* > 0$ est stable par rapport à de petites perturbations sur les fréquences de $A_1,\ldots,A_N$ mais devient instable lorsque A_{N+1} est introduit en petite quantité, alors il existe un et un seul équilibre stable p^{**} dans le système entier. A ce point, la viabilité moyenne $w(p)$ atteint un maximum global par rapport aux vecteurs de fréquence $p = (p_1,\ldots,p_{N+1})$.

Démonstration du théorème 4.1. Étant donné la conclusion du théorème 2.1, il suffit de considérer pour le modèle de détermination de sexe (2.1) un équilibre symétrique $p^* = (p_1^*,\ldots,p_N^*,0)$ satisfaisant aux hypothèses du théorème 4.2 avec $w(p^*) < 1/2$ (le cas $w(p^*) > 1/2$ est analogue par symétrie entre les sexes). Lorsque l'allèle A_{N+1} envahit l'équilibre p^* qui était stable jusqu'alors par rapport aux allèles représentés à l'équilibre, deux situations seulement peuvent se produire sous les hypothèses II: ou bien l'équilibre p^{**} du théorème 4.2 satisfait $w(p^{**}) < 1/2$ et alors

$$(4.19) \qquad\qquad w(p^*) < w(p^{**}) < 1/2$$

et p^{**} est le seul équilibre stable dans le système entier, ou bien $w(p^{**}) > 1/2$ et alors

$$(4.20) \qquad\qquad w(p^*) < 1/2 < w(p^{**})$$

et il existe une surface d'équilibres de parité séparant p^* et p^{**} avec aucun équilibre stable entre p^* et cette surface (voir (2.28) et figure 4.3). Donc dans les deux cas, s'il y a convergence, il y aura convergence vers un équilibre où $w(p)$ est plus près de $1/2$. Ceci complète la démonstration du théorème 4.1 et précise la signification d'un équilibre "attaignable".

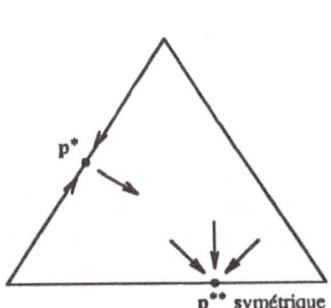
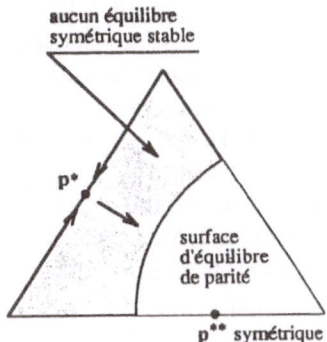

Figure 4.3. Configurations d'équilibre possibles pour le modèle V lorsqu'un allèle mutant envahit un équilibre symétrique p* stable jusque là.

Définition. Un équilibre "attaignable" est un équilibre vers lequel on peut avoir convergence s'il y a effectivement convergence.

Cas 3. f_{ij} = f(m_{ij}) *pour tout i,j où f est une fonction décroissante quelconque.*

Nous avons étudié ce cas par des simulations numériques. En supposant une relation fonctionnelle de la forme f_{ij} = f(m_{ij}) avec $0 < m_{ij} < 1$ pour les paramètres de viabilité m_{ij} et f_{ij} associés au génotype $A_i A_j$ pour tout i,j chez les mâles et les femelles, respectivement, nous avons introduit successivement de nouveaux allèles choisis au hasard (c'est-à-dire, des allèles avec des paramètres choisis au hasard) dans le système (4.2). A chaque fois, nous avons itéré les équations de récurrence (4.2) jusqu'à l'atteinte d'un état d'équilibre avant d'introduire l'allèle suivant en très petite quantité.

Nous avons considéré deux grandes classes de fonctions décroissantes f (voir figure 4.3): les *fonctions convexes* (dont les dérivées sont décroissantes) et *les fonctions concaves* (dont les dérivées sont croissantes).

Les détails de la simulation et les résultats numériques sont présentés dans Lessard (1988). Les conclusions générales sont les suivantes:

(1) Dans la plupart des cas convexes, le produit m • f tend rapidement vers sa valeur optimale d'un équilibre au suivant bien qu'il ne l'atteigne jamais vraiment.

(2) Dans la plupart des cas concaves, il y a plus de changements d'équilibre et plus d'allèles représentés à l'équilibre. De plus, on remarque que le produit m • f augmente toujours d'un équilibre au suivant si l'équilibre de départ est symétrique. Cependant, *bien que rare et petite, une décroissance de* m • f *d'un équilibre au suivant est possible.* En général, les phénomènes observés dans les cas concaves sont plus prononcés lorsque la concavité est faible. Les mêmes phénomènes sont d'ailleurs aussi susceptibles d'être observés dans des cas convexes lorsque la convexité est faible.

Le cas linéaire (cas 2) a aussi été simulé. Comme prévu, le produit m • f atteint rapidement sa valeur optimale à l'équilibre et l'introduction de tout nouvel allèle en petite quantité a seulement alors comme effet de conduire la population à un équilibre tout près du premier et appartenant à la même surface d'équilibres, donc de ramener la population à un équilibre où le produit m • f est optimal.

Comme nous l'avons vu, dans les cas non-linéaires le produit m • f n'est pas nécessairement croissant d'un équilibre au suivant. La tendance générale est cependant à la croissance car les décroissances sont rares et petites. A long terme, la population se rapprochera donc de la valeur optimale du produit m • f mais sans vraiment l'atteindre en général car cela exigerait des relations particulières entre les paramètres de viabilité qui sont improbables quoique possibles.

Cas 4. $0 < m_{ij}, f_{ij} < 1$ pour tout i,j .

En choisissant les paramètres de viabilité complètement au hasard entre 0 et 1 , on a observé aucune décroissance du produit m • f d'un équilibre au suivant dans nos simulations ce qui laisse supposer que les décroissances sont encore plus rares que ne le suggèrent nos résultats dans le cas 3. Les relations fonctionnelles entre les paramètres de viabilité susceptibles de mener à des décroissances sont donc en fait très improbables.

Appendices

4.A. DÉMONSTRATION DU LEMME 4.1

(i) $\Rightarrow$ *(ii)*. On suppose que $\hat{m} \cdot \hat{f}$ est le maximum de m • f pour (m,f) dans l'enveloppe convexe de S dénotée $\mathcal{E}(S)$. On veut montrer que tout (m,f) dans S appartient alors au demi-plan

$$(4.21) \qquad \hat{f}(m-\hat{m}) + \hat{m}(f-\hat{f}) \leq 0 .$$

En effet tout point sur la ligne $t(m,f) + (1-t)(\hat{m},\hat{f})$ pour $0 \leq t \leq 1$ est dans $\mathcal{E}(S)$ si (m,f) est dans S , et le produit des composantes le long de cette ligne a le membre de gauche de (4.21) comme dérivée par rapport à t à t = 0 . Mais cette dérivée est nécessairement ≤ 0 si $(\hat{m}, \hat{f})$ est un point de maximum par rapport à $\mathcal{E}(S)$.

(ii) $\Rightarrow$ *(i)*. Inversement si l'inégalité (4.21) est valide pour tout (m,f) dans S , elle l'est aussi pour tout (m,f) dans $\mathcal{E}(S)$ par linéarité. Alors, en utilisant le fait qu'une moyenne géométrique n'excède jamais une moyenne arithmétique, on obtient les inégalités

$$(4.22) \qquad \left(\frac{m \cdot f}{\hat{m} \cdot \hat{f}}\right)^{1/2} \leq \frac{m}{2\hat{m}} + \frac{f}{2\hat{f}} \leq 1$$

qui garantissent que $m \cdot f \leq \hat{m} \cdot \hat{f}$ pour tout (m,f) dans $\mathcal{E}(S)$.

Unicité de $(\hat{m},\hat{f})$. Remarquons qu'on a des égalités dans (4.22) si et seulement si $m/\hat{m} = f/\hat{f}$ et $m \cdot f = \hat{m} \cdot \hat{f}$. Ceci est possible seulement dans le cas où $m = \hat{m}$ et $f = \hat{f}$.

(i) $\Rightarrow$ *(iii)*. Dans ce cas on utilise les inégalités

$$(4.23) \qquad \frac{\hat{m}}{2\tilde{m}} + \frac{\hat{f}}{2\tilde{f}} \geq \left(\frac{\hat{m} \cdot \hat{f}}{\tilde{m} \cdot \tilde{f}}\right)^{1/2} \geq 1$$

pour tout $(\tilde{m},\tilde{f})$ dans $\mathcal{E}(S)$ avec égalité si et seulement si $(\tilde{m},\tilde{f}) = (\hat{m},\hat{f})$.

(iii) $\Rightarrow$ *(ii)*. On a alors

$$(4.24a) \qquad \frac{\hat{m}}{2[\hat{m}+t(m-\hat{m})]} + \frac{\hat{f}}{2[\hat{f}+t(f-\hat{f})]} \geq 1 \, ,$$

c'est-à-dire,

$$(4.24b) \qquad \frac{m-\hat{m}}{2[\hat{m}+t(m-\hat{m})]} + \frac{f-\hat{f}}{2[\hat{f}+t(f-\hat{f})]} \leq 0 \, ,$$

pour tout (m,f) dans $\mathcal{E}(S)$ et $0 < t < 1$. Ceci entraîne (4.21) en faisant tendre t vers 0.

4.B. DÉMONSTRATION DU THÉORÈME 4.2

Le cas $N = 1$ est trivial car le seul équilibre stable est celui qui correspond au maximum global de $w(p)$ par rapport aux vecteurs de fréquence $p = (p_1,p_2)$. A partir de maintenant, on va donc supposer $N \geq 2$.

La stabilité interne de $p^* = (p_1^*,\ldots,p_N^*,0)$ avec $p_i^* > 0$ pour $i = 1,\ldots,N$ garantit que la fonction quadratique

$$(4.25) \qquad w(p) = \sum_{i,j=1}^{N+1} m_{ij}p_i p_j$$

atteint un maximum global à p^* par rapport aux vecteurs de fréquence $p = (p_1,\ldots,p_N,0)$. Mais l'instabilité de p^* lorsque l'allèle A_{N+1} est introduit implique que $w(p)$ n'est pas à un maximum local à p^* par rapport aux vecteurs de fréquence $p = (p_1,\ldots,p_N,p_{N+1})$ avec $p_{N+1} > 0$.

a) paraboles

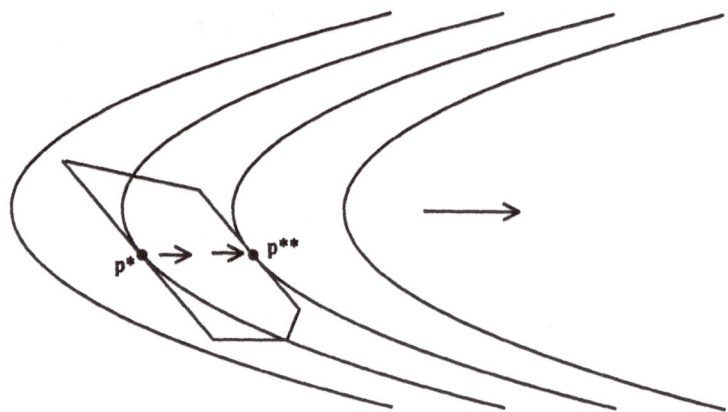

un et un seul maximum local (global) à p**

b) paraboles

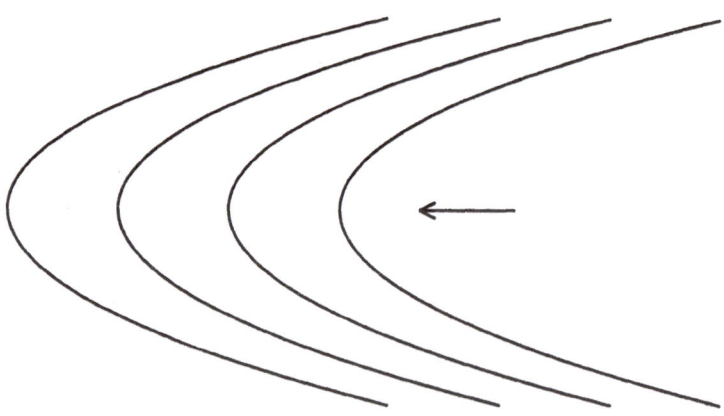

impossible

Figure 4.4.1.

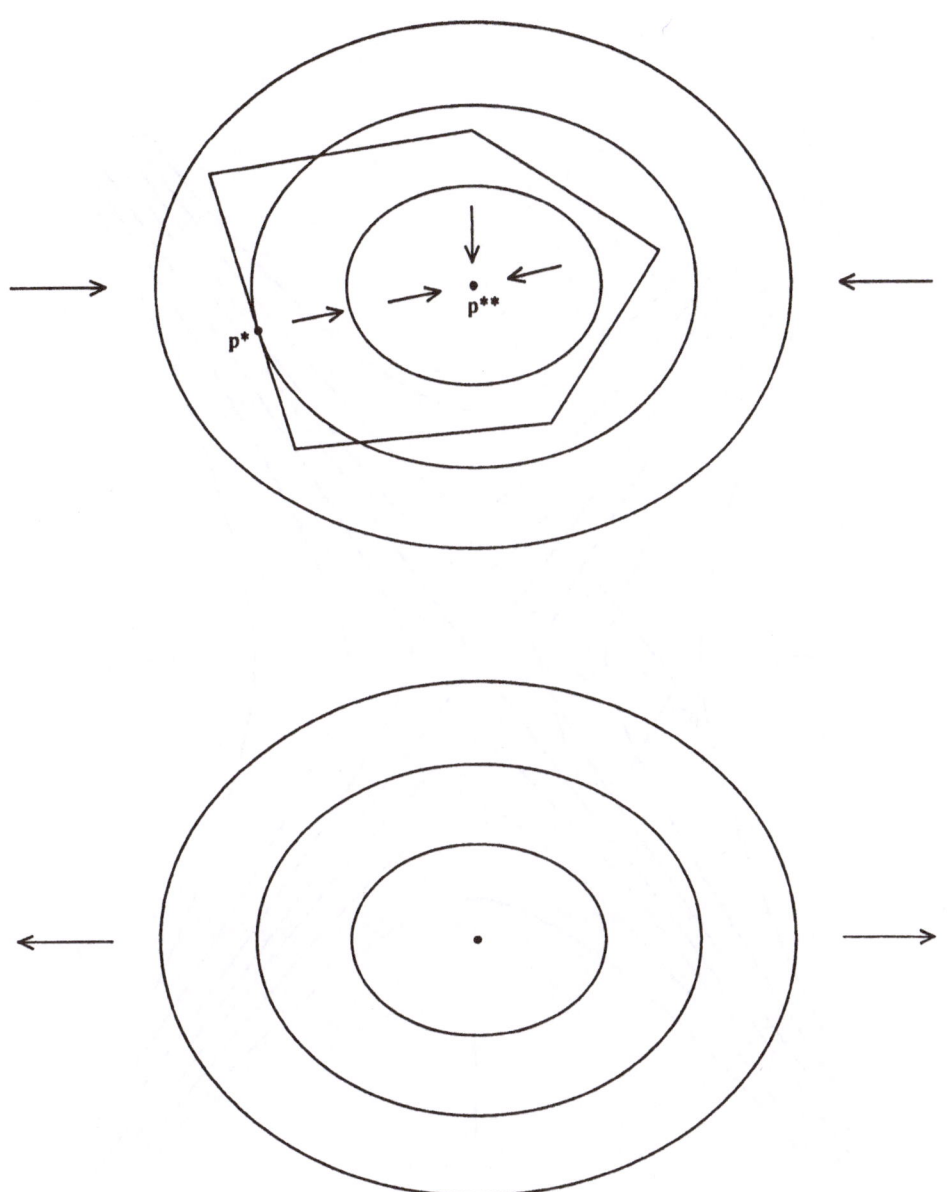

Figure 4.4.2.

e) hyperboles

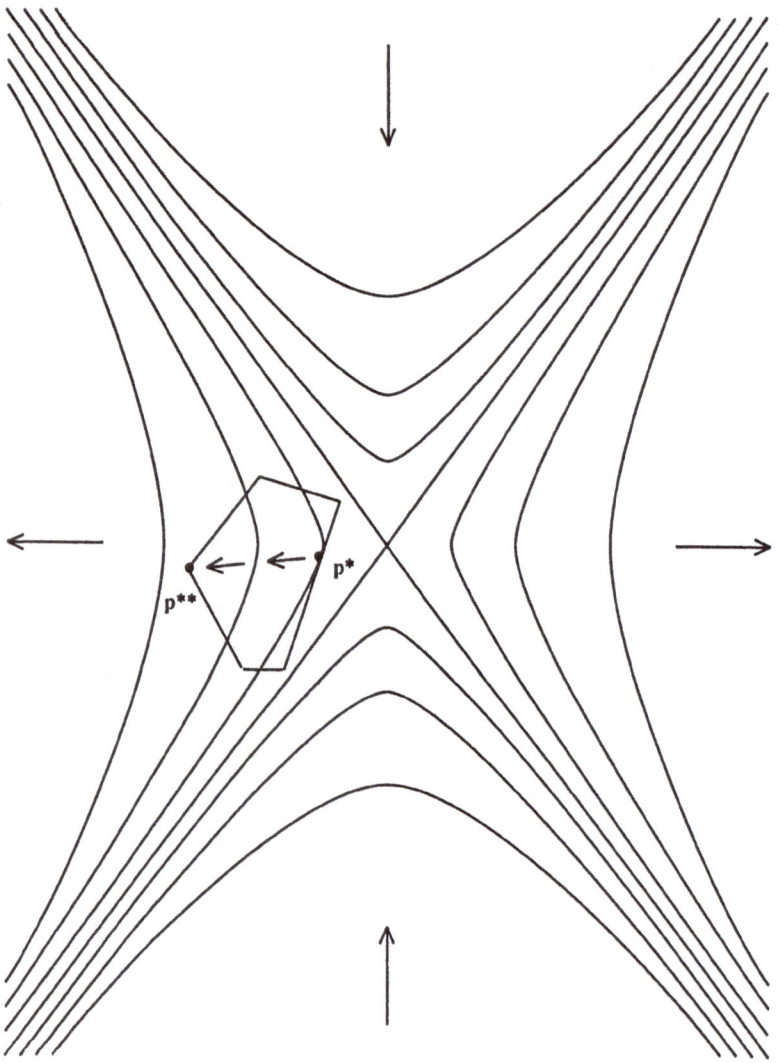

Figure 4.4.3.
Figure 4.4.1-4.4.3. Courbes de niveau de w(p) de (4.25) pour **p** dans un polyèdre convexe avec un point **p*** à l'intérieur d'un coté qu est un point de maximum local de w(p) restreint à ce coté. (→ indique la direction de croissance des courbes de niveau.)

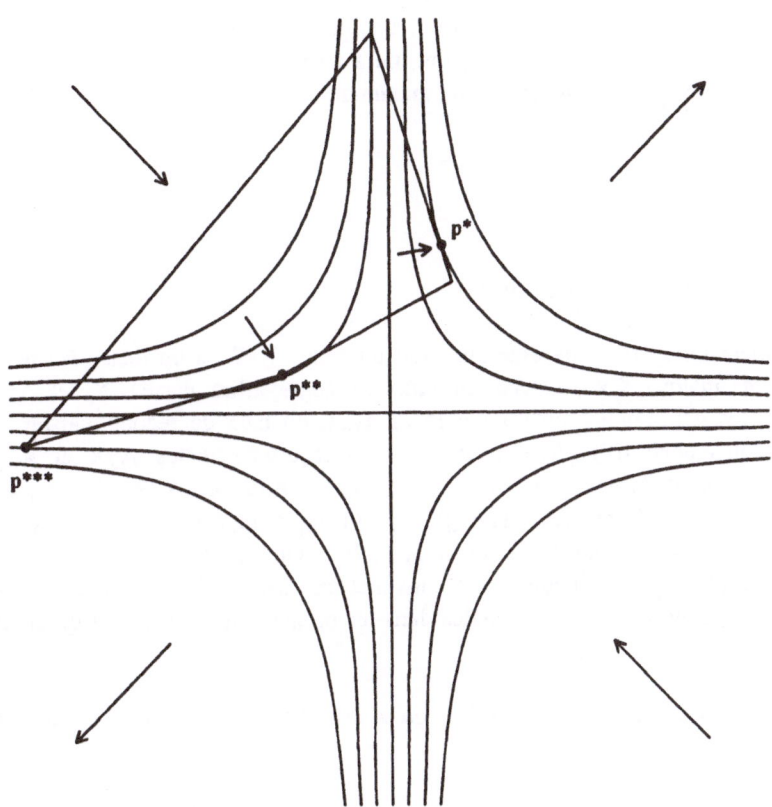

Figure 4.5. Exemple de trois maxima locaux de w(p) de (4.25) pour **p** dans un polyèdre convexe avec au moins un maximum local à l'intérieur d'un côté du polyèdre (selon un exemple de Cannings et Vickers).

Supposons maintenant que **p**** et **p***** sont deux équilibres stables, et donc correspondent eux à des maxima locaux de w(p) par rapport aux vecteurs de fréquence **p** à (N+1) composantes. On considère alors l'intersection du plan engendré par **p***, **p**** et **p***** avec le simplexe de ces vecteurs de fréquence **p** . Cet ensemble, qu'on dénote Γ , est nécessairement un polyèdre convexe et **p*** est situé à l'intérieur de l'un de ses côtés. De plus, à partir de **p***, les valeurs de w(p) doivent croître au moins initialement lorsqu'on se dirige vers l'intérieur de Γ mais décroître lorsqu'on se déplace de part et d'autre de **p*** sur le côté de Γ contenant **p*** .

Mais d'autre part, les courbes de niveau de la fonction quadratique w(p) restreinte au plan à lequel appartient Γ sont ou bien des paraboles (des droites dans le cas dégénéré), ou bien des ellipses, ou bien des hyperboles. Les seuls cas compatibles avec les valeurs de w(p) dans le voisinage de **p*** par rapport à Γ sont illustrés dans la figure 4.4. On voit que, dans chaque cas, il peut exister un et un seul maximum local (en fait global) de $w(\mathbf{p})$ par rapport à Γ. On obtient donc une contradiction avec l'existence de deux maxima locaux à $\mathbf{p^{**}}$ et $\mathbf{p^{***}}$. On en conclut qu'il peut exister un et un seul équilibre stable dans le système entier. A ce point, on a un maximum global de $w(\mathbf{p})$ par rapport aux vecteurs de fréquence $\mathbf{p}$ car un tel maximum doit correspondre à un équilibre stable.

5. Facteurs perturbateurs

5.1. LINKAGE AVEC D'AUTRES GÈNES

On considère un système de détermination du sexe X-Y à un locus, les femelles étant XX et les mâles XY, avec un taux de ségrégation entre X et Y chez les hétérozygotes déterminé à un second locus ayant un taux de recombinaison r avec le premier (voir modèle III'). Ainsi les mâles ayant reçu les gamètes A_i/X et A_j/Y de leurs parents transmettront les gamètes A_i/X, A_j/X, A_i/Y et A_j/Y à leurs descendants immédiats avec les fréquences $(1-m_{ij})(1-r)$, $(1-m_{ij})r$, $m_{ij}r$ et $m_{ij}(1-r)$, respectivement $(0 < m_{ij} = m_{ji} < 1$ pour tout $i,j = 1,...,N)$. On fait l'hypothèse générique que la matrice $\|m_{ij}\|_{i,j=1,...,N}$ et toutes ses sous-matrices principales sont non-singulières. On dénote la proportion moyenne de mâles dans les progénitures par m et on suppose $0 < r < 1/2$.

Remarque. Dans le cas $r = 1/2$, on a évolution vers $m = 1/2$ dans le sens du théorème 4.1.

Pour les variables p_i, q_i et z_i représentant les fréquences de A_i/X parmi les gamètes transmis par les femelles, A_i/Y parmi les gamètes porteurs du Y transmis par les mâles et A_i/X parmi les gamètes porteurs du X transmis par les mâles, respectivement $(i = 1,...,N)$, on obtient sous les hypothèses habituelles les équations de

$$(5.1a) \qquad p_i' = \frac{p_i + z_i}{2}, \quad i = 1,...,N,$$

$$(5.1b) \qquad q_i' = \frac{1}{m}\left[(1-r)q_i \sum_{j=1}^{N} m_{ij}p_j + rp_i \sum_{j=1}^{N} m_{ij}q_j\right], \quad i = 1,...,N,$$

$$(5.1c) \qquad z_i' = \frac{1}{1-m}\left[rq_i \sum_{j=1}^{N} (1-m_{ij})p_j + (1-r)p_i \sum_{j=1}^{N} (1-m_{ij})q_j\right], \quad i = 1,...,N,$$

où m est la proportion moyenne de mâles dans les progénitures, c'est-à-dire,

(5.2)
$$m = \sum_{i,j=1}^{N} m_{ij} p_i q_j \, .$$

Une classe importante d'équilibres de (5.1) est obtenue en faisant $p_i = q_i = z_i$ pour $i = 1,...,N$. Les équilibres "symétriques" ainsi obtenus sont appelés *équilibres de Hardy-Weinberg-Robbins* (HWR). A un tel équilibre, la fréquence de A_i pour tout i est la même chez tous les types de gamètes (femelles ou mâles, porteurs du X ou du Y). On dénote cette fréquence par p_i^* ($i = 1,...,N$) et la proportion de mâles correspondante à l'équilibre par m^*, c'est-à-dire,

(5.3)
$$m^* = \sum_{i,j=1}^{N} m_{ij} p_i^* p_j^* \, .$$

On remarque qu'un équilibre de HWR $p^* = (p_1^*,...,p_N^*)$ pour (5.1) est un équilibre pour le modèle de viabilité $M = \|m_{ij}\|_{i,j=1,...,N}$ défini par (2.5), et vice versa. De plus si un allèle A_{N+1} est introduit en petite quantité dans la population près de l'équilibre p^*, alors la proportion marginale de mâles dans les progénitures de mâles mutants sera

(5.4)
$$m_{N+1}^* = \sum_{i=1}^{N} m_{i,N+1} p_i^* \, .$$

On suppose toujours à partir de maintenant $m_{N+1}^* \neq m^*$.

La conclusion la plus surprenante au sujet des équilibres de HWR de (5.1) est sans doute la contradiction qui semble exister entre la stabilité interne et la stabilité externe dans le cas où il y a linkage ($0<r<1/2$) entre le locus qui détermine le sexe et celui qui détermine le sex ratio dans la progéniture. Dans ce qui suit la stabilité et l'instabilité sont sous-entendues à des taux géométriques.

Résultat. Au moins pour $(3 - \sqrt{3})/4 < r < 1/2$, un équilibre de HWR de (5.1) avec $m^* = 1/2$ est toujours stable par rapport aux allèles existants (stabilité interne) mais est envahi par n'importe quel allèle mutant (instabilité externe).

Le résultat exact est le suivant:

Théorème 5.1. (Lessard, 1987)[17] (a) *Un équilibre de HWR p^* de (5.1) avec proportion moyenne de mâles à l'équilibre $m^* < 1/2$ est stable intérieurement si et seulement si*

(5.5)
$$r_1(m^*) < r < r_2(m^*)$$

où

(5.6) $r_1(m^*) \to r_0 \leq \dfrac{3 - \sqrt{3}}{4}$ et $r_2(m^*) \to \dfrac{1}{2}$ $lorsque$ $m^* \to \dfrac{1}{2}$.

(b) *Si un allèle mutant* A_{N+1} *avec* m^*_{N+1} *comme proportion marginale de mâles à l'équilibre est alors introduit en petite quantité, l'équilibre* **p*** *de (a) devient instable extérieurement si et seulement si*

(5.7)   *ou bien* $m^*_{N+1} > m^*$

(5.8a)   *ou bien* $m^*_{N+1} < m^*$

(5.8b)   *et alors* $r < \dfrac{m^* - m^*_{N+1}}{1 - 2m^*_{N+1}} \to \dfrac{1}{2}$ *lorsque* $m^* \to \dfrac{1}{2}$.

(c) *Dans le cas où* $m^* > 1/2$, *il suffit de renverser les inégalités* (5.7) *et* (5.8a).

Le théorème 5.1 nous dit qu'un équilibre de HWR associé à un sex ratio plus près de 1/2 a plus de chances d'être stable intérieurement mais a aussi plus de chances d'être déstabilisé par un allèle mutant. A la limite, il peut même y avoir un opposition totale entre la stabilité interne et la stabilité externe d'après l'énoncé du résultat 5.1. Cette contradiction apparente peut être expliquée de la façon suivante: les modificateurs du sex ratio dans la progéniture des mâles ont toujours tendance a amener la population vers un sex ratio plus près de 1/2 comme dans le cas $r = 1/2$ mais le linkage avec le locus X-Y fait en sorte qu'un nouvel allèle peut envahir la population quelque soit son effet. Le théorème 5.1 met en évidence ce fait.

5.2. INTERACTIONS ENTRE INDIVIDUS

On considère le cas où le génotype à un locus détermine lequel de deux phénotypes sera exprimé, disons le phénotype 1 avec probabilité w_{ij} et 2 avec probabilité $1-w_{ij}$ pour un individu de génotype A_iA_j $(i,j = 1,...,N)$. Ces phénotypes ont des valeurs sélectives différentes chez les mâles et les femelles, disons F_1 et F_2 chez les femelles et M_1 et M_2 chez les mâles. Ces valeurs sélectives sont supposées positives et peuvent dépendre de la représentation des deux phénotypes dans la population.

Si p_i et q_i sont les fréquences de l'allèle A_i chez les femelles et les mâles, respectivement $(i = 1,...,N)$, alors les fréquences des phénotypes 1 et 2 chez les zygotes de la génération suivante dans une population infinie panmictique seront

(5.9) $w = \displaystyle\sum_{i,j=1}^{N} w_{ij}p_iq_j$

et $1 - w$, respectivement. Les valeurs sélectives du génotype A_iA_j chez les zygotes mâles et femelles, respectivement, seront des fonctions de w données par

(5.10a) $m_{ij}(w) = w_{ij}M_1(w) + (1-w_{ij})M_2(w)$

et

(5.10b) $f_{ij}(w) = w_{ij}F_1(w) + (1-w_{ij})F_2(w)$

pour $i,j = 1,...,N$. Les équations de récurrence pour les fréquences alléliques p_i et q_i sont alors les équations (4.2) avec les valeurs sélectives ci-dessus.

On vérifie facilement qu'à l'équilibre on doit avoir

(5.11) $p_iQ(w) = q_iQ(w)$ pour $i = 1,...,N$,

où

$$(5.12) \qquad Q(w) = \left[\frac{F_1(w)}{2F(w)} + \frac{M_1(w)}{2M(w)}\right] - \left[\frac{F_2(w)}{2F(w)} + \frac{M_2(w)}{2M(w)}\right],$$

les fonctions $F(w)$ et $M(w)$ représentant les valeurs sélectives moyennes chez les femelles et les mâles, respectivement, c'est-à-dire

(5.13) $F(w) = wF_1(w) + (1-w)F_2(w)$ et $M(w) = wM_1(w) + (1-w)M_2(w)$.

La fonction $Q(w)$ représente la différence des valeurs reproductives des phénotypes 1 et 2 selon leurs fréquences w et $1-w$, respectivement, d'après la formule de Shaw-Mohler (voir section 1). A l'équilibre, on doit avoir

(1) $Q(w) = 0$, *ou sinon* (2) $p_i = q_i$ *pour* $i = 1,...,N$.

Les équilibres satisfaisant la condition (1) sont appelés *équilibres phénotypiques* et ceux satisfaisant la condition (2) *équilibre génotypiques*. Génériquement, on peut supposer que ces deux conditions ne sont jamais satisfaites simultanément à l'équilibre ce qui exclut la possibilité d'un équilibre qui serait à la fois phénotypique et génotypique. En particulier, ceci entraîne que $M_1(w) \neq M_2(w)$ et $F_1(w) \neq F_2(w)$ à tout équilibre phénotypique.

Au sujet des équilibres phénotypiques, on remarque d'abord que $Q(w) = 0$ si et seulement si

$$(5.14) \qquad \frac{F_1(w)}{2F(w)} + \frac{M_1(w)}{2M(w)} = \frac{F_2(w)}{2F(w)} + \frac{M_2(w)}{2M(w)} = 1,$$

car

$$(5.15) \qquad w\left[\frac{F_1(w)}{2F(w)} + \frac{M_1(w)}{2M(w)}\right] + (1-w)\left[\frac{F_2(w)}{2F(w)} + \frac{M_2(w)}{2M(w)}\right] = 1 .$$

Un équilibre phénotypique du modèle (5.10) correspond donc à un équilibre de parité du modèle (2.1) avec matrice

$$(5.16) \qquad M = \frac{[M_1(w)-M_2(w)]W+M_2(w)U}{2M(w)}$$

où $W = \|w_{ij}\|_{i,j=1,\dots,N}$ est la *matrice de détermination phénotypique* et $Q(w) = 0$.

D'autre part, les équilibres génotypiques du modèle (5.10) correspondent aux équilibres symétriques du modèle (4.2) avec paramètres de viabilité $m_{ij}(w^*)/[2M(w^*)]$ et $f_{ij}(w^*)/[2F(w^*)]$ chez les mâles et les femelles, respectivement $(i,j = 1,\dots,N)$, où w^* est la valeur de w à l'équilibre. Comme pour le modèle (2.1), un tel équilibre correspond à un équilibre pour le modèle de viabilité (2.5) avec W comme matrice de viabilité. De plus il est stable si et seulement s'il est stable pour le modèle de viabilité (2.5) avec paramètres de viabilité

$$(5.17) \qquad \frac{m_{ij}(w^*)}{2M(w^*)} + \frac{f_{ij}(w^*)}{2F(w^*)} = Q(w^*)w_{ij} + \left[\frac{M_2(w^*)}{2M(w^*)} + \frac{F_2(w^*)}{2F(w^*)}\right]$$

pour $i,j = 1,\dots,N$. Enfin remarquons qu'un équilibre génotypique correspond aussi à un équilibre symétrique du modèle (5.16) pour tout w satisfaisant $Q(w) = 0$. Nous sommes maintenant en mesure de conclure:

Théorème 5.2. (Lessard, 1986)[18] *Soit* $W = \|w_{ij}\|_{i,j=1,\dots,N}$ *une matrice de détermination phénotypique satisfaisant aux hypothèses II. On considère le modèle (5.10) et la fonction Q de (5.12) tels que Q(w) = 0 à tout équilibre phénotypique et Q(w) ≠ 0 à tout équilibre génotypique*
(a) *Si* $Q(\tilde{w}) = 0$, *alors il existe un équilibre phénotypique avec* $w = \tilde{w}$ *si et seulement s'il existe deux équilibres génotypiques avec* $w = w^*$ *et* $w = w^{**}$ *tels que* $w^* < \tilde{w} < w^{**}$. *Dans ce cas il existe une surface d'équilibres phénotypiques séparant les deux équilibres génotypiques dans le sens de (2.28).*
(b) *Un équilibre génotypique avec* $w = w^*$ *est stable si et seulement s'il est stable pour le modèle de viabilité (2.5) avec matrice de viabilité* W *si* $Q(w^*) > 0$ *et* $U - W$ *si* $Q(w^*) < 0$.
(c) *Si* $\tilde{w}$ *est un zéro de plus à moins de Q (c'est-à-dire,* $Q(\tilde{w}) = 0$ *et la dérivée de Q à* $\tilde{w}$ *est < 0), alors on a évolution vers l'équilibre phénotypique correspondant à* $\tilde{w}$ *au moins faiblement dans le sens du théorème 4.2 et localement lorsque* w *est suffisamment près de* $\tilde{w}$.
(d) *De même on a évolution au moins faiblement et localement vers les états* $w = 0$ *et* $w = 1$ *lorsque* $Q(0) < 0$ *et* $Q(1) > 0$, *respectivement.*

Notes et commentaires sur la section 1

(1) Nous recommandons au lecteur l'excellent livre de Bull (1983) pour une étude détaillée des systèmes de détermination du sexe et de leur évolution.

(2) Les modèles I et II ont été analysés à divers niveaux de généralité et de formalisme (deux ou plusieurs modificateurs du taux de ségrégation, différences de viabilité ou de fertilité et/ou virilité, analyses locales ou globales, etc.) par Edwards (1961), Hamilton (1967), Cannings (1967), Thomson et Feldman (1975), Curtsinger et Feldman (1980), Karlin et Lessard (1986) et Lessard (1987). Nous présentons dans les sections 1.4 et 1.5 des analyses globales dans le cas de plusieurs modificateurs qu'on peut trouver dans Karlin et Lessard (1986) sous des hypothèses légèrement plus faibles dans le cas du modèle II. Il est cependant possible, sinon probable, que ces résultats soient connus depuis fort longtemps.

(3) Selon l'inégalité (1.2), la proportion de mâles dans la population est une fonction de Lyapounov pour le système (1.1). Les équilibres étant isolés, ceci garantit la convergence vers un équilibre à partir de n'importe quel point de départ. Dans le cas présent, la fonction de Lyapounov est linéaire et la convergence vers un équilibre peut être déduite facilement. Pour un exposé général sur les fonctions de Lyapounov pour les systèmes dynamiques à temps discret, nous recommandons l'article de LaSalle (1977).

(4) Cette similitude d'interprétation pour les théorèmes 1.1 et 1.2 a été relevée par Freddy Christiansen lors de la tenue du Séminaire.

(5) Les arguments intuitifs de Fisher on été successivement formalisés par un certain nombre d'auteurs. Parmi ceux-ci, mentionnons Shaw et Mohler (1953), Bodmer et Edwards (1960), Edwards (1962), MacArthur (1965) et Eshel (1975). Les formalisations les plus récentes et les plus rigoureuses sont l'objet de la présente étude.

(6) Chez l'homme, il y a environ 104 garçons pour 100 filles à la naissance et probablement plus à la conception car la mortinatalité est plus élevée chez les garçons. La mortalité est d'ailleurs plus élevée chez les hommes que chez les femmes à tout âge.

(7) Le théorème 1.3 est un cas particulier du théorème fondamental de la sélection naturelle en remarquant que le modèle haploïde (1.5) correspond à un modèle diploïde "multiplicatif". Cette correspondance est connue depuis longtemps au moins pour les modèles haploïdes sans distinction de sexes (voir, par exemple, Karlin, 1978). Plus récemment, une telle correspondance a été exploitée par Lessard (1984) pour des modèles haploïdes à deux phénotypes avec distinction de sexes et Gregorius (1982) pour des populations avec des phases haploïde et diploïde.

(8) Les premières versions du théorème fondamental de la sélection naturelle sont dues à R.A. Fisher et S. Wright. La croissance de la viabilité moyenne d'une génération à la

suivante implique que les équilibres stables correspondent aux maxima locaux de la viabilité moyenne.   La convergence vers un point d'équilibre à partir de n'importe quel état initial a été démontrée sous des conditions très générales même lorsque les équilibres forment des courbes ou des surfaces sur lesquelles la viabilité moyenne est constante (voir Lyubich et al., 1980).

Notes et commentaires sur la section 2

(9)   L'analogie des structures d'équilibre pour les modèles III, IV, IV', V et V' , ou seulement certains d'entre eux, a été relevée par Uyenoyama et Bengtsson (1979), Eshel et Feldman (1982a,b) et Karlin et Lessard (1986).   Pour le modèle V' , on fixe habituellement la proportion des individus haploïdes et on considère une différence de virilité entre les mâles haploïdes et diploïdes.

(10)   Les deux types d'équilibre, symétriques et de parité, ont été identifiés par Eshel et Feldman (1982a,b) qui ont donné la condition de stabilité externe des équilibres symétriques.   La condition de stabilité interne est un résultat particulier d'une analyse faite par Karlin (1978) prolongeant celle de Kingman (1961b) pour un modèle général de viabilité avec distinction de sexes.   Tel qu'énoncé et démontré, le théorème 2.1 a été présenté pour la première fois dans Karlin et Lessard (1983, 1984).

(11)   La suffisance du critère pour l'existence d'équilibres de parité comme la démonstration du théorème 2.2 est dans Karlin et Lessard (1984).   La nécessité du critère a été présentée dans Lessard (1986).

Notes et commentaires sur la section 3

(12)   Eshel (1975) a étudié le Modèle V dans le cas de deux allèles et a déduit les structures d'équilibre par des analyses de stabilité locale.

(13)   Le théorème 3.1 est seulement énoncé dans Karlin et Lessard (1986) à partir d'une affirmation dans Karlin (1972).   Nous donnons dans l'appendice de la section 3 une démonstration complète.   Selgrade et Ziehe (1988) ont donné une autre démonstration en montrant de plus que sur les courbes séparant les domaines de convergence des équilibres stables il y a convergence vers un équilibre instable.

Notes est commentaires sur la section 4

(14) Le lemme 4.1 est une formulation rigoureuse de propriétés de la formule de Shaw-Mohler (voir (1.7)) qui furent proposées pour la première fois par MacArthur (1965) et

Charnov (1982). Une démonstration complète du lemme 4.1 a été donnée dans Lessard (1988).

(15) Le théorème 4.1 confirme une conjecture d'Eshel et Feldman (1982a) qui, après avoir obtenu la condition de stabilité externe des équilibres symétriques de (2.1), affirmèrent que la population devrait tendre à se rapprocher toujours davantage de $m = 1/2$.

(16) Si p^* est stable (intérieurement et extérieurement), alors il peut exister plus qu'un autre équilibre stable contrairement à l'énoncé initial dans Karlin (1978). Ceci a été relevé par Cannings et Vickers qui ont aussi donné une démonstration du théorème 4.2 lors de la tenue du Séminaire. Un exemple donné par Cannings et Vickers est le suivant (voir figure 4.5):

$$M = \begin{pmatrix} 1 & 14 & 18 & 6 \\ 14 & 10 & 9 & 10.2 \\ 18 & 9 & 6 & 11 \\ 6 & 10.2 & 11 & 10 \end{pmatrix}$$

avec les équilibres stables

$$p^* = \begin{pmatrix} 1/3 \\ 1/3 \\ 1/3 \\ 0 \end{pmatrix}, \ p^{**} = \begin{pmatrix} 0 \\ 1/2 \\ 0 \\ 1/2 \end{pmatrix}, \ p^{***} = \begin{pmatrix} 0 \\ 0 \\ 1/6 \\ 5/6 \end{pmatrix}.$$

Notes et commentaires sur la section 5

(17) Le modèle (5.1) a été étudié par de nombreux auteurs dans le cas $N = 2$ (Edwards, 1961; Hamilton, 1967; Thomson et Feldman, 1975; Curtsinger et Feldman, 1980). Dans le cas général, la condition de stabilité externe (5.7) et (5.8) et la condition de stabilité interne (5.5) ont été déduites pour la première fois par Lessard (1987) qui a montré de plus qu'il y a naissance (par bifurcation) d'équilibres ponctuels qui ne sont pas de HWR lorsque r croise $r_2(m^*)$ et de cycles invariants lorsque r croise $r_1(m^*)$, ce qui confirme des simulations faites par Maffi et Jayakar (1981) dans le cas $N = 2$. Mais contrairement à ce qui était affirmé dans Lessard (1987), il peut exister des équilibres stables qui ne sont pas de HWR même lorsque r est très petit (Feldman et Otto, 1989).

(18) Le modèle (5.10) sans distinction de sexes (c'est-à-dire, avec $F_1(w) = M_1(w)$ et $F_2(w) = M_2(w)$) a été analysé avec toujours plus de rigueur et de généralité par Maynard

Smith (1981), Eshel (1982) et Lessard (1984), qui a démontré entre autres que les équilibres phénotypiques correspondant à des zéros de plus à moins de la fonction Q sont stables, la convergence étant même globale dans le cas linéaire.

Bibliographie

Bell, A.E. (1954), A gene in *Drosophila melanogaster* that produces all male progeny, *Genetics* **39**, 958-959.

Bodmer, W.F. and Edwards, A.W.F. (1960), Natural selection and the sex ratio, *Ann. Hum. Genet.* **24**, 239-244.

Bull, J.J. (1983), *Evolution of Sex Determining Mechanisms*, Benjamin-Cummings, Menlo Park, California.

Cannings, C. (1967), Equilibrium, convergence and stability at a sex-linked locus under natural selection, *Genetics* **56**, 613-618.

Charlesworth, D. and Charlesworth, B. (1981), Allocation of resources to male and female functions in hermaphrodites, *Biol. J. Linn. Soc.* **14**, 57-74.

Charnov, E.L. (1982), *The Theory of Sex Allocation*, Princeton University Press, Princeton.

Charnov, E.L., Maynard Smith, J. and Bull. J.J. (1976), Why be an hermaphrodite? *Nature* **263**, 125-126.

Crow, J.F. and Kimura, M. (1970), *An Introduction to Population Genetics Theory*, Harper and Row, New York.

Curtsinger, J.W. and Feldman, M.W. (1980), Experimental and theoretical analysis of the "sex ratio" polymorphism in *Drosophila pseudoobscura, Genetics* **94**, 445-466.

Edwards, A.W.F. (1961), The population genetics of "sex-ratio" in *Drosophila pseudoobscura, Heredity* **16**, 291-304.

Edwards, A.W.F. (1962), Genetics and the human sex ratio, *Adv. Genet.* **11**, 239-276.

Eshel, I. (1975), Selection on sex ratio and the evolution of sex determination, *Heredity* **34**, 351-361.

Eshel, I. (1982), Evolutionary stable strategies and viability selection in Mendelian populations, *Theoret. Population Biol.* **22**, 204-217.

Eshel, I. and Feldman, M.W. (1982a), On evolutionary genetic stability of the sex ratio, *Theoret. Population Biol.* **21**, 430-439.

Eshel, I. and Feldman, M.W. (1982b), On the evolution of sex determination and the sex ratio in haplodiploid populations, *Theoret. Population Biol.* **21**, 440-450.

Eshel, I. and Feldman, M.W. (1984), Initial increase of new mutants and some continuity properties of ESS in two-locus systems, *Amer. Natur.* **124**, 631-640.

Ewens, W. (1979), *Mathematical Population Genetics*, Springer-Verlag, Heidelberg.

Faulhaber, S.H. (1967), An abnormal sex ratio in *Drosophila simulans*, *Genetics* **56**, 189-213.

Feldman, M.W. and Otto, S.P. (1989), More on recombination and selection in the modifier theory of sex-ratio distortion, *Theoret. Population Biol.* (to appear).

Fisher, R.A. (1930), *The Genetical Theory of Natural Selection*, Oxford University Press, Oxford.

Fredga, K., Gropp, A., Winking, H. and Frank, F. (1976), Fertile XX and XY type females in the wood lemming *Myopus schisticolor*, *Nature (London)* **261**, 225-227.

Gantmacher, F.R. (1959), *The Theory of Matrices*, Vols I and II, Chelsea, New York.

Gershenson, S. (1928), A new sex ratio abnormality in *Drosophila obscura*, *Genetics* **13**, 488-507.

Gregorius, H.R. (1982), Selection in diplo-haplonts, *Theoret. Population Biol.* **21**, 289-300.

Hamilton, W.D. (1967), Extraordinary sex ratios, *Science* **156**, 477-488.

Hickey, W.A. and Craig, G.B. (1966), Genetic distortion of sex ratio in a mosquito, *Aedes aegypti*, *Genetics* **53**, 1177-1196.

Kallman, K.D. (1973), The sex-determining mechanisms of the platyfish, *Xiphophorus maculatus*, in: *Genetics and Mutagenesis of Fish* (J.H. Schroder, ed.), Springer Verlag, New York, 19-28.

Karlin, S. (1972), Some mathematical models of population genetics, *Amer. Math. Monthly*, **79**, 699-739.

Karlin, S. (1978), Theoretical aspects of multilocus selection balance I, in: *Mathematical Biology, Part II: Populations and Communities* (S.A. Levin, Ed.), MAA Studies in Mathematics, Vol. 16, Washington, D.C., 503-587.

Karlin, S. and Lessard, S. (1983), On the optimal sex ratio, *Proc. Nat. Acad. Sci. USA* **80**, 5931-5935.

Karlin, S. and Lessard, S. (1984), On the optimal sex ratio: A stability analysis based on a characterization for one-locus multiallele viability models, *J. Math. Biol.* **20**, 15-38.

Karlin, S. and Lessard, S. (1986), *Theoretical Studies on Sex Ratio Evolution*, Princeton University Press, Princeton, New Jersey.

Kingman, J.F.C. (1961a), A matrix inequality, *Quart. J. Math.* 12, 78-80.

Kingman, J.F.C. (1961b), A mathematical problem in population genetics, *Proc. Camb. Phil. Soc.* 57, 574-582.

LaSalle, J.P. (1977), Stability theory for difference equations, in: *Studies in Ordinary Differential Equations* (J. Hale, ed.), MAA Studies in Mathematics, Vol. 14, Math. Association of America, Washington, D.C., 1-31.

Lessard, S. (1984), Evolutionary dynamics in frequency-dependent two-phenotype models, *Theoret. Population Biol.* 25, 210-234.

Lessard, S. (1986), Evolutionary principles for general frequency-dependent two-phenotype models in sexual populations, *J. Theoret. Biol.* 119, 329-344.

Lessard, S. (1987), The role of recombination and selection in the modifier theory of sex-ratio distortion, *Theoret. Population Biol.* 31, 339-358.

Lessard, S. (1988), Ressource allocation in Mendelian populations: Further in ESS theory, in: *Mathematical Evolutionary Theory* (M.W. Feldman, ed.), Princeton University Press, Princeton, 207-246.

Lyubich, Yu. I., Maistrovskii, G.D. and Ol'khovskii, Yu. G. (1980), Selection-induced convergence to equilibrium in a single-locus autosomal population, *Problems Inform. Transmission* 16, 66-75.

MacArthur, R.H. (1965), Ecological consequences of natural selection, in: *Theoretical and Mathematical Biology* (T.H. Waterman and H. Morowitz, eds.), Blaisdell, New York, 388-397.

Maffi, G. and Jayakar, S.D. (1981), A two-locus model for polymorphism for sex-linked meiotic drive modifiers with possible applications to *Aedes aegypti, Theoret. Population Biol.* 19, 19-36.

Maynard Smith, J. (1981), Will sexual population evolve to an ESS?, *Amer. Natur.* 117, 1015-1018.

Newton, M.E., Wood, R.J. and Southern, D.I. (1978), Cytological mapping of the M and D loci in the mosquito, *Aedes aegypti (L.), Genetica* 48, 137-143.

Selgrade, J.F. and Ziehe, M. (1988), Convergence to equilibrium in a genetic model with differential viability between the sexes, *J. Math. Biol.* 25 (1987), 477-490.

Shaw, R.F. and Mohler, J.D. (1953), The selective advantage of the sex ratio, *Amer. Natur.* 87, 337-342.

Stalker, H.D. (1961), The genetic systems modifying meiotic drive in *Drosophila paramelanica, Genetics* 46, 177-202.

Sturtevant, A.H. and Dobszhansky, T. (1936), Geographical distribution and cytology of "sex ratio" in *Drosophila pseudoobscura* and related species, *Genetics* 21, 473-490.

Thomson, G.J. and Feldman, M.W. (1975), Population genetics of modifiers of meiotic drive. IV. On the evolution of sex-ratio distortion, *Theoret. Population Biol.* 8, 202-211.

Uyenoyama, M.K. and Bengtsson, B.O. (1979), Towards a genetic theory for the evolution of the sex ratio, *Genetics* 93, 721-736.

Whiting, P.W. (1943), Multiple alleles in complementary sex determination of *Habrobracon*, *Genetics* 28, 365-382.

TOPICS IN EVOLUTIONARY ECOLOGY

Simon Levin
Section of Ecology and Systematics,
Center for Environmental Research,
Center for Applied Mathematics
Cornell University
Ithaca, NY 14853, USA
and
Carlos Castillo-Chavez
Center for Applied Mathematics,
Biometrics Unit
Cornell University
Ithaca, NY 14853, USA

ABSTRACT. Various topics in evolutionary ecology are discussed, ranging from evolution as optimization, the joint evolution of avoidance and tolerance of toxins, dispersal and dormancy as adaptations in variable environments, seed dispersal models, and diffuse coevolution of chemical defenses and detoxification mechanisms.

1. Introduction

Mathematical models have played a central role in evolutionary theory at least since the pioneering work of Fisher, Wright and Haldane. Two approaches have been predominant: mechanistic and reductionistic ones that assume considerable detail about genetic mechanisms, and phenotypically-based ones that assume quantitative inheritance, or that suppress genetic detail entirely. Despite the successes of both approaches, they have in general failed to deal adequately with some of the central problems of evolutionary ecology – those in which strong nonlinear feedbacks result from intra-specific or inter-specific frequency dependence. Thus, the need for new ideas and new approaches is as pressing as ever.

These notes, which are based on a series of lectures given at the Université de Montréal by the first author, begin with a discussion of evolution as optimization, and the constraints that arise due to history, stochasticity, and frequency dependence. Life-history evolution in variable environments is treated through a consideration of dispersal and dormancy strategies, in which the importance of frequency dependence is illustrated. Finally, consideration is given to coevolution in host-parasite and plant-herbivore systems,

S. Lessard (ed.), Mathematical and Statistical Developments of Evolutionary Theory, 327–358.
© *1990 by Kluwer Academic Publishers.*

with special attention to the coevolution of virulence and resistance in situations of tight and diffuse interactions.

2.   Evolution as optimization

To a great extent, evolutionary ecology deals with the study of adaptations. The adaptationist approach views evolution as a problem solver, providing solutions that are in some sense "optimal" in a given environment.   In short, evolution optimizes some nebulous quantity, the organism's "fitness".   Herein lie the principal problems with the adaptationist approach: the identification of the end result (the definition of fitness) and the determination of the "purpose" of the genetic algorithm.   As Gould (1977) argues, "...evolution has no purpose. Individuals struggle to increase the representation of their genes in future generations, and that is all."   Lewontin (1977) clearly states that "Adaptation, for Darwin, was a process of becoming rather than a state of final optimality." Jacob (1977) expounds this view very eloquently when he points out that the process of evolution, being more similar to the work of a tinkerer rather than that of a master craftsman, is constrained by past history, chance, and the mechanics of the evolutionary process itself.

Sewall Wright, Ronald Fisher, and J.B.S. Haldane were among the most prominent figures in the development of the modern synthesis of population genetics and evolutionary theory. Among their contributions were the development of a mathematical framework for population genetics and the exploration of the population-level consequences of natural selection and other evolutionary processes.   The paradigm developed through Fisher's Fundamental Theorem of Natural Selection and Wright's Adaptive Landscape – that through natural selection, fitness will gradually improve at a rate proportional to the remaining genic variance, and that the process can be viewed as hill-climbing – has become one of the most powerful in evolutionary theory, and provided a mathematical justification for the rise in status of optimization theory.

However, the usefulness of this paradigm always has been questioned.   Recently, Provine (1986) has argued that the underlying concept of the multi-dimensional adaptive landscape is a useless and misleading metaphor, and indeed that Wright's own presentation of the notion has itself gone through substantive evolution. Levin (1978, 1983a), somewhat more positive on the subject, nonetheless has argued that it can mislead: "the conclusion that populations evolve towards maximization of mean fitness is easily vitiated, and the worst culprit is frequency dependence."

Despite the above objections, the paradigm that emerges from these approaches still provides one of the most important theoretical concepts in evolutionary theory. In some special simple cases it provides the correct picture (see Levin 1978; Ewens, 1979).   In more complex situations, it provides a possible starting point for future extensions. For example, various authors have shown that variants on the metaphor are still valid for traits that are density-dependent, although multiple loci and frequency dependence introduce more fundamental problems. Although versions of the Fisher theorem can be developed under weak frequency dependence or weak epistasis (Ewens 1969a,b; Nagylaki 1976), in general the undulating landscapes that arise under frequency dependence and coevolution mandate entirely new approaches.

We begin by discussing the problems that arise when landscapes are rugged and very high-dimensional, and then move on to the central problems of evolutionary ecology – frequency dependence and coevolution.

TOWARDS A GENERAL THEORY OF ADAPTIVE WALKS

The problems of high-dimensionality and ruggedness have been examined by Kauffman and Levin (1987), and their discussion forms our starting point.   In particular, consideration of the hill-climbing metaphor as a heuristic solution algorithm is shown to be fraught with problems – false peaks, multiple pathways, etc.   Computer simulations of simple cases reinforce our understanding of the importance of stochasticity and history.

The mutational process plays a crucial role in the generation of genetic variability. In this section we assume that we are dealing exclusively with point mutations that switch, insert, or delete single nucleotide bases. This is a reasonable starting point, because other types of mutations can be thought of as mechanisms that produce many mutations of the above type simultaneously. Kauffman and Levin (1987) construct a genotypic space by assuming that each genotype is surrounded by 1-mutant neighbors; that is, a single mutational alteration will transform the genotype under consideration into a neighboring type. We have then a space in which each point denotes a particular genotype and has as its immediate neighbors genotypes that differ by a single mutation. Note that the topology on this space is given by the mutational "move" generator that specifies the allowable transformations, i.e., that specifies which entities can mutate in one step to one another. Furthermore, observe that in this case the process is symmetric and reversible.   This restriction can be relaxed.

Kauffman and Levin (1987) define a mapping from this genotypic space to the appropriate phenotypic space, and specify the fitness associated with each attribute (phenotype). The discrete distribution of fitness values across the genotypic lattice will be referred to as the fitness landscape. If, for example, we restrict ourselves to a haploid organism with DNA genotypes of length 100,000 nucleotides, then each position in the DNA sequence can be occupied by 4 alternative bases, and each genome has 300,000 1-mutant neighbors in the space of haploid genotypes. Hence, each genotype is surrounded by huge numbers of 1-mutant neighbors with (possibly) different fitness values. In this scenario, to the extent that the metaphor of the previous section is valid, adaptive evolution can be thought of as an uphill walk via 1-step fitter variants until a local or a global optimum is reached.   This gradual adaptive climbing through mutation and selection provides the simplest trial and error optimization method, and is mimicked by the development of heuristic methods in combinatorial optimization (e.g., Lin and Kernighan 1973) and neural computing (Hopfield 1982).

What are the constraints that arise from such an approach to improvement? If the notion of neighborhood is extended to include k point changes (k = 1,2,3,...), how many local optima are there in the space with respect to k-mutants? If adaptive movement is allowed only through fitter neighbors, what is the expected number of improved variants passed on the way to a local optimum? How long are adaptive walks that use this optimization algorithm? In how many ways can adaptive walks branch at each uphill step? How many local optima are there available for an arbitrary initial genotype? What role does the initial fitness play in the availability of local optima? What is the probability of attaining a global maximum? What is the correlation structure in a fitness landscape?

Consider the expected character of adaptive walks in (uncorrelated) spaces where the fitness value of each genotype is drawn at random from some fixed underlying distribution. Furthermore, replace the actual fitness values assigned to the genotypes in the space by their rank orders. The least fit has rank 1, the most fit has rank T. We assume that there are no tie values and that the fitness values are distributed uniformly on the real line. For concreteness consider (as in Kauffman and Levin 1987) a space of length-N peptides that can use only two amino acids. Hence we can represent each peptide as a binary string of length N. The space of peptides is the N-dimensional Boolean hypercube. There are 2^N strings, and we assign order fitnesses from 1 to 2^N at random without replacement to each of the points in the N-dimensional Boolean hypercube. The probability that a vertex is a local maximum is given by

(1) $$P_m = 1/(N+1) ,$$

and the expected number of local optima with respect to 1-mutant neighbors, M_1, is

(2) $$M_1 = \frac{2^N}{N+1} .$$

For 1- and 2-step mutant neighbors, the expected number of local optima, M_2, is

(3) $$M_2 = \frac{2^{N+1}}{2+N(N+1)} ,$$

and for k-step mutants, the expected number of local optima M_k is

(4) $$2^{N+k} / \sum_{j=0}^{k} \binom{N}{j} .$$

Hence, in an uncorrelated fitness landscape, the number of local (1-step) optima is of the same order of magnitude as the number of possible peptides. If the peptides use B amino acids, then any peptide of length N has $D = (B-1)N$ 1-mutant neighbors, and M_1 is now given by

(5) $$M_1 = \frac{B^N}{D+1} = \frac{B^N}{N(B-1)+1} .$$

As before, the number of local optima increases exponentially in N.

Kauffman and Levin (1987) show that the probability that an entity in this space is a local optimum is low if its rank is low, and rises rapidly when the rank increases. They further show that an upper bound for the average walk length R is $R = \log_2(D-1)$, and find that for greedy walks (those that always choose the best improvement) in an uncorrelated space the average walk length is less than 2. Gillespie has independently derived this result recently (Gillespie, personal communication), and Weinberger (1988) has confirmed it by developing more accurate estimations of walk length. In addition, using the fact that the number of alternative pathways towards increased fitness values decreases linearly with rank order, Kauffman and Levin (1987) calculate an upper bound for the expected number of local optima B accessible from the lowest rank entity:

$$(6) \qquad\qquad\qquad B = D^{(\log_2 D-1)/2}.$$

Hence, only a tiny fraction of all local optima are accessible from any entity on adaptive walks via 1-mutant fitter variants in uncorrelated landscapes.   Walks in correlated landscapes in general will be longer.

For some implications of the results of this baseline case to the length of adaptive walks in the immune system, and to branching phylogenies in biological evolution, see Kauffman and Levin (1987) and Kauffman et al. (1988). Gillespie (1983, 1984) uses a variant of this model on the molecular clock hypothesis to show that burst-like evolution fits better with a selectionist theory that with a neutral theory.

Kauffman and Levin (1987) examine these results further by applying this evolutionary algorithm to optimization problems such as the traveling salesman problem (Lin and Kernighan 1973, Johnson and Papadimitrou 1985).   They demonstrate the tendency of the scheme to get hung up on false peaks, and the importance of stochastic phenomena, especially early in evolution. They furthermore show that the most efficient optimization occurs for intermediate levels of mutation: low levels of mutation rapidly lock the system in to false peaks, whereas high levels do not take advantage of local information and progress already made.   Furthermore, optimization is significantly enhanced when the process occasionally can go downhill and traverse valleys.   This approach, which allows one to get free from false peaks, involves "simulated annealing" in heuristic combinatorial optimization, and can arise from a number of genetic mechanisms (shifting balance, genotypic variance, outcrossing, etc.).

EVOLUTION IN VARYING ENVIRONMENTS

The problems associated with the adaptationist approach include: the definition of the putative quantity to be maximized; the determination of what is heritable; the high dimensionality, which leads to large numbers of optima; pleiotropy, linkage and epistasis; temporal variation in fitness; frequency and density dependence; and coevolutionary interactions (among populations and with the environment).   The last two classes of problems are perhaps the central ones in understanding natural communities and ecosystems.   Essentially, whatever is being optimized is a tradeoff against different environments, which the species defines and alters as it evolves. One of the most important sets of constraints, and one of the least explored in theory, arises from the tradeoffs involving different phenotypic aspects. These may involve pleiotropy or interactions among loci, but most often involve different genotypes being favored in different parts of a heterogeneous environment. Castillo-Chavez et al. (1989) explore the tradeoffs between the evolution of habitat selection and physiological adaptation in a heterogeneous environment. They start by developing a two-locus model that considers a panmictic population in which prereproductive individuals are mobile enough to move among patches. Alleles at one locus code for the absence or presence of physiological adaptation to detrimental patches, and alleles at the second locus code for the absence or presence of behavior that cause avoidance to detrimental patches. It is further assumed that the effects of alleles controlling physiology and behavior are additive and that fitnesses are frequency independent.

Table 1: Fitnesses of the various genotypes.

The fitness (w) of each genotype is dependent upon whether it is in the toxic (T) or non-toxic (UT) environment. The fitnesses (w) in the environment (L) are $w_{L(RR)}$, $w_{L(rr)}$, and $(w_{L(RR)}+w_{L(rr)})/2$ for individuals with RR, rr, and Rr respectively, where L is either (T) or (UT). X and Y are defined in the text.

$$w_{(RRAA)} = Xw_{T(RR)} + (1-X)w_{UT(RR)}$$

$$w_{(RRAa)} = \left(\frac{X+Y}{2}\right)w_{T(RR)} + \left(1 - \frac{X+Y}{2}\right)w_{UT(RR)}$$

$$w_{(RRaa)} = Yw_{T(RR)} + (1-Y)w_{UT(RR)}$$

$$w_{(RrAA)} = X\left[\frac{w_{T(RR)}+w_{T(rr)}}{2}\right] + (1-X)\left[\frac{w_{UT(RR)}+w_{UT(rr)}}{2}\right]$$

$$w_{(RrAa)} = \left(\frac{X+Y}{2}\right)\left[\frac{w_{T(RR)}+w_{T(rr)}}{2}\right] + \left(1 - \frac{X+Y}{2}\right)\left[\frac{w_{UT(RR)}+w_{UT(rr)}}{2}\right]$$

$$w_{(Rraa)} = Y\left[\frac{w_{T(RR)}+w_{T(rr)}}{2}\right] + (1-Y)\left[\frac{w_{UT(RR)}+w_{UT(rr)}}{2}\right]$$

$$w_{(rrAA)} = Xw_{T(rr)} + (1-X)w_{UT(rr)}$$

$$w_{(rrAa)} = \left(\frac{X+Y}{2}\right)w_{T(rr)} + \left(1 - \frac{X+Y}{2}\right)w_{UT(rr)}$$

$$w_{(rraa)} = Yw_{T(rr)} + (1-Y)w_{UT(rr)}$$

More specifically, two semi-dominant alleles at a single locus, allele R (resistant) and allele r (susceptible), determine the degree of physiological resistance. Repellency is governed by two semi-dominant alleles at the second locus: allele A codes for a high degree of avoidance, while allele a codes for lower avoidance. Ten genotypes are possible. To impose a price on avoidance, Castillo-Chavez et al. (1989) allow the fitnesses (w) to depend on whether the individual is found in a chemically treated (T) or untreated (UT) environment. Define the probabilities that a particular individual will be found in a given environment as:

(7)
$$\text{Prob(individual AA is in T)} = X,$$
$$\text{Prob(individual aa is in T)} = Y,$$
$$\text{Prob(individual Aa is in T)} = (X+Y)/2.$$

The Rr genotype's fitness in a particular environment is defined to be the arithmetic mean of the corresponding (identical at the complementary locus) homozygous genotypes in the same environment. The overall expected fitness for any genotype is the arithmetic mean of its expected fitnesses, W_T and W_{UT}, respectively in toxic and nontoxic environments, weighted by the probabilities specified by (7) (the overall fitnesses are summarized in Table 1, from Castillo-Chavez et al. 1989).

Introduce the notation $w_{(RRAA)} = \alpha$, $w_{(RRaa)} = \beta$, $w_{(rrAA)} = \gamma$, $w_{(rraa)} = \delta$, for the fitnesses of the double homozygotes, and let $x_1(t)$, $x_2(t)$, $x_3(t)$, and $x_4(t)$ denote the frequencies of the chromosomal types RA, Ra, rA, and ra, respectively. Then the frequencies of these chromosomal types in successive generations are related by the iterative scheme (Felsenstein 1965):

(8)
$$x_i' = \frac{w_i}{\overline{w}(x)} x_i + e_i k D \frac{w_H}{\overline{w}(x)}, \quad i = 1,2,3,4,$$

where $e_1 = e_4 = -e_2 = -e_3 = -1$. The prime denotes the succeeding generation, D denotes the linkage disequilibrium coefficient, k the recombination fraction between the two loci and $w_H = (\alpha+\beta+\gamma+\delta)/4$ the double heterozygote fitness. w_i is the mean fitness associated with allele x_i, and $\overline{w}$ is the mean fitness in the population.

The most general outcome of this scheme is fixation for a single gametic type. For example, if in the presence of a toxin the type (ra) that can make no response is the least fit $(\delta < \alpha,\beta,\gamma)$, then the plant species will evolve either a physiological or behavioral response, and may evolve both if the double homozygote possessing both (resistance and avoidance) is the most fit double homozygote $(\alpha > \beta,\gamma)$. On the other hand, if in contrast, $\alpha < \beta,\gamma$, then a bistable situation may result in which either resistance or avoidance can evolve, depending on initial gene frequencies. In this case, which implies a burden to having both features when one will do, the most fit double homozygote will not necessarily prevail. The initial conditions, the recombination rate, and the values of α and γ can influence recombination. The complexity of the situation is illustrated in Figure 1 (from Castillo-Chavez et al. 1989).

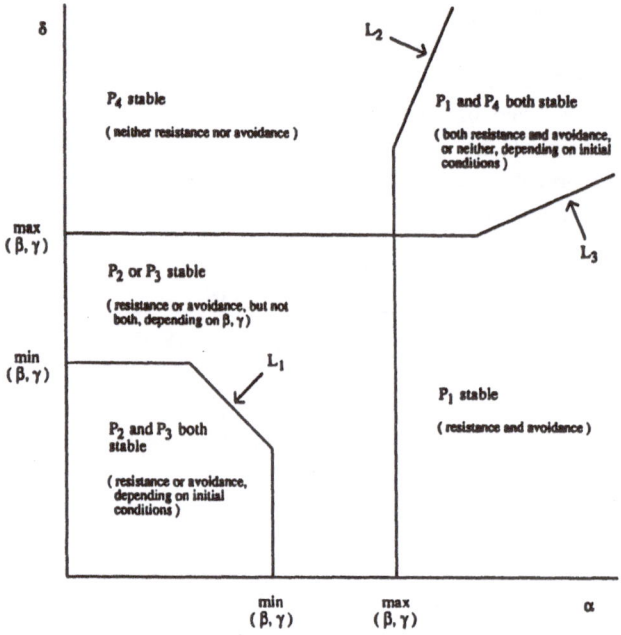

Figure 1. Stability regions for various equilibria.

$$L_1 : \alpha + \delta + \max(\beta,\gamma) = \frac{3+k}{1-k} \min(\beta,\gamma) \,,$$

$$L_2 : \beta + \gamma + \delta = \frac{3+k}{1-k} \alpha \,, \quad L_3 : \beta + \gamma + \alpha = \frac{3+k}{1-k} \delta \,.$$

The problem of joint selection for behavioral and physiological traits is exemplary of more general problems concerned with multiple genetic responses to single selective factors (e.g., Cohan 1984). Problems such as this frustrate the application of simple optimization approaches, because interest must be on the diversity of environments that might be confronted. Nonetheless, there are situations in which the objections to the adaptationist programme are more fundamental because of the importance of nonlinear feedbacks through frequency dependence and coevolution. We begin to treat these in the next section.

3.   Dispersal

Among the central problems in ecology are the statistical description of movement and the understanding of population distributions in terms of individual behavior. Questions regarding issues as diverse as the evolution of life history traits or the spread of genetically engineered organisms are crucially tied to our understanding of the dispersal patterns of plants and animals. One of the most fascinating challenges in evolutionary ecology is to determine the role that the spatial and temporal structure of the environment plays in the dispersal of individuals. These questions were raised years ago by Skellam (1951) and Hutchinson (1951). The evolutionary aspects of dispersal have received much less attention because of the complexities introduced by frequency dependence (Levin 1987).

In 1977, Hamilton and May asked: What are the advantages of dispersal for annual plants living in a renewable and stable environment? If there is a cost to dispersal, and if the habitat is uniformly good (or bad), then why disperse at all? The use of naive optimization arguments in such situations would dictate against dispersal, since there is cost without apparent gain. If, however, we consider the frequency dependence that is implicit when different genotypes are in competition, then the answer is quite different. Dispersers outcompete nondispersers. Furthermore, by applying the concept of evolutionarily stable strategies, Hamilton and May found that the best possible strategy within their model is to disperse with a probability equal to the reciprocal of 1 plus the probability of loss during dispersal. Hence, even if 90% of the dispersers are lost before reaching an appropriate site, the evolutionarily stable strategy (ESS) still is to disperse 52.6% of the seeds.

The Hamilton and May approach is elegant in its clear and simple demonstration of the basic need to consider frequency dependence. To examine the evolution of dispersal, however, one must consider a more general class of environments and strategies. Indeed, dispersal is just one possible evolutionary response to local unpredictability (broadly understood) and takes its place in a spectrum that includes dispersal, dormancy, diapause, iteroparity, and vegetative spread. In what follows we focus on two particular strategies: dispersal and dormancy.

Intuitively, dispersal and dormancy can be thought of as alternative strategies for individuals that have to deal with the spatial and temporal variability of the environment. In the previous section, we discussed an analogous problem: in the face of a toxic environment, two alternative strategies are physiological resistance and behavioral avoidance. It was seen that there are tradeoffs among these. Hence, similarly, one expects that the evolved level of one factor (dispersal or resistance) is a function of the other (dormancy or avoidance).

To gain understanding of the differences and similarities between dormancy and dispersal, and with the objective of determining the conditions needed for a strategy to dominate, Levin et al. (1984) developed a simple model of population growth in varying environments. In what follows, the effects of the spatial and temporal structure of the environment, as well as the relative cost of both strategies, are discussed in the context of the Levin, Cohen, Hastings model. Details can be found in Cohen and Levin (1987).

A seed population of annual plants in a patchy environment is considered. For patch j, the basic growth equation before germination, for the seed population, is given by:

$$(9) \qquad S_{t+1}^{j} = S_{t}^{j}[GY_{t}^{j}(1-D)+(1-G)V] + \frac{ADG}{L} \sum_{i=1}^{L} Y_{t}^{i}S_{t}^{i} .$$

We assume that an equation of this type applies for each genotype (= phenotype). In the above equation, G denotes the constant annual germination fraction and D denotes the constant dispersal fraction of seeds. Only the parameters G and D are genotype dependent; all others are assumed to be the same for all genotypes (although the approach could be extended to examination of the evolution of those parameters as well). A denotes the fraction of dispersing seeds that are successful at reaching a safe habitat, and V denotes the survival of those nongerminating seeds that remain dormant. Since the seeds are dispersed uniformly over all L patches, we take the summation over L. Y denotes a density-dependent yield function that is assumed to have the form $Y(Z) = K/(Z+b)$, where Z denotes the total density of all competing types in a given patch, and K is a random variable that denotes the total seed yield of the patch. For more general growth functions, see Levin et al. (1984). K is assumed to be independently distributed among patches; however, within a given patch, K has several possibilities: it may vary independently among years, or it may show a positive or a negative temporal correlation. D^* [G^*] , the evolutionarily stable strategy for dispersal [germination], is defined by the condition that its genotype, once established, cannot be invaded by any rare mutant playing a different strategy. In what follows, the details of the genetic system are ignored and alternative strategies are assumed to involve competing asexual clones.

Numerical simulations show that D^* is an increasing function of the germination fraction, G, if the latter is held fixed, and is a decreasing function of V (the survival of nongerminating seeds). On the other hand, Ellner (1985) shows that in the absence of dispersal, G^* is given implicitly by

$$1/V = \text{Expectation}(S_t/S_{t+1}),$$

and the numerical simulations seem to agree with Ellner's result as D^* approaches zero.

In the more general case, simulations indicate that G^* is an increasing function of dispersal (D), and of the effectiveness of dispersal (A). Furthermore, A and D seem to affect optimal germination mostly through the factor $F = AD/(1-D)$, which represents the seed's effective dispersal fraction, and G^* approaches zero as V approaches unity.

When dispersal and dormancy both are subject to selection, the optimal strategy is obtained as the intersection of the curves $D^*(G)$ and $G^*(D)$. In the absence of temporal correlation in environmental variation, this optimum appears to be a stable equilibrium. If, however, the environment is cyclical, then this internal equilibrium is unstable and there are two competing boundary equilibria. Simulations show that coexistence among these boundary equilibria is possible, but that more generally one of the strategies outcompetes the other. More specifically, the conclusions of the numerical simulations as reported in Cohen and Levin (1987) are:

(1) The optimal dispersal decreases as the level of dormancy $H = (1-G)V$ increases.

(2) The optimal dormancy level decreases as the level of dispersal $F = AD/(1-D)$ increases.

(3) The ratio between dispersal and dormancy in the joint optimal strategy is affected by the ratio between the effectiveness of dispersal A and the survival of dormant seeds V. Therefore, the distribution of dispersal and dormancy among plant families or species

from the same environment should be negatively correlated. This agrees with observations (Ellner and Shmida, unpublished, Venable and Lawlor 1980).

(4) In environments that vary periodically, there is no single joint optimal strategy with intermediate levels of dispersal and dormancy. Cohen and Levin (1987) used simulations to investigate stability and found that the only stable equilibria were at the boundaries $G^* = 1$ or $D^* = 0$. It was found, however, that two such boundary strategies could coexist in a stable frequency dependent equilibrium, which would be the eventual evolutionary equilibrium reached among many competing mutants with a wide range of dispersal and dormancy levels. For further details see Cohen and Levin (1987). Further investigations, to be published (Cohen and Levin, submitted), have focused on the influences of temporal and spatial correlation patterns and environmental variability.

From the above summary we can see that the tradeoffs between dispersal and dormancy are somewhat analogous to the tradeoffs between resistance and behavior previously discussed. It is rare to find populations that select both strategies, since this will add the burden of having both features when one will do.

The most important conclusion of these investigations is the essential nature of the concept of ESS when frequency dependence is involved. We cannot compare one strategy against another unless we put them in competition, and attempts to approach such problems from the view point of optimization theory generally give incorrect answers.

RANDOM WALK MODELS OF DISPERSAL

Having made the evolutionary case for the existence of dispersal – to escape local environmental deterioration, to reduce sib competition, to average the negative consequences of unpredictability, and to explore new habitats – we make a detour to ask about the observed patterns of dispersal. How far and how rapidly do organisms disperse?

The classical models of movement (e.g., Skellam 1951, Okubo 1980) are based on random walk models. Random walk models are derived from the assumption that individuals move in a series of discrete steps, the direction of each step being determined by probabilities totally specified by positional information (but see Kareiva and Shigesada (1983) for a discussion of correlated walks). The application of such models to populations of organisms (or molecules) does not require that the basic assumptions be valid for the actual movements of individuals, but rather that other details of how the individual moves be irrelevant to the patterns of spread of populations.

The simplest one-dimensional random walk model can be motivated by the following experiment (see Okubo 1980, Levin 1986). Assume that an organism is located at the origin of the real line; that at discrete times $k\tau$, it jumps either forward (right) or backward (left) λ units; and that either event has probability $1/2$. If m and n are integers, then the probability that at time $n\tau$ the organism is at position $m\lambda$ after its latest jump is given by the general term of the Bernoulli distribution:

$$(10) \qquad \text{Prob} = \left(\frac{1}{2}\right)^n \frac{n!}{\left(\frac{n+m}{2}\right)! \left(\frac{n-m}{2}\right)!} \cdot$$

As n increases, this converges to the Gaussian distribution given by

$$(11) \qquad Ce^{-m/2}$$

where

$$(12) \qquad C = \sqrt{\frac{2}{\pi n}} \cdot$$

If we let $x = \lambda m$ and $t = \tau n$, then the Gaussian distribution is given by

$$(13) \qquad C \exp\left(-\frac{x^2}{4t} \cdot \frac{2\tau}{\lambda^2}\right).$$

This tends to

$$(14) \qquad \rho(x,t) = C \exp\left(-\frac{x^2}{4Dt}\right), \text{ where } C = \frac{1}{2\sqrt{\pi Dt}},$$

provided that λ and τ shrink to zero in a way that the limit

$$(15) \qquad \lim_{\lambda,\tau \to 0} \frac{\lambda^2}{\tau} = 2D$$

exists. D is known as the diffusion coefficient. For generalizations to higher dimensions see Okubo (1980) or Lin and Segel (1974). As Okubo (1980) points out, the diffusion approximation is valid only on scales that involve a great many individual steps.

Observe that the population is normally distributed for $t > 0$, and has variance $2Dt$, which increases linearly with time; and that this distribution satisfies the diffusion or heat equation . Note further that the diffusion equation more generally describes the spread of a diffusing population with any distribution (that is, not just the normal distribution that would result if all individuals began at the same point in space and time). Furthermore, it can be shown that the variance $V(t)$ has the general form:

$$(16) \qquad V = V_0 + 2Dt ;$$

that is, the variance increases linearly with time from its initial value V_0.

Kareiva (1983), using data on the foraging movements of phytophagous insects, estimated D from the slope of the regression of V on t. He used his estimate of D to generate a series of probability distributions for the spread of insects, and to compare them with actual observations.   Agreement was excellent in many cases, but in some instances the habitat-dependent diffusion model,

(17)
$$\frac{\partial P}{\partial t} = \frac{\partial^2}{\partial x^2} \cdot (D(x)P)$$

provided a better fit. He concluded that the basic diffusion model was an excellent starting point, but that modifications of this basic formalism are necessary to take into consideration the substantial habitat variability that organisms often experience.

If growth and spread occur simultaneously, then the diffusion model gives way to

(18)
$$\frac{\partial P}{\partial t} = D\frac{\partial^2 P}{\partial x^2} + F(P,x,t) \ ,$$

where $F(P,x,t)$ denotes local population growth. If $F(P,x,t) = rP(1-P)$, then we arrive at the simplest model introduced by Fisher (1937) to describe the rate of advance of advantageous alleles, given that selection is operating on two alleles at a single autosomal locus.   A more general, cubic, form is necessary when there is partial or complete dominance, and can lead to fundamentally different results. P in this context denotes the frequency of the advantageous allele.   The correct approach to the population genetics problem is to imbed this within a fuller treatment of genotype frequencies (see Aronson and Weinberger 1975, Hoppensteadt 1975, Hadeler and Rothe 1975).   However, the basic insights that emerge from Fisher's model, at least regarding rates of spread, are essentially the same (see Hadeler 1976).

Fisher's fundamental insight, based on such models, lies in his estimate of the asymptotic speed of advance of a wave front. Fisher's conjecture – that an advancing wave would relax asymptotically to a front with this characteristic speed – was formalized by Kolmogorov et al. (1937), who considered the general equation

(19)
$$\frac{\partial P}{\partial t} = D\frac{\partial^2 P}{\partial x^2} + f(P) \ ,$$

where

(20)
$$f(0) = f(1) = 0 , \ f > 0 \ \text{ on } (0,1)$$

and

(21)
$$f'(0) > f'(P) \ \text{ on } [0,1] \ .$$

By looking for traveling waves (non-negative solutions of the form)

(22)
$$P = H(x-ct), \ c > 0 \ ,$$

Kolmogorov et al. (1937) proved the existence of monotone wave solutions for all wave speeds greater than or equal to the critical speed

(23) $$c^* = 2\sqrt{D \cdot f'(0)} \ .$$

There are no such solutions for $c < c^*$; furthermore, if P is initially given by a Heaviside distribution, then the wave corresponding to $c = c^*$ is attracting (see Hadeler 1976). For a complete mathematical treatment, the reader is referred to Bramson's (1983) monograph and to Fife (1984).

Skellam (1951) applied models of this type to the study of species invading new habitats, and Aronson and Weinberger (1975) have used systems of equations of this type in population genetics. Kendall (1965), Hadeler (1984), and other investigators have extended them to the study of the spread of epidemics. Recent applications are provided by Lubina and Levin (1988) and Andow et al. (1989). In many cases, the agreement between theory and experiment is excellent; in others, the assumption that movement is the result of numerous small steps clearly leads to the wrong answers, and more general redistribution kernels are necessary (see for example Mollison 1977).

ADVECTION-DIFFUSION MODELS OF DISPERSAL

The consideration of population rates of spread is predicated on assumptions concerning the movements of individuals. The spread of plant populations occurs via seeds and pollen. We therefore conclude the section on dispersal with a brief presentation of an advection-diffusion description for the wind dispersal of seeds and pollen (Okubo and Levin 1989). The shape of the *dispersal curve*, that is, the curve relating the number of dispersed seeds to distance from source, varies depending upon the speed of descent (the "settling" velocity), the height of release, wind speed and turbulence, and specific morphological adaptations for dispersal (Augspurger and Franson 1987). Typically, it falls off with large distances; but because of the effects of wind, it achieves its apex at some distance away from a point source. On the other hand, for a distributed source, we have a different situation, as the peak usually occurs at or close to the boundary of the source region.

To understand what factors control the forms of such dispersal curves, Okubo and Levin (1989) consider diffusive and advective forces with regard to properties of the propagules and height of release. As a first approximation, they do not take into account the influence of the parent plant on microscale air movements (Niklas 1984), and do not allow seeds to move once they strike the ground.

Dispersal curves with phenomenological derivations have been used widely; examples include the inverse power law (Gregory 1968), and the negative exponential (Frampton et al. 1942, Kiyosowa and Shiyomi 1972). These curves do not deal with transients, being confined to the asymptotic distribution of seeds, spores, or pollen from point releases, or the time-averaged solutions for continuous point sources. More importantly, they involve curve-fitting, and do not allow predictions to be made based on physical parameters such as wind velocity, turbulence, seed weight or height of release. More details can be found in Gregory and Read (1949) and Minogue (1986).

The inverse power law is given by

$$(24) \qquad y = as^{-b} ,$$

where s denotes the distance from source, y the probability distribution associated with dispersal, and a and b are constants. It transforms to a straight line on a log-log plot, making parameter estimation simpler; b is dimensionless, and hence it provides an advantage when one is dealing with studies on different scales.

The log-linear (negative exponential) model has the shape

$$(25) \qquad y = ae^{-bs} ,$$

which transforms to linear on a semi-log plot. Note that for this model y remains finite as s tends to zero. Each of these models has advantages (see Gregory 1968, McCartney and Bainbridge 1984, Fitt and McCartney 1986). However (Okubo and Levin 1989), they do not allow extrapolation from one solution to another based on independently measured physical parameters, and provide no understanding of the underlying mechanisms.

GAUSSIAN PLUME MODELS

Gaussian plume models have been used primarily for the description of the dispersion of air pollutants from smokestacks, but they also have been applied to spore dispersal (see for example Gregory et al. 1961, Fitt and McCartney 1986).

The Gaussian plume method (Csanady 1973; Hanna et al. 1982) uses Sutton's (1947) steady-state solution for a special type of the diffusion equation. The assumptions are: reflection at the surface of the earth, constant wind speed u in the x-direction at source height, and a continuous point source at height H above the ground. In addition, diffusion in the x direction is neglected relative to advection. Furthermore, it is assumed that particles are deposited at the surface of the earth at horizontal position (x,y) at the rate

$$(26) \qquad D = S(x,y,0)V_d ,$$

where V_d is the deposition velocity (Chamberlain 1975).

Using the reflection boundary conditions, one obtains the solution

$$(27) \qquad S(x,y,z) = n(x) \frac{\exp(-y^2/2\sigma_y^2)}{2\pi\bar{u}\sigma_z\sigma_y} \left[\exp\left(\frac{-(H-z)^2}{2\sigma_z^2} \right) \right] ,$$

where n = n(x) is the effective *source* strength at distance x, and the standard deviations σ_z, σ_y are function of x. (See Pasquill and Smith 1983, p. 333.) Dependence of n on x allows for losses due to deposition (Horst 1977). This model assumes that we are dealing with very light particles, and hence it does not take gravity into consideration. For heavy particles, the *tilted plume* model is obtained by replacing the effective height H of the plume by $H-xW_s/u$, where W_s is the settling velocity of seeds. This extends the plume model to the situation when particulates have a non-trivial settling velocity (see e.g. Csanady 1973).

Under simplifying assumptions (see Okubo and Levin 1989), it is found that the rate of deposition at the ground is given by

$$(28) \qquad D = S(x,y,0)W_s,$$

where $W_s = V_d$. From this, Okubo and Levin (1989) determine an expression for the concentration of seeds at the ground level:

$$(29) \qquad D = Q(x,y) = \frac{n(x)W_s}{2\pi\bar{u}\sigma_y\sigma_z} \exp\left\{\frac{y^2}{2\sigma_y^2} + \frac{(H-W_s x/\bar{u})^2}{2\sigma_z^2}\right\}.$$

The crosswind-integrated deposition rate (CWID) is obtained by integrating across the direction of the wind:

$$(30) \qquad \text{CWID} = \int_{-\infty}^{\infty} Q(x,y)dy = Q(x) = \frac{nW_s}{\sqrt{2\pi}\,\bar{u}\,\sigma_z} \exp\left\{\frac{-(H-W_s x/\bar{u})^2}{2\sigma_z^2}\right\}.$$

Ignore the decay in $n(x)$ [set $n(x)$ = constant], and set

$$(31) \qquad \sigma_z^2 = 2Ax/\bar{u},$$

where A is the vertical diffusivity. This is motivated by the fact that, under pure diffusion, variance increases at the rate $2At$, and by the fact that the time to reach position x is x/u. The distribution is skewed; the maximum is less than or equal to the mean, and is given by

$$(32) \qquad x_m = \frac{\bar{u}H}{W_s} \left[\{1+(A/HW_s)^2\}^{\frac{1}{2}} - (A/HW_s)\right],$$

which agrees with the mean in the absence of vertical diffusivity (A = 0).

Define

$$(33) \qquad 2AH = W^* \text{ (vertical mixing velocity)},$$

and rewrite (32) as

$$(34) \qquad \lambda\frac{x_m}{H} = \frac{\bar{u}}{W_s},$$

where

(35)
$$\lambda = (1+(W*/2W_s)^2)^{\frac{1}{2}} + W*/2W_s .$$

For small values of $W*/W_s$ (heavy seeds)

(36)
$$\lambda \approx 1 + W*/2W_s \approx 1 ,$$

and so $x_m \sim Hu/W_s$; whereas, for large values of $W*/2W_s$ (light seeds),

(37)
$$\lambda \approx W*/W_s \gg 1 ,$$

and so $x_m \sim Hu/W*$.

In Okubo and Levin (1989), the above model is extended to incorporate more precisely the dynamics of advective and diffusive movements in both the horizontal and vertical directions, and correct boundary conditions at the earth's surface. Horizontal advection is determined by mean wind speeds, while the vertical advective force is gravitational. These and other assumptions are made to determine the equations governing the dispersal of seeds or pollen from an isolated plant or tree. The major change in the calculation of the mode is that formula (35) is replaced by

(35a)
$$\lambda = 1 + W*/W_s .$$

Actual dispersion relationships obtained with data in 15 studies are compared with model predictions (Okubo and Levin 1989).

These models, borrowed from the atmospheric diffusion literature, allow establishment of a framework for organizing data concerning the relationship of dispersion distances to environmental and species-specific parameters, and to such other parameters as height of release. This presents us with an improved situation because the conventional models are phenomenological, and hence do not provide a basis for extrapolation from one environment to another. Further details and a more elaborate discussion can be found in Okubo and Levin (1989).

4. Tight and diffuse coevolution

As mentioned previously, the study of coevolutionary interactions is one of the central problems in evolutionary biology. In this section, following now conventional usage, we make a clear distinction between *tight* coevolution, involving a few closely linked species, and *diffuse* coevolution, in which the influences are spread over many species.

Many models of tight coevolutionary interactions fall within the framework of explicit genetic models (e.g., Levin and Udovic 1977). This approach is useful when the bases for inheritance are well understood, and when the number of loci involved is small. However, for parasite-host systems, most such classical models ignore the central ecological and epidemiological interactions that are faced by intimately interacting species. Incorporating the nature of these interactions into models is critical if we are to understand phenomena such as the evolution of virulence and disease, because classical models do not take account of the truly tight interdependence of the host and parasite.

The study of diffuse coevolutionary interactions, involving a multitude of species, demands a different perspective and a different approach. To this end, we discuss some preliminary work on the evolution of chemical defenses.

TIGHT COEVOLUTION: MODELS OF HOST-PARASITE COEVOLUTION

One of the best examples of the successful application of explicit genetic models for tight coevolutionary interactions involves the gene-for-gene systems of cereal plants and flax and their fungal pathogens (rusts). In these systems, specific genes for host resistance may be attached to specific genes for parasite virulence; and hence there is strong selection for specific characters (e.g., Feeny 1975, Janzen 1980).

The study of the cereal-rust interactions builds on the experimental work of Flor (1955, 1956) and the theoretical work of Mode (1958, 1960, 1961). It is a common characteristic of these models to omit the epidemiological details and formulate the probability of association between parasites and hosts in terms of a mass-action law. For a review of the literature on cereal-rust interactions, see Levin (1983b).

To be more specific, we briefly describe the treatment of this problem by Lewis (1981a, 1981b). In Lewis (1981a), the host is assumed to be a diallelic diploid and the pathogen, a diallelic haploid. Pathogen fitnesses, for each host-parasite association, are described in the table below. The fitness w of the host in each pair is 1 minus the pathogen fitness.

Table 2

Host Genotype

		AA	A a	a a
Pathogen Genotype	B	α	β	γ
	b	γ	β	α

It is assumed that the frequencies of the particular associations are proportional to the products of the corresponding associated types. If, in addition, x denotes the fixed probability that a host is parasitized, then we arrive at the following model (due to Lewis):

$$(38) \qquad \rho' = \rho \, \frac{\rho W_{AA} + (1-\rho) W_{Aa}}{\rho(\rho W_{AA} + (1-\rho) W_{Aa}) + (1-\rho)(\rho W_{Aa} + (1-\rho) W_{aa})}$$

for the host. Here

$$W_{AA} = 1 - x + x[q(1-\alpha)+(1-q)(1-\gamma)] = 1 - x[q\alpha+(1-q)\gamma]$$

(39) $\quad W_{Aa} = 1 - x + x[q(1-\beta)+(1-q)(1-\beta)] = 1 - x[\beta]$

$$W_{aa} = 1 - x + x[q(1-\gamma)+(1-q)(1-\alpha)] = 1 - x[q\gamma+(1-q)\alpha]$$

and

(40)
$$q' = q \frac{V_B}{qV_B+(1-q)V_b}$$

for the pathogen, in which

(41)
$$V_B = p^2\alpha + 2p(1-p)\beta + (1-p)^2\gamma .$$

and

(42)
$$V_b = p^2\gamma + 2p(1-p)\beta + (1-p)^2\alpha .$$

Here, p and q are, respectively, the allelic frequencies of allele A in the host and allele B in the pathogen in the present generation, while p' and q' denote the frequencies of A and B in the next generation.

From the symmetry of the system, it follows that is has an internal polymorphic equilibrium at $p = q = 0.5$. However, this equilibrium is stable if and only if

(43)
$$\beta < \frac{\alpha+\gamma}{2}\sqrt{1-2\frac{\alpha-\gamma^2}{(\alpha+\gamma)^2}} < \frac{\alpha+\gamma}{2}$$

From the viewpoint of the host population, this condition is stronger than marginal overdominance, which is always a necessary condition for the stability of polymorphic equilibria in the absence of frequency dependence (see Levin and Udovic 1977). In addition, oscillatory solutions have been found when β is increased above the threshold specified above, in agreement with fluctuations observed in many simplified agricultural systems.

It follows that when resistance is dominant and virulence recessive, as in many cereal-rust systems, stable polymorphisms cannot be established through such models, since marginal overdominance is impossible. How then is stability realized in host-parasite systems? The most likely explanation is through explicit or implicit frequency dependence, which stabilizes such interactions (see Gillespie 1975).

Frequency dependence of some form is inescapable when one is interested in the evolution of virulence. As Anderson and May (1982b) point out, most standard textbooks take the dogmatic approach that parasite evolution is towards less and less virulent pathogen strains, with commensalism the inevitable end point. The situation, however, is not this simple. As Levin (1983a) states: "Evolution in parasite populations represents an interplay between conflicting factors: within an individual host, the race is to the swift and evolution will favor those with the highest rate of reproduction, which is likely to mean those with higher virulence. But the parasite population is a shifting mosaic of demes associated with individual hosts, and the capacity for profligate growth may doom one's host to a shorter life expectancy and reduce the contribution to the larger (mega-) population. Depending on the balance between these factors, some evolution towards attenuation might be expected among parasites, but this attenuation may be checked far short of commensalism (Levin and Pimentel 1981, Anderson and May 1982a, Bremermann and Pickering 1982)."

The most famous example of loss of reduced virulence occurred in the (European) rabbit-myxoma system in Australia. The myxoma virus was introduced to control the rabbit population, which had denuded the landscape (Fenner and Ratcliffe 1965); hence, loss of virulence may lead to loss of control. To examine this system, Levin (1983a, see also Levin and Pimentel 1981) built upon classical epidemiological models to arrive at the model:

$$\frac{dS}{dt} = (r_0 S + r_1 I_1 + r_2 I_2 + r_3 I_3) - bS - \beta_1 SI_1 - \beta_2 SI_2 + v_1 I_1 + v_2 I_2$$

$$\frac{dI_1}{dt} = \beta_1 SI_1 - (b+\alpha_1)I_1 - v_1 I_1 + w_2 I_3 - \gamma_2 \beta_2 I_1 I_2$$

(44)

$$\frac{dI_2}{dt} = \beta_2 SI_2 - (b+\alpha_2)I_2 - v_2 I_2 + w_1 I_3 - \gamma_1 \beta_1 I_1 I_2$$

$$\frac{dI_3}{dt} = (\gamma_1 \beta_1 + \gamma_2 \beta_2)I_1 I_2 - (w_1 + w_2)I_3 - (b+\alpha_3)I_3 \ .$$

Here, S denotes susceptible hosts; I_1, I_2 denotes hosts infected with strain 1 and 2 respectively; and I_3 denotes hosts infected with both strains. The parameters r_i represent the birth rates; v_i, w_i represent the recovery rates; b and $b + \alpha_i$ represent the death rates; β_i denote the transmission rates; and γ_i denote the secondary infection rates. Such a framework allows explicit consideration of evolutionarily stable strategies, while recognizing the importance of the host-parasite interaction.

Possible outcomes of this model include: competitive exclusion of either viral type, their stable coexistence, or unbounded behavior. A positive polymorphic equilibrium satisfies the following conditions

$$I_2 = \frac{\beta_1}{(\gamma_2\beta_2 - w_2 Q)}\left(S - \frac{b + \alpha_1 + v_1}{\beta_1}\right)$$

(45)
$$I_1 = \frac{\beta_2}{(w_1 Q - \gamma_1\beta_1)}\left(\frac{b + \alpha_2 + v_2}{\beta_2} - S\right)$$

$$I_3 = QI_1 I_2 = \frac{\gamma_1\beta_1 + \gamma_2\beta_2}{w_1 + w_2 + b + \gamma_3} I_1 I_2 .$$

Note that in order for I_1 and I_2 to be positive, we need that:

(46)
$$\frac{b + \gamma_2 + v_2}{\beta_2} > S > \frac{b + \gamma_1 + v_1}{\beta_1}$$

and

(47)
$$\gamma_2\beta_2 w_1 > \gamma_1\beta_2(w_2 + b + \gamma_3) .$$

The most important questions involve an explanation of which viral strains survive and why, and of how the virus can be coupled with other control measures to lead to effective control of the virus. We note that the results of the above modelling exercise only scratch the surface of the complicated questions regarding the evolution of virulence in the parasite and of resistance in the host. For the myxoma-rabbit system, other factors have to be considered: seasonality and multiple modes of transmission as well as the role played by this pathogen in regulating the host population.

Dwyer et al. (submitted) emphasize the importance of directing attention to an analysis of the myxoma-*Oryctolagus* interaction. The system is of fundamental theoretical and applied importance. If myxomatosis evolves to the point that control is lost, the rabbit population again may become a serious pest. Because the underlying processes occur on a variety of temporal and spatial scales, mathematical models are critical in dissecting the complex system, and in identifying underlying mechanisms. Dwyer et al. (submitted) develop a simulation model that incorporates many of the aspects left out by simple models. Preliminary investigations seem to show that the spatial structure of the population plays a very important role in the observed coexistence of intermediate types.

What can be said concerning the evolution of other viral diseases? Influenza, to be discussed in the next paragraphs, provides one particularly interesting example, because changes in a few surface antigens led to the proliferation of a variety of strains, and to the potential for reappearance of strains previously lost. Thus, periodic or other recurrent behavior is to be expected in influenza, and such behavior indeed is observed (see Liu and Levin 1989, Hethcote and Levin 1989). For AIDS, that modern scourge, the hope that might be raised by contemplation of the myxoma story is short-lived. By the standards of myxoma, AIDS is already attenuated, in that infected individuals live a very long time.

Thus, selection for reduced virulence is not a potent force at all, and the rapidly changing AIDS virus is more likely to evolve in the direction of increased virulence.

In the influenza-man system, attention is focused on the potential for cross-immunity (a measure of reduced susceptibility to related strains of type A influenza) to facilitate oscillations and coexistence of strains (Castillo-Chavez et al. 1988, 1989). Recent work shows the existence of long-lasting cross-immunity between related strains (i.e., variants of the same subtype) in human influenza (Couch and Kasel 1983). Cross-immunity implies that the presence of one strain of the virus can reduce the pool of susceptible individuals for co-circulating strains, introducing a form of exploitation competition (Catillo-Chavez et al. 1988, 1989).

Castillo-Chavez et al. (1989) present models to elucidate the recently observed co-circulation of related strains, by extending the classical epidemiological approaches to allow for immunological interactions between strains (cross-immunity). For a homogeneous population, they introduce the diagram

(48)

$$
\begin{array}{ccccc}
X & \Rightarrow & Y_1 & \Rightarrow & Z_1 \\
\Downarrow & & & & \Downarrow \\
Y_2 & & & & V_2 \\
\Downarrow & & & & \Downarrow \\
Z_2 & \Rightarrow & V_1 & \Rightarrow & W
\end{array}
$$

In the above system, the population has been divided into 8 classes: X (fraction susceptible), Y_i (fraction infected by strain i), Z_i (fraction recovered from the other strain), V_i (fraction infected by strain i after recovery from the other strain), and W (recovered from both strains). Castillo-Chavez et al. (1988, 1989) assume that the population is homogeneously mixing, and that the usual bilinear incidence function describes transmission. They then formulate the following two-stain epidemiological model:

$$
(49) \qquad X'(t) = [\beta_1(Y_1+V_1)+\beta_2(Y_2+V_2)-\mu]X + \mu ,
$$

$$
(50) \qquad Y_i'(t) = \beta_i(Y_i+V_i)X - (\gamma_i+\mu)Y_i ,
$$

$$
(51) \qquad Z_i'(t) = \gamma_i Y_i - [\sigma_j\beta_j(Y_j+V_j)+\mu]Z_i ,
$$

$$
(52) \qquad V_i'(t) = \sigma_i\beta_i(Y_i+V_i)Z_j - (\gamma_i+\mu)V_i ,
$$

(53)
$$W'(t) = \gamma_1 V_1 + \gamma_2 V_2 - \mu W \, ,$$

where

(54)
$$i = j, j = 2 \quad \text{or} \quad i = 2, j = 1 \, .$$

In addition, β_i denotes the transmission coefficient of strain i. σ_i denotes the *susceptibility factor* (where $j = 3 - i$); that is, σ_i is a measure of the relative susceptibility of types Z_i and X in terms of their acquisition of strain j. Usually, but not always, σ_i is between 0 and 1. Furthermore, γ_i denotes the recovery rate form strain i, and μ denotes the (constant) natural mortality rate. Thus, the model is flexible enough to cover the range of possibilities, from closely related strains to distinct subtypes.

Mathematical analysis and numerical simulations indicate that the above system cannot produce sustained oscillations. However, slowly-damped and hence biologically important oscillations are generated as a result of the cross-immunity.

When a heterogeneous host population is considered (age-structured population), then the model above is replaced by:

(55)
$$\frac{\partial x(a,t)}{\partial a} + \frac{\partial x(a,t)}{\partial t} = -(\lambda_1(t)b(a) + \lambda_2(t)b(a) + \mu(a))x(a,t) \, ,$$

(56)
$$\frac{\partial y_i(a,t)}{\partial a} + \frac{\partial y_i(a,t)}{\partial t} = \lambda_i(t)b(a)x(a,t) - (\gamma_i + \mu(a,t)) \, , \quad i = 1,2$$

(57)
$$\frac{\partial z_i(a,t)}{\partial a} + \frac{\partial z_i(a,t)}{\partial t} = \gamma_i y_i(a,t) - \sigma_i \lambda_j(t)b(a)z_i(a,t) - \mu(a)z_i(a,t) \, , \quad i = 1,2$$

(58)
$$\frac{\partial v_i(a,t)}{\partial a} + \frac{\partial v_i(a,t)}{\partial t} = \sigma_i \lambda_i(t)b(a)z_j(a,t) - (\gamma_i + \mu(a))v_i(a,t) \, , \quad i = 1,2$$

(59)
$$\frac{\partial w(a,t)}{\partial a} + \frac{\partial w(a,t)}{\partial t} = (\gamma_1 + \gamma_2 - \mu(a))w(a,t) \, ,$$

(60)
$$\lambda_i(t) = \beta_i \int_0^\infty b(a')[y_i(a',t) + v_i(a',t)]da' \, ,$$

(61)
$$x(0,t) = \rho \, , \quad y_i(0,t) = 0 \, , \quad z_i(0,t) = 0 \, , \quad v_i(0,t) = 0 \, , \quad w(0,t) = 0 \, ,$$

$$x(a,0) = x_0(a) , \quad y_i(a,0) = y_{0i}(a) , \quad z_i(a,0) = z_{0i}(a) ,$$

(62)

$$v_i(a,0) = v_{0i}(a) , \quad w(0,t) = w_0(a) .$$

Furthermore,

(63)
$$\rho = \left[\int_0^\infty e^{-M(a)} da \right]^{-1} \quad \text{where} \quad M(a) = \int_0^a \mu(a) d\alpha .$$

Here $x(a,t)$, $y_i(a,t)$, $z_i(a,t)$, $v_i(a,t)$, and $w(a,t)$ denote the densities of the individuals in each class previously defined, and a is an independent variable that denotes the age of an individual. $b(a)$ represents the age-specific contact rate, λ_i denotes the instantaneous force of infection, β_i denotes the transmission scaling factor, $m(a)$ is the age-specific mortality rate, and σ_i denotes the (constant) recovery rate. In this case Castillo-Chavez et al. (1989) and Andreasen (1989) have suggested that sustained oscillations are possible, due to the interaction between cross-immunity and age structure.

DIFFUSE COEVOLUTION

The problem of diffuse coevolution is probably ecologically more important than tight coevolution, but is much less understood and rarely modeled. As Levin (1983b) remarks: "...many problems of interest in the evolution of ecological communities are much more diffuse, involving many species with varying degrees of relationship to one another. Problems of this sort arise in the consideration of the chemical defenses of plants in response to insects and other pests (Feeny 1982), for often these do not have the finely tuned species-for-species relationship already discussed for the cereals and their rusts. Similar problems occur in predator-prey systems, which are by nature less specific than the host-parasite relationships; in competition theory; and regarding the evolution of the vertebrate immune system". In this section, we describe some early and tentative attempts to approach such problems.

   In what follows we provide a very brief introduction to preliminary investigations by Levin, Segel, and Adler (unpubl.), who have begun to develop a framework within which to examine the patterns of diffuse coevolution in plant-herbivore communities. They point out that "the lack of evidence for the tight coevolution between pairs of species may inappropriately direct attention away from the obvious coevolution of defensive chemicals and mechanisms for detoxifying them." Given the impracticability of a reductionistic approach that includes the detailed genetics of every species, they focus on macroscopic variables such as the number and frequency distribution of different kinds of chemicals, and on other community-level descriptors. Assume that the plants possess toxins that, if unneutralized, prevent herbivores from consuming them. First, consider an oversimplified situation in which there is a pool of N chemical defenses such that each plant has exactly n . In addition, each herbivore is able to detoxify m of these toxins. Furthermore, assume that the particular group of defensive chemicals or detoxifying agents is drawn at random by the plant or the herbivore from the pool; it is further assumed that the plant repels or resists the herbivore if and only if it has at least one defensive chemical that the herbivore cannot counteract.

If $L(m,n)$ denotes the probability that a plant with n defensive chemicals will "lose" in an encounter with a herbivore capable of detoxifying m substances, then clearly

$$(64) \qquad\qquad\qquad L(m,n) = 0 \text{ if } m < n .$$

Since, for $m \geq n$, $L(m,n)$ denotes the probability that m elements chosen at random from a set of N elements lie within a particular subset of size n, then

$$(65) \qquad L(m,n) \; = \frac{m(m-1)...(m-n+1)}{N(N-1)...(N-n+1)} \; = \binom{N-n}{m-n} \Big/ \binom{N}{m}, \text{ if } m \geq n .$$

The above expressions allow calculation of the probabilities of various results of an individual encounter between a plant and a herbivore. Levin, Segel, and Adler (ms.) investigate the implications of these assumptions in the development or evolution of a community. The first question that they ask is: What will be the fate of a rare mutant or migrant that appears in the community? In order to answer this question, one must assign benefits and costs to the winners and losers during a given encounter. Furthermore, costs have to be assigned for a given level of chemical defense or detoxifying ability.

The simplest assumptions are that the new migrant (or mutant) cannot interbreed with the resident types, and that the number of encounters per unit time remains constant. More specifically, the total cost per plant per unit time associated with herbivory is

$$(66) \qquad\qquad\qquad k L(m,n) + cn ,$$

where the constant k is the product of the number of encounters per unit time and the cost per loss. The constant c gives the cost per unit time of keeping a single chemical defense. From here we easily conclude that the invasion of an (m,n) community by a plant with $n+1$ chemical defenses is possible provided that

$$(67) \qquad\qquad\qquad \theta_p \; = \frac{c}{k} < L(m,n) - L(m,n+1) .$$

If we replace the linear cost function cn by a function $f(n)$ of n, then we obtain:

$$(68) \qquad\qquad\qquad \frac{\mu(n)}{k} < L(m,n) - L(m,n+1) ,$$

where $\mu(n) = f(n+1) - f(n)$ is the marginal cost of adding an additional defensive chemical to the n that already are present.

The net gain associated with herbivory per herbivore per unit time is, in the simplest case,

$$(69) \qquad\qquad\qquad e L(m,n) - bm .$$

An (m,n) community can be invaded by a rare herbivore with $m+1$ detoxifying agents if and only if

$$(70) \qquad \theta_h = \frac{b}{e} < L(m+1,n) - L(m,n) .$$

The linear cost function bm similarly can be generalized. Note that similar considerations show that inequalities (67) and (70) are also the conditions respectively that an $(m,n+1)$ community cannot be invaded by a type-n plant and an $(m+1,n)$ community cannot be invaded by a type-m herbivore. From this, Levin, Segel, and Adler (ms), endeavor to build a tapestry of increasing realism, initially expanding their investigations to the situation in which a distribution of phenotypes (containing different numbers of chemicals) exist within the community, and from there to consideration of the spectrum of available chemicals, and the distribution of phenotypes in this aspect space. At this point, the treatment makes contact with the earlier work of Levin and Segel (1984) on pattern diversity in aspect space.

These investigations of diffuse coevolution are very preliminary. The objective – to develop macroscopic descriptors at the community and ecosystem level – are essential to the development of interfaces between population biology and ecosystem science. Imaginative approaches to such problems represent one of the unmet challenges of evolutionary theory.

Acknowledgments

Simon Levin gratefully acknowledges support during the preparation of these notes from All Souls College and the Centre for Mathematical Biology at the University of Oxford and from the Science and Engineering Research Council of Great Britain (Grant No. GRT/D/13573). Simon Levin also acknowledges support from the National Science Foundation, Grant No. BSR-8806202. Carlos Castillo-Chavez has been partially supported by the Center for Applied Mathematics and the Office of the Provost at Cornell University, as well as by a Ford Foundation Postdoctoral Fellowship for Minorities.

References

Anderson, R.M. and May, R.M. (1982a), Coevolution of hosts and parasites, *Parasitology* 85, 411-426.

Anderson, R.M. and May, R.M. (eds. 1982b), *Population Biology of Infectious Diseases, Dahlem Konferenzen*, Springer-Verlag, Berlin.

Andow, D.A., Kareiva, P.M., Levin, S.A. and Okubo, A. (1989), Spread of invading organisms: patterns of spread, in: *Evolution of Insect Pests: The Pattern of Variations* (K.C. Kim, ed.), John Wiley, New York (in press).

Andreasen, V. (1989), Multiple scales in the dynamics of infectious diseases, in: *Mathematical Approaches to Ecological and Environmental Problem Solving* (C. Castillo-Chavez, S.A. Levin and C. Shoemaker, eds.), Lecture Notes in Biomathematics, Springer-Verlag, Heidelberg (in press).

Aronson, D.G. and Weinberger, H.F. (1975), Nonlinear diffusion in population genetics, combustion, and nerve propagation, in: *Partial Differential Equations and Related Topics* (J. Goldstein, ed.), Lecture Notes in Mathematics 445, Springer-Verlag, Heidelberg, 5-49.

Augspurger, C.K. and Franson, S.E. (1987), Wind dispersal of artificial fruit varying in mass, area, and morphology, *Ecology* **68**, 27-42.

Bramson, M. (1983), *Convergence of Solutions of the Kolmogorov Equation to Travelling Waves*, Mem. Amer. Math. Soc. 285, Amer. Math. Soc., Providence, R.I.

Bremermann, H.J. and Pickering, J. (1982), A game theoretical model of parasite virulence, *J. Theoret. Biol.* **100**, 411-426.

Castillo-Chavez, C., Levin, S.A. and Gould, F. (1988), Physiological and behavioral adaptation to varying environments: a mathematical model, *Evolution* **42**(5), 986-994.

Castillo-Chavez, C., Hethcote, H.W., Andreason, V., Levin, S.A. and Liu, W.-m. (1988), Cross-immunity in the dynamics of homogeneous and heterogeneous populations, in: *Mathematical Ecology. Proc. of the Autumn Course Research Seminars, Trieste 1986* (L. Gross, T. Hallam and S. Levin, eds.), World Scientific Publishing Co., Singapore, 303-316.

Castillo-Chavez, C., Hethcote, H.W., Andreasen, V., Levin, S.A. and Liu, W-m. (1989), Epidemiological models with age structure and cross-immunity, *J. Math. Biol.* (in press).

Chamberlain, A.C. (1975), The movement of particles in plant communities, in: *Vegetation and the Atmosphere, Vol. 1* (J.L. Monteith, ed.), Academic Press, New York, 155-203.

Cohan, F. (1984), Genetic divergence under uniform selection I. Similarity among populations of *Drosophila melanogaster* in their responses to artificial selection for modifiers of ci^D , *Evolution* **38**, 55-71.

Cohen, D. and Levin, S.A. (1987), The interaction between dispersal and dormancy strategies in varying and heterogeneous environments, in: *Mathematical Topics in Population Biology, Morphogenesis and Neurosciences, Proc. Kyoto 1985* (E. Teramoto and M. Yagamuti, eds.), Springer-Verlag, Heidelberg, 110-122.

Cohen, D. and Levin, S.A. (1989), Dispersal in patchy environments: the effects of temporal and spatial structure. Submitted.

Couch, R.B. and Kasel, J.A. (1983), Immunity to influenza in man, *Ann. Rev. Microbiol.* **37**, 529-549.

Csanady, G.T. (1973), *Turbulent Diffusion in the Environment*, D. Reidel, Boston.

Dwyer, G.M., Levin, S.A. and Buttel, L. (1989), Myxomatosis and the European rabbit, *Oryctolagus cuniculus*: preliminary analysis of a model system. Submitted.

Ellner, S.P. (1985), ESS germination strategies in randomly varying environments, I. Logistic-type models, *Theoret. Population Biol.* **28**, 50-79.

Ewens, W.J. (1969a), A generalized fundamental theorem of natural selection, *Genetics* **63**, 531-537.

Ewens, W.J. (1969b), Mean fitness increases when fitnesses are additive, *Nature* **221**, 1076.

Ewens, W.J. (1979), *Mathematical Population Genetics*, Springer-Verlag, New York.

Feeny, P. (1975), Biochemical coevolution between plants and their insect herbivores, in: *Coevolution of Animals and Plants* (L.E. Gilbert and P.H. Raven, eds.), University of Texas Press, Austin and London, 3-19.

Feeny, P. (1982), Coevolution of plants and insects, in: *Current Themes in Tropical Sciences, 2: Natural Products for Innovative Pest Management*, (T.R. Odhiambo, ed.), Pergamon Press, Oxford, Chapter 11.

Felsenstein, J. (1965), The effect of linkage on directional selection, *Genetics* **52**, 349-363.

Fenner, F. and Ratcliffe, R.N. (1965), *Myxomatosis*, Cambridge University Press, London.

Fife, P.C. (1984), Current topics in reaction-diffusion systems, in: *Proceedings of NATO Conference on Nonequilibrium Phenomena in Physics and Related Fields* (M.G. Velarde, ed.), Plenum, New York.

Fisher, R.A. (1937), The wave of advance of advantageous genes, *Ann. Eugen. London*, **7**, 355-369.

Fitt, B.D.L. and H.A. McCartney, (1986), Spore dispersal in relation to epidemic models, in: *Plant Disease Epidemiology. 1.* (K.J. Leonard and W.E. Fry, eds.), Macmillan, New York, 311-345.

Flor, H.H. (1955), Host-parasite interacions in flax rust – its genetics and other implications, *Phytopathology* **45**, 680-685.

Flor, H.H. (1956), The complementary genic systems in flax and flax rust, *Advances in Genetics* **8**, 29-54.

Frampton, V.L., Linn, M.B., Hansing, E.D. (1942), The spread of virus diseases of the yellow type under field conditions, *Phytopathology* **32**, 799-808.

Gillespie, J.H. (1975), Natural selection for resistance to epidemics, *Ecology* **56**, 493-495.

Gillespie, J.H. (1983), A simple stochastic gene substitution model, *Theoret. Population Biol.* **23**(2), 202-215.

Gillespie, J.H. (1984), Molecular evolution over the mutational landscape, *Evolution* **38**(5), 1116-1129.

Gould, S.J. (1977), *Ever Since Darwin*, Norton, New York.

Gregory, P.H. (1968), Interpreting plant disease dispersal gradients, *Annual Review Phytopathology* **6**, 189-212.

Gregory, P.H., Longhurst, T.J. and Sreeramula, T. (1961), Dispersion and deposition of airborne *Lycopodium* and *Ganoderma* spores, *Annals of Applied Biology* **49**, 645-658.

Gregory, P.H. and Read, D.R. (1949), The spatial distribution of insect-borne plant-virus diseases, *Annals of Applied Biology* **36**, 475-482.

Hadeler, K.P. (1976), Nonlinear diffusion equations in biology, in: *Ordinary and Partial Differential Equations* (W.N. Everett and B.D. Sleeman, eds.), Lecture Notes in Biomathematics 564, Springer-Verlag, Heidelberg.

Hadeler, K.P. (1984), Spread and age structure in epidemic models, in: *Perspectives in Mathematics. Anniversary of Oberwolfach, 1984.* Birkhäuser-Verlag, Basel, 295-320.

Hadeler, K.P. and Rothe, E. (1975), Travelling fronts in nonlinear diffusion equations, *J. Math. Biol.* **2**, 251-263.

Hamilton, W.D. and May, R.M. (1977), Dispersal in stable habitats. *Nature* **269**(5629), 578-581.

Hanna, S.R., Briggs, G.A. and Hosker R.P., Jr. (1982), *Handbook on Atmospheric Diffusion*, DOE/TIC-11223 (DE82002045), Technical Information Center, U.S. Department of Energy.

Hethcote, H.W. and Levin, S.A. (1989), Periodicity in epidemiological models, in: Applied Mathematical Ecology (S.A. Levin, T.G. Hallam and L.J. Gross, eds.) Biomathematics 18, Springer-Verlag, Heidelberg (in press).

Hopfield, J.J. (1982), Neural networks and physical systems with emergent collective computational abilities, *Proc. Natl. Acad. Sci. USA* **79**, 2554-2558.

Hoppensteadt, F. (1975), *Mathematical Theories of Populations: Demographics, Genetics and Epidemics*, SIAM Reg. Conf. Series 20, Philadelphia, PA.

Horst, T.W. (1977), A surface depletion model for deposition from a Gaussian plume, *Atmospheric Environment* **11**, 41,46.

Hutchinson, G.E. (1951), Copepodology for the ornithologist, *Ecology* **32**, 571-577.

Jacob, F. (1977), Evolution and tinkering, *Science*, **196**, 1161-1166.

Janzen, D.H. (1980), When is it coevolution?, *Evolution* **34**, 611-612.

Johnson, D.S. and Papadimitriou, C.H. (1985), Computational complexity, in: *The Traveling Salesman Problem* (E.L. Lawler, J.K. Lenstra, A.H.G. Rinnooy Kan and D.B. Shmoys, eds.), Wiley, Chichester, 37-85.

Kareiva, P.M. (1983), Local movement in herbivorous insects: applying a passive diffusion model to mark-recapture field experiments, *Oecologia* **57**, 322-327.

Kareiva, P.M. and Shigesada, N. (1983), Analyzing insect movement as a correlated random walk, *Oecologia* **56**, 234-238.

Kauffman, S.A. and Levin, S.A. (1987), Towards a general theory of adaptive walks on rugged landscapes, *J. Theoret. Biol.* **128**, 11-45.

Kauffman, S.A., Weinberger, E.D. and Perelson, A.S. (1988), Maturation of the immune response via adaptive walks on affinity landscapes, in: *Theoretical Immunology, Part One* (A.S. Perelson, ed.), SFI Studies in the Sciences of Complexity. Addison Wesley Publishing Co.

Kendall, D.G. (1965), Mathematical models of the spread of infection, in: *Mathematics and Computer Science in Biology and Medicine*, H.M.S.O., London, 213-225.

Kiyosowa, S. and Shiyomi, M. (1972), A theoretical evaluation of the effect of mixing resistant variety with susceptible variety for controlling plant diseases, *Annals of the Phytopathological Society of Japan* **38**, 41-51.

Kolmogorov, A., Petrovskij, I. and Piskunov, N. (1937), Étude de l'équation de la diffusion avec croissance de la quantité de la matière et son application à un problème biologique, *Bull. Univ. Moscou Sér. Internation., Sec. A.* **1**(6), 1-25.

Levin, S.A. (1974), Dispersion and population interactions, *Amer. Natur.* **108**, 207-228.

Levin, S.A. (1978), On the evolution of ecological parameters, in: *Ecological Genetics: The Interface* (P.F. Brussard, ed.), Springer-Verlag, New-York, 3-26.

Levin, S.A. (1983a), Some approaches to the modelling of coevolutionary interactions, in: *Coevolution* (M. Nitecki, ed.), University of Chicago Press, Chicago, 21-65.

Levin, S.A. (1983b), Coevolution, in: *Population Biology* (H.I. Freedman and C. Strobeck, eds.), Lectures Notes in Biomathematics 52, Springer-Verlag, 328-334.

Levin, S.A. (1986), Random walk models of movement and their implications, in: *Mathematical Ecology, an Introduction* (T.G. Hallam and S.A. Levin, eds.), Springer-Verlag, Berlin, Heidelberg, 149-154.

Levin, S.A. (1987), Ecological and evolutionary aspects of dispersal, in: *Mathematical Topics in Population Biology, Morphogenesis and Neurosciences, Proc. Kyoto 1985* (E. Teramoto and M. Yagamuti, eds.), Springer-Verlag, Berlin, 80-87.

Levin, S.A., Cohen, D. and Hastings, A. (1984), Dispersal strategies in patchy environments, *Theoret. Population Biol.* 26(2), 165-191.

Levin, S.A. and Pimentel, D. (1981), Selection of intermediate rates of increase in parasite-host systems, *Amer. Natur.* 117, 308-315.

Levin, S.A. and Segel. L.A. (1984), Pattern generation in space and aspect, *SIAM Review* 27, 45-67.

Levin, S.A., Segel, L.A. and Adler, F., Diffuse coevolution in plant-herbvivore and plant-pathogen communities. Manuscript.

Levin, S.A. and Udovic, J.D. (1977), A mathematical model of coevolving populations, *Amer. Natur.* 111, 657-675.

Lewis, J.W. (1981a), On the coevolution of pathogen and hosts: I. General theory of discrete time coevolution, *J. Theoret. Biol.* 93, 927-951.

Lewis, J.W. (1981b), On the coevolution of pathogens and hosts: II. Selfing hosts and haploid pathogens, *J. Theoret. Biol.* 93, 953-985.

Lewontin, R.C. (1977), Adaptation, *Encyclopedia Einaudi   Turin* 1, 198-214.

Lin, C.C. and Segel, L.A. (1974), *Mathematics Applied to Deterministic Problems in the Natural Sciences*, Macmillan, New York.

Lin, S. and Kernighan, B.W. (1973) An effective heuristic algorithm for the traveling salesman problem, *Oper. Res.* 21, 498.

Liu, W-m. and Levin, S.A. (1989), Influenza and some related mathematical models, in: *Applied Mathematical Ecology*, (S.A. Levin, T.G. Hallam and L.J. Gross, eds.), Biomathematics 18, Springer-Verlag, Heidelberg (in press).

Lubina, J.A. and Levin, S.A. (1988), The spread of a reinvading species: range expansion in the California sea otter, *Amer. Natur.* 131(4), 526-543.

McCartney, H.A. and Bainbridge, A. (1984), Deposition gradients near to a point source in a barley crop, *Phytopathologische Zeitschrift* 109, 219-236.

Minogue, K.P. (1986), Disease gradients and the spread of disease, in: *Plant Disease Epidemiology*, (K.J. Leonard and W.E. Fry, eds.), Macmillan, New York, 285-310.

Mode, C.J. (1958), A mathematical model for the coevolution of obligate parasites and their hosts, *Evolution* 12, 158-165.

Mode, C.J. (1960), A model of a host-pathogen system with particular reference to the rust of cereal, in: *Biometrical Genetics*, Pergamon Press, New York, 84-96.

Mode, C. (1961), A generalized model of a host-pathogen system, *Biometrics* **17**, 386-404.

Mollison, D. (1977), Spatial contact models for ecological and epidemic spread, *J. Roy. Statist. Soc. Ser. B.* **39**, 283-326.

Nagylaki, T. (1976), The evolution of one- and two-locus systems, *Genetics* **83**, 583-600.

Niklas, K.J. (1984), The motion of windborne pollen grains around conifer ovulate cones: implications on wind pollination, *Amer. J. Botany* **71**, 356-374.

Okubo, A. (1980), *Diffusion and Ecological Problems: Mathematical Models*, Biomathematics 10, Springer-Verlag, New York.

Okubo, A. and Levin, S.A. (1989), A theoretical framework for the analysis of data on the wind dispersal of seeds and pollen, *Ecology* **70**(2), 329-338.

Pasquill, F. and Smith, F.B. (1983), *Atmospheric Diffusion*, Third Edition, Ellis Horwood Ltd., Chichester.

Provine, W.B. (1986), *Sewall Wright and Evolutionary Biology*, The University of Chicago Press, Chicago.

Skellam, J.G. (1951), Random dispersal in theoretical populations, *Biometrika* **38**, 196-218.

Sutton, O.G. (1947), The theoretical distribution of airborne pollution from chimneys, *Quarterly Journal of the Royal Meteorological Society* **73**, 426-436.

Venable, D.L. and Lawlor, L. (1980), Delayed germination and dispersal in desert annuals: escape in space and time, *Oecologia* **46**, 272-282.

Weinberger, E. (1988), A more rigorous derivation of some properties of uncorrelated fitness landscapes, *J. Theoret. Biol.* **134**, 125-129.

INDEX